Edik U. Rafailov, Maria Ana Cataluna, and Eugene A. Avrutin

Ultrafast Lasers Based on Quantum Dot Structures

Related Titles

Markel, V. A., George, T. F. (eds.)

Optics of Nanostructured Materials

568 pages
2008
E-Book
ISBN: 978-0-470-35349-3

Piprek, J. (ed.)

Nitride Semiconductor Devices: Principles and Simulation

519 pages with 220 figures and 53 tables
2007
Hardcover
ISBN: 978-3-527-40667-8

Champion, Y., Fecht, H.-J. (eds.)

Nano-Architectured and Nanostructured Materials
Fabrication, Control and Properties

166 pages with 101 figures and 16 tables
2004
Online Book Wiley Interscience
ISBN: 978-3-527-60601-6

Harrison, P.

Quantum Wells, Wires and Dots
Theoretical and Computational Physics

480 pages
2000
Hardcover
ISBN: 978-0-471-98495-5

Edik U. Rafailov, Maria Ana Cataluna, and Eugene A. Avrutin

Ultrafast Lasers Based on Quantum Dot Structures

Physics and Devices

WILEY-VCH Verlag GmbH & Co. KGaA

The Authors

Prof. Edik U. Rafailov
University of Dundee
Photonics and Nanoscience Gr.
Dundee, UK
e.u.rafailov@dundee.ac.uk

Dr. Maria Ana Cataluna
University of Dundee
Photonics and Nanoscience Gr.
Dundee, UK
M.A.Cataluna@dundee.ac.uk

Dr. Eugene A. Avrutin
University of York
Department of Electronics
York, UK
eaa2@ohm.york.ac.uk

All books published by **Wiley-VCH** are carefully produced. Nevertheless, authors, editors, and publisher do not warrant the information contained in these books, including this book, to be free of errors. Readers are advised to keep in mind that statements, data, illustrations, procedural details or other items may inadvertently be inaccurate.

Library of Congress Card No.: applied for

British Library Cataloguing-in-Publication Data
A catalogue record for this book is available from the British Library.

Bibliographic information published by the Deutsche Nationalbibliothek
The Deutsche Nationalbibliothek lists this publication in the Deutsche Nationalbibliografie; detailed bibliographic data are available on the Internet at http://dnb.d-nb.de.

© 2011 WILEY-VCH Verlag & Co. KGaA, Boschstr. 12, 69469 Weinheim, Germany

All rights reserved (including those of translation into other languages). No part of this book may be reproduced in any form – by photoprinting, microfilm, or any other means – nor transmitted or translated into a machine language without written permission from the publishers. Registered names, trademarks, etc. used in this book, even when not specifically marked as such, are not to be considered unprotected by law.

Cover Design   Adam Design, Weinheim
Typesetting   Thomson Digital, Noida, India
Printing and Binding   betz-druck GmbH, Darmstadt

Printed in the Federal Republic of Germany
Printed on acid-free paper

ISBN: 978-3-527-40928-0

Contents

Introduction *IX*
Acknowledgments *XI*

1 **Semiconductor Quantum Dots for Ultrafast Optoelectronics** *1*
1.1 The Role of Dimensionality in Semiconductor Materials *1*
1.2 Material Systems Used *4*
1.2.1 III–V Epitaxially Grown Quantum Dots *4*
1.2.2 QD-Doped Glasses *6*
1.2.3 Quantum Dashes *6*
1.3 Quantum Dots: Distinctive Properties for Ultrafast Devices *7*
1.3.1 Inhomogeneous Broadening *7*
1.3.2 Ultrafast Carrier Dynamics *9*

2 **Foundations of Quantum Dot Theory** *11*
2.1 Energy Structure and Matrix Elements *11*
2.2 Theoretical Approaches to Calculating Absorption and Gain in Quantum Dots *14*
2.3 Kinetic Theory of Quantum Dots *22*
2.4 Light–Matter Interactions in Quantum Dots *37*
2.5 The Nonlinearity Coefficient *51*

3 **Quantum Dots in Amplifiers of Ultrashort Pulses** *55*
3.1 Optical Amplifiers for High-Speed Applications: Requirements and Problems *55*
3.2 Quantum Dot Optical Amplifiers: Short-Pulse Operating Regime *62*
3.3 Quantum Dot Optical Amplifiers at High Bit Rates: Low Distortions and Patterning-Free Operation *63*
3.4 Nonlinear Operation and Limiting Function Using QD Optical Amplifiers *76*

Ultrafast Lasers Based on Quantum Dot Structures: Physics and Devices.
Edik U. Rafailov, Maria Ana Cataluna, and Eugene A. Avrutin
Copyright © 2011 WILEY-VCH Verlag GmbH & Co. KGaA, Weinheim
ISBN: 978-3-527-40928-0

4	**Quantum Dot Saturable Absorbers** *77*	
4.1	Foundations of Saturable Absorber Operation *77*	
4.2	The General Physical Principles of Saturable Absorption in Semiconductors *80*	
4.2.1	Physical Processes in a Saturable Absorber *80*	
4.2.2	Geometry of Saturable Absorber: SESAM versus Waveguide Absorber – The Cavity Enhancement of Saturable Absorption and the Standing Wave Factor in SESAMs *84*	
4.3	The Main Special Features of a Quantum Dot Saturable Absorber Operation *87*	
4.3.1	Bandwidth of QD SAs *88*	
4.3.2	Dynamics of Carrier Relaxation: Ultrafast Recovery of Absorption *88*	
4.3.3	Saturation Fluence *94*	
5	**Monolithic Quantum Dot Mode-Locked Lasers** *99*	
5.1	Introduction to Semiconductor Mode-Locked Lasers *99*	
5.1.1	Place of Semiconductor Mode-Locked Lasers Among Other Ultrashort Pulse Sources *99*	
5.1.2	Mode-Locking Techniques in Laser Diodes: The Main Principles *100*	
5.1.3	Passive Mode Locking: The Qualitative Picture, Physics, and Devices *101*	
5.2	Theoretical Models of Mode Locking in Semiconductor Lasers *103*	
5.2.1	Small-Signal Time Domain Models: Self-Consistent Pulse Profile *103*	
5.2.2	Large-Signal Time Domain Approach: Delay Differential Equations Model *109*	
5.2.3	Traveling Wave Models *120*	
5.2.4	Frequency and Time–Frequency Treatment of Mode Locking: Dynamic Modal Analysis *125*	
5.3	Main Predictions of Generic Mode-Locked Laser Models and their Implication for Quantum Dot Lasers *126*	
5.3.1	Laser Performance Depending on the Operating Point *126*	
5.3.2	Main Parameters that Affect Mode-Locked Laser Behavior *129*	
5.4	Specific Features of Quantum Dot Mode-Locked Lasers in Theory and Modeling *131*	
5.4.1	Delay Differential Equation Model for Quantum Dot Mode-Locked Lasers *132*	
5.4.2	Traveling Wave Modeling of Quantum Dot Mode-Locked Lasers: Effects of Multiple Levels and Inhomogeneous Broadening *141*	
5.4.3	Modal Analysis for QD Mode-Locked Lasers *153*	
5.5	Advantages of Quantum Dot Materials in Mode-Locked Laser Diodes *154*	
5.5.1	Advantages of QD Saturable Absorbers *154*	
5.5.2	Broad Gain Bandwidth *154*	
5.5.3	Low Threshold Current *155*	
5.5.4	Low Temperature Sensitivity *155*	

5.5.5	Suppressed Carrier Diffusion	*156*
5.5.6	Lower Level of Amplified Spontaneous Emission	*157*
5.5.7	Linewidth Enhancement Factor	*157*
5.6	Ultrashort Pulse Generation: Achievements and Strategies	*158*
5.6.1	Monolithic Mode-Locked Quantum Dot Lasers	*158*
5.6.2	Chirp Measurement and Pulse Compression	*161*
5.6.3	Toward Higher Power: Tapered Lasers	*164*
5.6.4	Toward Higher Repetition Rates	*165*
5.6.5	External Cavity QD Mode-Locked Lasers	*166*
5.7	Noise Characteristics of QD Mode-Locked Lasers	*167*
5.7.1	Timing Jitter	*167*
5.7.2	Pulse Repetition Rate Stability and Resilience to Optical Feedback	*170*
5.7.3	Performance Under Optical Injection	*172*
5.8	Performance of QD Mode-Locked Lasers at Elevated Temperature	*174*
5.8.1	Stable Mode Locking at Elevated Temperature	*174*
5.8.2	Pulse Duration Trends at Higher Temperatures	*175*
5.8.3	The Use of p-Doping in QD Mode-Locked Lasers	*176*
5.9	Exploiting Different Transitions for Pulse Generation	*176*
5.9.1	Mode Locking via Ground and Excited States	*176*
5.9.2	The Excited-State Transition as Tool for Novel Mode-Locking Regimes	*179*
5.10	Summary and Outlook	*180*
5.10.1	QD Mode-Locked Laser Diodes: New Functionalities	*180*
5.10.2	Future Directions	*181*
6	**Ultrashort Pulse Solid State Lasers Based on Quantum Dot Saturable Absorbers**	*183*
6.1	A Brief Historical Overview of Ultrashort-Pulse Generation	*183*
6.2	Macroscopic Parameters of Saturable Absorbers	*184*
6.3	QD SESAMs for Efficient Passive Mode Locking of Solid-State Lasers Emitted around 1 μm	*187*
6.4	QD SESAMs for Efficient Passive Mode Locking of Solid-State Lasers Emitted around 1.3 μm	*193*
6.5	QD SESAMs for the Passive Mode Locking of Fiber Lasers	*199*
6.6	Mode-Locked Semiconductor Disk Lasers Incorporating QD SESAMs	*201*
6.7	Optically Pumped Quantum Dot VECSELs	*204*
7	**Saturable Absorbers Based on QD-Doped Glasses**	*207*
7.1	II–VI Semiconductor Nanocrystals in Glass	*207*
7.2	IV–VI Semiconductor QD-Doped Glasses for Ultrashort-Pulse Generation from Solid-State Lasers	*209*
7.3	QD-Doped Glass Saturable Absorbers for Passive Mode Locking around 1.3 μm	*210*

7.4		Cr:YAG Laser Passively Mode Locked with a QD-Doped Glass Saturable Absorber *212*
7.5		PbS QD-Doped Glass Saturable Absorbers for Passive Mode Locking around 1 µm and Their Nonlinear Characteristics *214*
8		**Emerging Applications of Ultrafast Quantum Dot Lasers** *217*
8.1		Optical Communications *217*
8.2		Datacoms *219*
8.3		Biophotonics and Medical Applications *220*
8.4		Outlook *220*

References *223*

Index *241*

Introduction

Over the past three decades, laser physics has advanced dramatically owing to the efforts of scientists and technologists. Refined laser operation in continuous wave (cw) or quasi-cw (i.e., pumped by current pulses longer than the characteristic times of the laser dynamics) regimes has been accompanied by reliable techniques for the generation of periodic sequences of ultrashort optical pulses.

Mode locking is a technique that facilitates the generation of the shortest pulse durations and the highest repetition rates available from ultrafast lasers. The basis of this methodology lies in locking the phases of the longitudinal modes of a laser resonator. Summing all the mode frequencies with a fixed phase relationship results in a periodic sequence of intense ultrashort pulses with a repetition rate related inversely to the cavity round-trip time. The locking of the phases can be achieved either through an applied modulation of the losses or gain in the laser or by exploitation of nonlinearities such as the optical Kerr effect in some crystal-based laser systems. Mode locking is a well-established technique and routine generation of ultrashort pulses is now available from a wide range of lasers, with Kerr-lens mode-locked solid-state lasers (notably titanium-doped sapphire) producing the shortest pulses to date in the near-infrared spectral region.

Access to practical femtosecond lasers and related sources has opened up a range of applications from real-time monitoring of chemical reactions [1] to ultrahigh bit rate optical communications [2, 3], thereby enabling the realization of new concepts in ultrafast optical networking, signal processing, and information transmission [4]. Compact, mass-produced lasers having multigigahertz repetition rates are becoming key components in photonic switching devices, optical interconnects, and clock distribution in integrated circuits [5]; ultrafast electro-optical sampling [6]; and high-speed analogue-to-digital converters [7, 8]. The enormous impact of ultrafast optical sources has already been recognized in the form of two Nobel Prizes awarded to A. Zewail (1999) and T. Hansch (2005) for applications in femtochemistry and laser-based precision spectroscopy, respectively [9, 10].

To increase the applicability of ultrafast optical sources, a number of research groups have been exploring the alternatives to Ti:sapphire and similar lasers, which

Ultrafast Lasers Based on Quantum Dot Structures: Physics and Devices.
Edik U. Rafailov, Maria Ana Cataluna, and Eugene A. Avrutin
Copyright © 2011 WILEY-VCH Verlag GmbH & Co. KGaA, Weinheim
ISBN: 978-3-527-40928-0

tend to be rather bulky and inefficient. Semiconductor lasers represent one of the most exciting options for the generation of femtosecond optical pulses because these offer admirable potential for compactness and integrability as well as excellent prospects for direct electrical control. The mode locking of semiconductor diode lasers (at present, mainly based on quantum well heterostructures) by active and passive modulation of cavity losses is an established technique for the generation of picosecond and high repetition rate optical pulses in the near-infrared spectral range [11, 12]. In many cases, it is possible to incorporate the components required for passive mode locking directly into the device structure, thereby further simplifying the fabrication techniques. Progress in semiconductor laser development has led to the production of ultrashort and high repetition rate pulses from simple system configurations that, although not easily achieving the femtosecond pulse durations that are routinely available from the diode-pumped crystal-based lasers, are showing promise for efficient and controllable operation in ultrafast regimes from electrically pumped devices. Novel structures based on quantum dots have enhanced the characteristics of these lasers further and opened up new possibilities in ultrafast science and technology by enabling the generation and amplification of femtosecond optical pulses directly from laser diode devices, thus providing a cost-efficient pulse source [13].

Furthermore, quantum dot-based materials that demonstrate ultrabroadband absorption and ultrafast carrier dynamics are currently one of the most promising materials for the design of ultrafast saturable absorbers that have been employed in a wide range of laser systems, turning them into more compact, reliable, and inexpensive optical sources.

In this book, we summarize the progress and recent results in the development of efficient and compact ultrafast laser and amplifier devices based on quantum dot materials and structures. To better understand the potential of quantum dot devices, we start with a brief introduction to these materials and highlight the unique characteristics that make these low dimensional materials attractive for ultrafast applications (Chapter 1). In Chapter 2, we lay the foundations of quantum dot theory that are necessary for the following chapters. Chapters 3 and 4 will focus on the physics and state of the art in quantum dot-based optical amplifiers and saturable absorbers that are compatible with ultrashort pulses, respectively. Chapter 5 accounts for the main achievements in the generation of ultrashort pulses directly from quantum dot semiconductor diode lasers. In Chapter 6, an overview of some applications of quantum dot-based saturable absorbers in a number of mode-locked lasers is provided. Besides, a short overview of the recent progress in quantum dot vertical external cavity surface emitting lasers is also given. Chapter 7 will report on the progress of quantum dot in glass saturable absorbers. Finally, in Chapter 8, we conclude with an analysis of emerging future applications of ultrafast quantum dot lasers. The reader is expected to have reasonable knowledge of semiconductor physics and laser principles, with the particulars related to both the ultrafast pulse generation and propagation and the quantum dots presented in the relevant chapters.

Acknowledgments

Particular thanks go to E.V. Viktorov at Université Libre de Bruxelles for carefully reading most parts of the manuscript and making invaluable suggestions, not to mention extremely useful discussions. We are also grateful to I. Montrosset at Politecnico di Torino, A. Vladimirov at WIAS Berlin, and H. Simos at the University of Athens for useful discussions of some of the theoretical aspects of the book and to W. Sibbett and A. Lagatsky at St Andrews University for stimulating discussions on the experimental aspects. We are very grateful to V. Ustinov at Ioffe Institute (St Petersburg) and D. Livshits at Innolume GmbH (Dortmund) for providing us the samples used in our experimental studies and for fruitful discussion of experimental methodologies. Finally, we are grateful to all the authors (and publishers) who gave their consent to reproduce the figures representing their results in this book, and, last but not least, to all the staff at John Wiley & Sons, Inc. (particularly V. Molière and A. Tschörtner) for their helpfulness and patience.

The authors acknowledge the financial support from European Community's Seventh Framework Programme FAST-DOT under grant agreement 224338 and the UK Engineering and Physical Sciences Research Council (EPSRC).

M.A. Cataluna also acknowledges the financial support from the Royal Academy of Engineering/EPSCR Research Fellowship.

1
Semiconductor Quantum Dots for Ultrafast Optoelectronics

1.1
The Role of Dimensionality in Semiconductor Materials

The history of semiconductor lasers has been punctuated by dramatic revolutions. Several proposals of injection semiconductor lasers were studied between the late 1950s and the early 1960s, and the first demonstrations of p–n junction GaAs lasers followed in 1962 [14]. Until the 1980s, only bulk materials were used in semiconductor devices. Originally, these were homostructure devices, functionalized by a doping profile including a p–n junction in the same material. However, these lasers exhibited very low efficiency due to high optical and electrical losses. At that time, pioneers such as Alferov and Kroemer independently proposed a laser construction known as the double heterostructure and involving an active layer of a semiconductor with a relatively narrow bandgap layer surrounded by two injector layers of more broadband material [15]. Such a design offers several crucial advantages. First, it efficiently localizes the charge carriers of both signs (electrons and holes) within the active layer, by creating a potential well for both types of carriers. This potential well also leads to the possibility of achieving carrier densities in the active layer exceeding the doping level of the injector layers (the superinjection effect); thus, low-doped injectors can be used and the optical losses are significantly decreased compared to those in early homostructure lasers, which were by necessity highly doped. Second, as narrowband materials tend to have a higher refractive index, the electronic confinement is accompanied by the *optical* confinement, with an epitaxially determined waveguide formed in the transverse direction (perpendicular to the plane of the epitaxial layers). In the simplest version of a double heterostructure, the active layer doubles as the waveguide core; most structures currently used, including quantum dot (QD) lasers, use a *separate confinement* arrangement whereby the relatively thin active layer is embedded in the broader *optical confinement layer* (made of a broader bandgap material to localize the carriers in the active layer), which, in its turn, is sandwiched between two cladding (p- and n-emitter) layers. These have a still broader bandgap and thus, importantly, a lower refractive index than that of the optical confinement layer to provide the optical confinement/waveguiding.

The enhanced electronic and optical confinement (localization) in the double heterostructure drastically improved the operational characteristics of laser diodes, in particular the threshold current density (current per unit area) J_{th}, which decreased by as much as two orders of magnitude.

Another revolution followed when it was realized that the confinement of electrons to lower dimensional semiconductor structures translates into completely new optoelectronic properties, compared to bulk semiconductors. The obvious can therefore be asked: how small should this confinement be? To answer this question, one should recall the concept of the de Broglie wavelength of thermalized electrons λ_B, as shown in Equation 1.1:

$$\lambda_B = \frac{h}{p} = \frac{h}{\sqrt{2m^* E}}, \tag{1.1}$$

where h is Planck's constant, p is the electron momentum, m^* is the electron effective mass, and E is the energy.

In the case of III–V compound semiconductors (such as the AlGaAs and InGaAsP material systems), λ_B is typically on the order of tens of nanometers for carriers with typical thermal energies [16]. Therefore, if one of the dimensions of a semiconductor is comparable to, or less than λ_B, the electrons will be confined to two dimensions and so the energy–momentum relations will dramatically change because quantization effects start to take place in one of the dimensions. This is the case for charge carriers in a quantum well (QW) structure, which have been intensively studied since the 1970s and are now prevalent in semiconductor lasers and, to a lesser degree, in amplifiers.

The next stage in reducing dimensionality in a semiconductor active medium is the quantum wire (QWR), which is a one-dimensional confined structure in the sense that the carrier movement is free in one dimension and confined in two others. Quantum wires can be fabricated by direct lithographic methods, but possibly the most important one-dimensionally confined structure from the laser device perspective is the quantum dash (QDH) structure considered in more detail below.

Finally, a quantum dot is a nanostructure in which the electron and hole movement is confined in all the three dimensions. Quantum dots are thus tiny clusters of semiconductor material having all three dimensions of only a few nanometers.

The spatial confinement of the carriers in lower dimensional semiconductors leads to dramatically different energy–momentum relations in the directions of confinement, which results in completely new density of states, compared to the bulk case, as depicted in Figure 1.1.

As dimensionality decreases, the density of states is no longer continuous or quasi-continuous but becomes quantized. In the case of quantum dots, the charge carriers occupy only a restricted set of energy levels rather like the electrons in an atom, and for this reason, quantum dots are sometimes referred to as "artificial atoms." It is important to stress, however, that QDs actually contain hundreds of thousands of atoms.

For a given energy range, the number of carriers necessary to fill out these states reduces substantially as the dimensionality decreases, which implies that it becomes

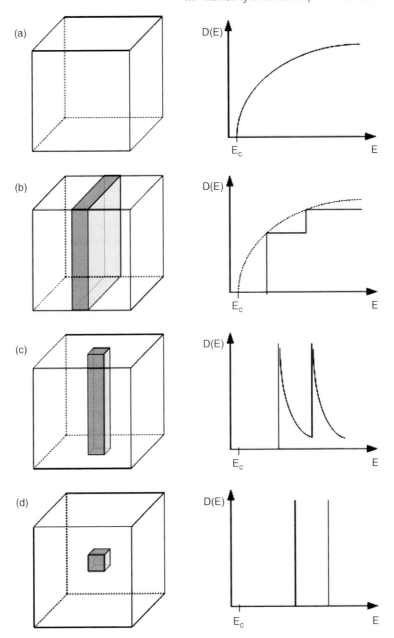

Figure 1.1 (*Right*) Schematic morphology and (*left*) density of states for charge carriers in semiconductor structures with different dimensionalities: (a) bulk, (b) quantum well, (c) quantum wire, and (d) quantum dot.

easier to achieve transparency and, eventually, the inversion of population needed to achieve optical gain and thus laser operation – with the resulting reduction of the threshold current density J_{th} of a laser. In fact, the reduction in J_{th} has been quite spectacular over the years, with sudden jumps whenever the dimensionality is decreased [15]. Low-dimensional lasers also exhibit reduced temperature sensitivity of J_{th}, and in the case of quantum dots, an infinite characteristic temperature has even been predicted.

The implications of a quantized density of states for QDs are immense and were identified as early as 1976 when a QD laser was first proposed [17]. A subsequent theoretical treatment was published in 1982 [18] in which it was predicted that the ideal three-dimensional localization of carriers would translate into a low-threshold, temperature-insensitive, and single-frequency laser. The latter two predictions relied, however, on the assumption of the dots being nearly identical and this having almost the same energy level separation (weak inhomogeneous broadening). While realization of such QD systems may yet become possible in future, all the dot structures grown so far have a significant size (and possibly composition) dispersion, leading to a significant inhomogeneous broadening of the emission line. Thus, some of the initial expectations from QD laser, such as the single-frequency lasing, had to be revised, while others, such as application for ultrafast optoelectronics, became more feasible. This will be discussed in more detail in Chapter 2, which will present a summary of quantum dot theory concentrating on the areas of interest to the subject of short pulse generation and amplification.

1.2
Material Systems Used

A variety of materials have been used in the past in the fabrication of quantum dots. In the following sections, we will describe the main QD material systems that have shown particular promise for ultrafast optoelectronic applications.

1.2.1
III–V Epitaxially Grown Quantum Dots

Since the pioneering work reported in 1985 [19], where the formation of InAs clusters in a GaAs matrix was demonstrated, a number of groups have synthesized and studied self-organized QD structures in a range of distinctive systems [20]. One of such groups of materials is based on III–V QDs, epitaxially grown on a semiconductor substrate. To date, the most promising results have been achieved through the spontaneous formation of three-dimensional islands during strained layer epitaxial growth – a process known as the Stranski–Krastanov mechanism. In this process, when a film is epitaxially grown over a substrate, the initial growth occurs layer by layer, but beyond a certain critical thickness, three-dimensional islands begin to form – the quantum dots. A continuous film, of quantum well thickness and usually assumed to have quantum well properties, lies underneath the dots and is called the wetting layer. A crucial requirement of this technique is that the lattice constant of the deposited material is larger than that of the substrate, so that the additional strain leads to the formation of clusters. This is the case of an InAs film (lattice constant of 6.06 Å) on a GaAs substrate (lattice constant of 5.64 Å) or on an InP substrate (lattice constant of 5.87 Å). Many different systems can be grown using this technique and these are not limited to group III–V constituents.

Despite being an extremely complex process, the Stranski–Krastanov mode is now widely used in the self-assembly of quantum dots. Epitaxial quantum dot materials can be grown using molecular beam epitaxy (MBE) and metal organic chemical vapor deposition (MOCVD). These techniques have been extensively used in the past decades to grow QW materials and therefore the growth of QDs benefited immensely from well-established procedures. This aspect is also advantageous for commercialization, as manufacturers do not need to invest in new epitaxy equipment to fabricate these structures.

Stranski–Krastanov-grown quantum dots typically have a pyramidal shape, with a base of 15–20 nm and a height of the order of ∼5 nm (it is possible that after overgrowth, this shape is modified). At present, the densities of quantum dots lie typically between 10^9 and 10^{12} cm^{-2}. The relatively sparse distribution of quantum dots results in a low value of gain. Thus, the levels of gain and optical confinement provided by a single layer of quantum dots may not be enough for the optimal performance of a laser. To circumvent this problem, quantum dot structures are routinely grown in stacks that allows an increase in the modal gain without increasing the internal optical mode loss [21]. Further optical confinement is enabled through the embedding of QD arrays within layers of higher refractive index and bandgap energy, therefore forming a heterostructure.

Figure 1.2a and b [22] shows a surface view of a system of self-organized QDs and a side view of a stack of three QD layers, respectively, to give an idea of dimensions, distances, and shapes of typical self-organized QDs in a practical device structure. In this case, the dot density in each layer is around 4×10^{10} cm^{-2}, with the lateral size of QDs of 15–18 nm. The apparent dot bases are almost square shaped and oriented along the ⟨100⟩ direction in the crystal, with a slight elongation along one of the ⟨110⟩

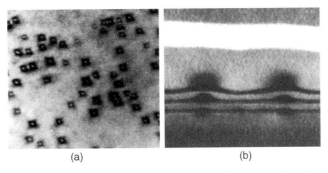

(a)   (b)

Figure 1.2 Scanning electron microscope image of a plan view (a) and side view (b) of a system of three-stacked QDs in a laser structure. Courtesy of D. Livshits, Innolume GmbH.

directions. Note also that the average size of the dots appears to increase from layer to layer, contributing to the inhomogeneous broadening as discussed below.

Materials based on InGaAs/InAs dots grown on a GaAs substrate have been investigated quite comprehensively in the past few years because their emission wavelengths can not only be tuned between 1.0 and 1.3 µm but can also be extended to 1.55 µm [20], through careful adjustment of the growth conditions. InGaAs/InAs QDs grown on InP substrate emit in the 1.4–1.9 µm wavelength range, thus enabling easy access to the optical communications band around 1.55 µm.

Epitaxially grown QD materials have been successfully deployed in the form of lasers, amplifiers, and saturable absorbers – and as such, they hold great promise for a complete optoelectronic integration of an array of distinct devices on the same wafer.

1.2.2
QD-Doped Glasses

QD-doped glasses represent the second key group of quantum dot materials that are of significant interest in ultrafast physics. These materials consist of semiconductor nanoparticles (such as PbS, PbSe, and CdTe) incorporated into a variety of transparent dielectric matrices, with excitonic absorption peaks in the spectral range of about 0.5–2.5 µm [23–25]. One of the main advantages of QD-doped glasses is that they are much cheaper and easier to produce than their epitaxially grown counterparts. QD-doped glasses have not yet realized their full potential as gain media in that efficient laser emission has not been observed to date. Moreover, they are not yet practical within the context of fabricated electrically injected devices. This material system is thus being investigated mainly as an absorber medium. QD-doped glasses have, in fact, been successfully used as ultrafast saturable absorbers and optical switches because they exhibit similar ultrafast properties to III–V QDs in heterostructures.

1.2.3
Quantum Dashes

Quantum dashes are often mentioned alongside quantum dots and are produced in a similar technological process, typically by growing InAs (notionally a few monolayers) on top of thin AlGaInAs/InP layers – such QDHs emit in the 1.55 µm wavelength band. However, the energy structure of quantum dashes is very different from those of quantum dots. Like QDs, QDHs have a vertical dimension of the order of 3–4 nm and a lateral dimension of 10–20 nm. However, unlike a QD, a QDH is elongated in the other lateral dimension, its typical length being on the order of fractions of a micron, which is considerably larger than the electron and hole de Broglie wavelengths. Therefore, two of the three components of the carrier wave vector in a QDH are quantized, but the third one is quasi-continuous, meaning that instead of discrete energy levels, like QDs, a single QDH has quasi-continuous energy bands for both electrons and holes, like bulk and QW materials, although the

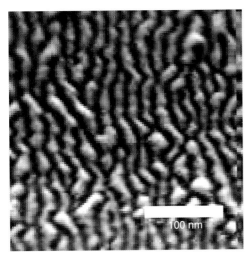

Figure 1.3 Scanning electron microscope image of a plan view of a sample of nonovergrown InAs quantum dashes. Reproduced with permission from Ref. [26].

density of states for electrons and holes has sharp peaks, reminiscent of the delta functions of QDs. In theoretical models, the properties of QDHs are quite well described [26] as similar to those of quantum wire assemblies with fluctuating transverse/lateral dimensions. This approach is not absolutely accurate because, unlike quantum wires, QDHs are self-assembled with a high density (hence there is a possibility of tunneling between them), and their geometry is more irregular than that of ideal quantum wires (an example is shown in Figure 1.3), but it has been shown in several papers that this irregularity does not significantly affect the QDH properties [26].

1.3
Quantum Dots: Distinctive Properties for Ultrafast Devices

1.3.1
Inhomogeneous Broadening

The main motivation behind the idea of a QD laser was to conceive a design for a low-threshold, single-frequency, and temperature-insensitive laser given the quantum nature of the density of states. Interestingly, practical devices exhibit the predicted outstandingly low thresholds [27, 28], but the spectral bandwidths of such lasers are significantly broader than those of conventional quantum well lasers [29]. This characteristic is attributable to *inhomogeneous broadening* with its dependence on the dot size distributions. Indeed, because of their self-organized growth, QDs exhibit a Gaussian distribution of sizes, with a corresponding, nearly Gaussian in the first approximation, distribution of emission frequencies. Also, fluctuations in the

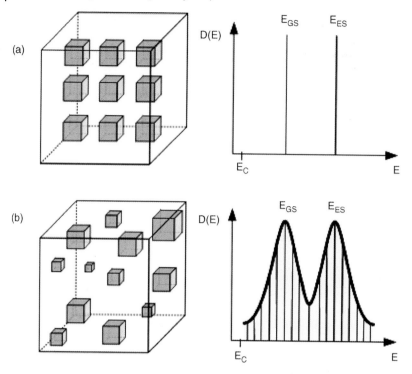

Figure 1.4 (Right) Schematic morphology and (left) density of states for charge carriers in (a) an ideal quantum dot system and (b) a real quantum dot system, where inhomogeneous broadening is illustrated.

elastic strain in different parts of the wafer will affect the level of energy [30]. The effects of inhomogeneous broadening on the density of states are schematically illustrated in Figure 1.4. The recent results from a number of research groups, including our own, indicate that because of this specific characteristic these structures could be designed to offer some advantages in ultrafast science and technology. This is because a very wide bandwidth is available for the generation, propagation, and amplification of ultrashort pulses, which can be tuned across a broad spectral latitude. Moreover, well-established growth technologies for such structures allow this broadening to be controlled and tailored by selecting QD layers with suitable size distributions [31, 32]. However, to date, only a small fraction of these possibilities have been realized, and optical pulses generated by quantum dot sources are one to two orders of magnitude longer than the inverse width of the inhomogeneously broadened gain spectrum. It is also important to stress that a highly inhomogeneously broadened gain also encompasses a number of disadvantages because it partially defeats the purpose of a reduced dimensionality, by broadening the density of states. Indeed, the fluctuation in the size of the QDs increases the transparency current and reduces the modal and differential gain [33, 34].

As for epitaxially grown QD layers, it has been shown that the variation in the dot size in QD-doped glasses leads to a change in the spectral location of the first excitonic absorption peak, giving a possibility of continuous absorption tuning to a substantial spectral extent [35].

1.3.2
Ultrafast Carrier Dynamics

In the initial studies of QD-based materials, it was thought that their charge carrier dynamics would be significantly slower than those of quantum well materials due to a phonon bottleneck effect [36, 37]. This effect was predicted on the basis that due to the discrete energy levels, electrons would not be able to relax via phonon interaction because it would not be possible to match the phonon energy; this expected limitation became known as *the phonon bottleneck*. Interestingly, experiments have demonstrated quite the opposite. As a consequence of access to a number of recombination paths for the carriers, QD structures exhibit ultrafast recovery times, under both absorption and gain conditions [38].

In several such assessments, the absorber dynamics of surface and waveguided QD structures were investigated by using a pump–probe technique (see, for example, Refs [39, 40]). This evaluation showed the existence of at least two distinct time constants for the recovery of the absorption. A fast recovery of around 1 ps is followed by a slower recovery process that extends to 100 ps (Figure 1.5). The fast recovery time is particularly useful for enabling saturable absorbers to mode lock lasers at high

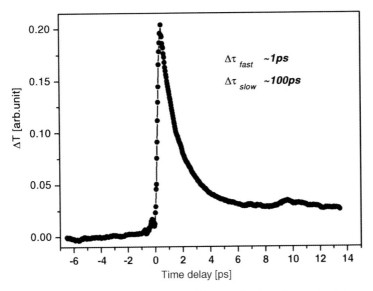

Figure 1.5 Pump–probe measurements of the carrier lifetime of a quantum dot waveguide device. $\Delta\tau_{fast}$ and $\Delta\tau_{slow}$ are fast and slow recovery times, respectively, and ΔT corresponds to the temporal changes in transmission.

repetition rates, where the absorption recovery should occur within the round-trip period of the cavity.

QDs suspended in glasses also exhibit fast carrier dynamics. Phonon bottleneck (see Section 2.3) has not been observed in these materials because of the existence of other relaxation channels for the carriers. These channels can be surface/defect states, electron–hole interaction in some materials [41], or multiphonon emission [42]. Their bleaching relaxation kinetics also exhibits a biexponential character, with fast and slow components, where the fast component decreases with reduction in the QD radius [35]. A review of all these dynamic processes can be found in Ref. [43].

2
Foundations of Quantum Dot Theory

A large number of theoretical papers on quantum dot structures, specifically lasers, have been published in the past decade or so. Here, we shall concentrate on those features and results that are the most relevant for the ultrafast behavior and short-pulse generation. Apologies are extended to colleagues whose work could not be included.

2.1
Energy Structure and Matrix Elements

As mentioned above, Stranski–Krastanov-grown quantum dots typically have a pyramidal shape (at least approximate), with quantization dimensions. Besides the dots, the active layer also contains one or several quantum wells (QWs), in the form of either the wetting layer or the technological QW in a dots-in-a-well (DWELL) system; below, we shall refer to the wells as wetting layer for definiteness.

It is impossible to solve Schroedinger equation for electrons and holes analytically in the three-dimensional pyramidal potential, so energy structure and interlevel transition matrix elements of quantum dots can be determined only numerically.

A typical result of such calculations, which agrees with experimental observations as discussed above, is that radiative transitions contributing to optical properties of epitaxial quantum dots typically involve electrons in at least two energy levels, referred to as the *ground level* (GL), located about 100 meV below the bandgap of the wetting layer, and the *excited level* (EL) or, more accurately, the *first* excited level, as shown in Figure 2.1. According to simulations, the ground state is a singlet state, with a degeneracy (entirely due to spin) of $\varrho_G = 2$. The excited state has a degeneracy, including spin, of $\varrho_E = 4$. A second excited level above the first excited level, with a degeneracy $\varrho_{E2} = 6$, is predicted by some theoretical calculations (see, for example, Ref. [44]); however, its effect on the optical properties of dots is less crucial. In most of the discussions below, we shall refer to just "the excited level," meaning the first, or lowest, excited level. The reason for different degeneracy of levels as usual in quantum mechanics is the different symmetry of the wave functions of the ground and excited levels [45, 46], which also determine the highest overlap between the wave functions of the electrons and holes in the same state.

Ultrafast Lasers Based on Quantum Dot Structures: Physics and Devices.
Edik U. Rafailov, Maria Ana Cataluna, and Eugene A. Avrutin
Copyright © 2011 WILEY-VCH Verlag GmbH & Co. KGaA, Weinheim
ISBN: 978-3-527-40928-0

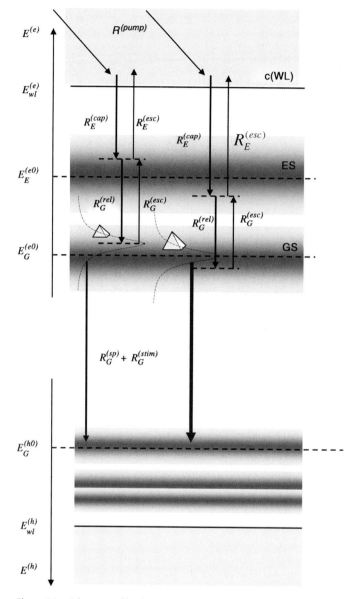

Figure 2.1 Schematic of level structure and the main electron kinetic processes in the most established dynamic model of a QD laser or amplifier. Stimulated transitions are assumed to affect the ground state only, and direct capture from the WL to the ground state is not included.

The energy structure for heavier holes may include a larger number of levels (up to 20 [47]), which is sometimes approximated as a continuum of energies.

Wave functions for electron and hole states allow the matrix elements of the optical transition to be calculated.

In the first approximation, for an infinitely deep potential, optical absorption or emission is possible only for the transition between a certain electron level and the hole level with the same quantum number. In practice, this rule is relaxed due to the finite potential depth and the mixing between heavy and light hole states, so that transitions between ground-level electrons and holes in excited levels are allowed [47]; however, ground- to ground-state transitions are still the most intense; the same may be expected for excited- to excited-state transitions.

A schematic of level structure involved in interpreting the QD dynamics is shown in Figure 2.1. Microscopic calculations of the energy levels and wave functions in the QDs, as well as the transition matrix elements constructed using these wave functions, have been reported in a large number of studies (see earlier monographs on quantum dots for an overview). The general tendency, as can be expected, is that, for dots of the same shape, the levels become deeper with the dot dimension. More precisely, the level position, counted from the dot material band edge, increases in a superlinear fashion as the dot becomes smaller (as in QWs, where the level position counted from the bandgap of the well material is roughly inversely proportional to the well thickness *squared*). An example of electron and hole ground energy levels, calculated for pyramidal InAs/GaAs dots as functions of the pyramid base (replotted after Ref. [33] using exponential fits proposed in that paper), is shown in Figure 2.2.

Microscopically calculated energy levels and transition matrix elements of quantum dots have also been directly used in some simulations on ultrafast dynamics [44]. With a better understanding of QD properties, such calculations may be hoped to gain full predictive capability. For the time being, their practical usefulness is limited by lack of knowledge of the dot shape, strain, and composition, particularly as, in order to take into account inhomogeneous broadening described in the previous

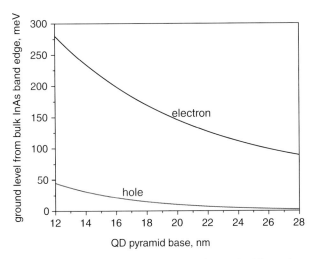

Figure 2.2 Calculated positions of the ground energy level for an electron and a hole in a pyramidal InAs/GaAs QD as functions of the pyramid base, measured from the InAs band edges. Adapted from Ref. [33].

chapter, it is desirable to know exactly the variation in shape and size of dots in the layer. Therefore, in a large number of studies related to practical devices reported to date, theoretical as well as mixed theoretical–experimental, including analysis of the ultrafast dynamics, both the level positions and the matrix elements are not calculated microscopically but used as phenomenological parameters, ideally extracted from experiments (and the matrix element is often taken, in the first approximation, as independent of dot size). Below, we shall assume that they are known, one way or the other.

2.2
Theoretical Approaches to Calculating Absorption and Gain in Quantum Dots

For understanding the ultrafast behavior of quantum dot devices, some model of absorption and gain in such media is required, to include both the dynamics of various populations in the dots and the dependence of gain on these populations. A large number of models have been presented in the literature; most of them share a common approach, but almost all differ in detail; below, we shall concentrate on what we believe is an optimal theoretical formalism for the analysis of ultrafast devices and also mention some other approaches.

In the simplest model, the broadening of transitions is not considered, and the dots may be described by a total volume density $N_D = n_l N_D^{(2d)}/d_{act}$, where n_l is the number of dot layers; d_{act} is the total thickness of the active layer composed of these dot layers, the wetting layers they sit on, and the spacer layers between them; and $N_D^{(2d)}$ is the two-dimensional density of dots per layer that is usually quoted in experimental papers. The total numbers of states (per unit volume) available for electrons on ground and excited states are $\varrho_G N_D$ and $\varrho_E N_D$, respectively, where ϱ_G and ϱ_E are the degeneracy numbers of the corresponding states (the subscript G denoting the ground state and E the excited state). As mentioned above, in self-organized quantum dots, the degeneracy, including spin, is $\varrho_G = 2$ and $\varrho_E = 4$. By assigning each dot in the ground state a resonant absorption cross section σ_G, which is proportional to the squared matrix element of electron–hole transition and thus related to spontaneous recombination time in the dot layer, we can write the simplest estimate for the absorption coefficient (at some spectral point, say, the spectral maximum) of a sample containing a quantum dot density N_D as

$$a_G \approx \sigma_G \varrho_G N_D. \qquad (2.1)$$

The equation is approximate because it ignores the absorption spectrum, the dot size, and shape spread and does not take into account Coulomb interaction between an electron and a hole in the dot. In addition, in the presence of electron–hole pairs in some of the dots, the absorption is modified, as in other materials, to reflect the blocking of the transition from occupied hole states into the full electron states (dynamic Moss–Burstein effect). The electron population of each of the dot levels – ground or excited – can be described by an occupation probability (otherwise called simply the occupancy or distribution function) f_G for the ground level and f_E for the excited level. Then, using "Einstein relations" that state the equality of matrix elements of stimulated emission and absorption, we can write, in the first approximation,

$$a_G \approx \sigma_G(1-2f_G)\varrho_g N_D. \qquad (2.2)$$

The factor 2 reflects the fact that the presence of the electron–hole pair fills both an electron and a hole state and assumes that their occupation function is identical; this is known as the *excitonic approach*. A more accurate theory will be discussed below.

When the bracketed factor in (2.2) becomes negative, *population inversion* is achieved in the sample, and it is customary to introduce a *(material) gain* coefficient of a dot-containing sample $g = -a$. In the simple model above, in the case of a ground-state transition, the gain coefficient is written as a simple linear function of the ground-state occupancy:

$$g_G \approx \sigma_G(2f_G-1)\varrho_g N_D = \sigma_G(2N_G-\varrho_g N_D). \qquad (2.3)$$

Here, $N_G = f_G \varrho_G N_D$ is the number of electrons in the ground level, and the cross section σ_G is referred to, in the context of gain modeling, as the gain cross section (linear gain coefficient) that is proportional to the squared matrix element of electron–hole transition and thus related to spontaneous recombination time in the dot layer. Neglecting the Coulomb interaction between the electron and the hole, the gain cross section in (2.3) is the same as the absorption cross section introduced in (2.1) and (2.2).

An implicit assumption made in Equations 2.2 and 2.3, as mentioned above, is that the dots with electrons on the ground level are neutral, with the occupancy of the electron state equal to that of the hole state. A more accurate model, used by some authors, is one that does not impose dot neutrality (although the whole system including the dots and the wetting layer may be neutral); in this case,

$$g_G = \sigma_G\left(f_G^{(e)} + f_G^{(h)} - 1\right)\varrho_g N_D = \sigma_G\left(N_G^{(e)} + N_G^{(h)} - \varrho_g N_D\right) \qquad (2.4)$$

with different occupancies $f_G^{(e)}$ and $f_G^{(h)}$ introduced for electrons and holes. An expression similar to (2.3) or (2.4) can also be written for the gain due to excited-state transitions.

The simple approach outlined above has proven quite successful in explaining – at least qualitatively – the dynamics of quantum dot lasers and amplifiers, including fast pulse operating regime of an amplifier [48] and mode-locking operation of a laser [49], and appears to be adequate in most cases when explicit account for *spectral hole burning* effects is not required. However, it is recognized that the simple formulas (2.3) and (2.4) are quite crude in describing the properties of realistic QDs. First, strictly speaking, the use of the averaged occupancy is in itself an approximation, which will be discussed in more detail in the discussion of the kinetics of dots. The second, and arguably the main, limitation of formulas (2.3) and (2.4) is that they do not make any explicit provision for inhomogeneous broadening of the transition level due to size (and also strain and shape) dispersion of realistic dots. They can be seen as either neglecting the inhomogeneous broadening or considering it implicitly, with the occupancies $f_G^{(e)}$ and $f_G^{(h)}$ being assumed to be averaged over all dot sizes. The former, however, is not justified for the available self-organized QD devices since the inhomogeneous broadening of electron levels is known to be on the order of 50 meV and thus comparable to the ground-to-excited-state energy separation and to exceed homogeneous broadening by a factor of several times to an order of magnitude. The latter, on the other hand, means that the simple linear dependence (2.4) of the gain on the electron/hole density (or occupancy) is only an approximation that, at the very least, needs to be validated, and in some cases may have to be substituted, by a more rigorous calculation.

Such a calculation, generalizing (2.3) or (2.4) for a more accurate treatment of the spectral properties of a QD laser or amplifier, involves an explicit introduction of inhomogeneous broadening, which leads to an energy distribution of levels. The form of this distribution, strictly speaking, is not trivial. If the random distribution of dots in size (the length of the dot base) is assumed to be the main source of energy distribution, then, due to the strong nonlinear dependence of the electron- and hole-level position on size, a Gaussian distribution in size will result in a *non-Gaussian* distribution in the level energy. If, however, the energy distribution is narrow enough so that the size dependence of the level position can be approximately taken as linear, the energy distribution can be taken in the standard Gaussian form. Thus, for example, for electrons and holes in the ground level, the energy $E_G^{e,h}$ in an individual dot can take a range of values around the average value $E_G^{(e,h)0}$, depending on the dot size, strain, and shape, with the distribution function (probability of an energy level existing per unit energy interval) given by

$$D_{inh}^{(e)}\left(E_G^{(e)}\right) = \frac{1}{\sqrt{\pi}\Delta E_{inh}^{(eG)}} \exp\left[-\left(\frac{E_G^{(e)} - E_G^{(e0)}}{\Delta E_{inh}^{(eG)}}\right)^2\right]. \quad (2.5)$$

A similar expression, with a distribution width $\Delta E_{inh}^{(hG)}$, can be written for ground-level hole energies, and so in terms of the electron–hole energy separation ($E_G^{(e-h)}$) we can also assume

$$D_{inh}\left(E_G^{(e-h)}\right) = \frac{1}{\sqrt{\pi}\Delta E_{inh}^{(G)}} \exp\left[-\left(\frac{E_G^{(e-h)} - E_G^{(e-h,0)}}{\Delta E_{inh}^{(G)}}\right)^2\right], \quad (2.6)$$

where the broadening $\Delta E_{inh}^{(G)}$ is due to both the electron- and the hole-level broadening, and if the level energy distribution is narrow enough for (2.5) to hold for both electron and hole energies, the inhomogeneous broadening of transitions is given by [33]

$$\Delta E_{inh}^{(G)} = \Delta E_{inh}^{(eG)} + \Delta E_{inh}^{(hG)}. \quad (2.7)$$

Note that the contribution from the electron level is likely to be more important than that of heavier holes, as can be deduced from Figure 2.2. The parameter $\Delta E_{inh}^{(G)}$ can be estimated from an experimental luminescence or gain spectrum of the dot layer. Depending on the material, number of dot layers, and growth conditions, values ranging in the interval $\Delta E_{inh}^{(G)} \sim 20$–$60$ meV around the notional ground-state

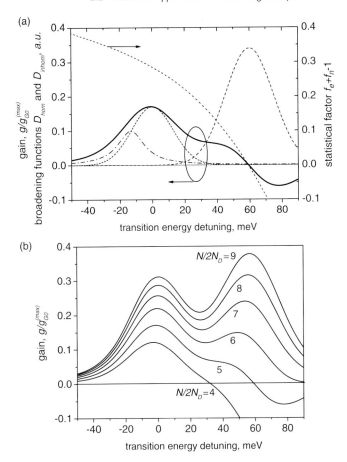

Figure 2.3 (a) A typical calculated gain spectrum (solid line) in a typical quantum dot laser, illustrating the role of homogeneous (dash-dotted line) and inhomogeneous (dashed line) broadening and their typical strengths, as well as the role of carrier statistics (short dashed line). Total (2D) carrier density $N = N^{(e)} = N^{(h)} = 10 N_D$. (b) Evolution of the gain spectrum with an increase in carrier density. Separate quasi-equilibrium statistics ($T = 300$ K) is assumed for electrons and holes (no excitonic approximation). Direct contribution of the wetting layer to the gain/absorption spectrum is not included.

electron- and hole-level separation $E_G^{(e-h,0)}$ have been quoted in the literature, giving the distribution width of ground-state transition energies at half maximum $2\sqrt{\ln 2}\Delta E_{\text{inh}}^{(G)} \sim 30\text{–}90$ meV. Excited-state transitions are believed to be slightly, but not significantly, more broadened. As illustrated in Figure 2.3, the function D_{inh} plays a similar role in determining the QD gain to that of the reduced density of states in the expression for gain in a bulk, QW, or QDH material, although the physical origin of these functions is different. Indeed, the reduced density of states is due to a continuum of energy states available to carriers at the same point in the physical space in the semiconductor, whereas the inhomogeneous broadening D_{inh} relates to a size spread in dots that are located at *different* points in physical space. The applicability limit of the introduction of D_{inh} (and indeed of the notion of material gain) is that a sample of the structure treated in the theory should be "microscopically large," that is, contain a number of dots $\gg 1$, so that the local statistical distribution D_{inh} can be introduced and be physically meaningful. For example, given a dot area density of 10^{10} cm^{-2}, 10 layers of dots in the structure, and a stripe width of 10 μm, a longitudinal slice of a "thickness" of 0.1 μm (the length scale that features, say, in the effects of longitudinal hole burning that may affect the operation of a mode-locked laser, as discussed later, and is likely to be the smallest size of a QD-containing sample needed in a dynamic model) will contain about 10^3 dots. Thus, the introduction of a statistical inhomogeneous broadening function D_{inh} – and of the material gain coefficient G itself – will still be reasonably justified.

The *homogeneous* broadening of a radiative transition within an individual dot is caused by any interaction/collision with either phonons or other charge carriers, either localized inside the dot or surrounding it, which destroy the quantum coherence in an electron–hole pair. In the simplest approximation, these collisions are described by a single characteristic time $T_2 \sim 100$ fs (Markoffian relaxation approximation). Then, a density matrix approach for a two-level system gives a Lorentzian line broadening shape

$$D_{hom}\left(\hbar\omega, E_G^{(e-h)}\right) = \frac{1}{\pi} \frac{\Delta E_{hom}}{\left(\hbar\omega - E_G^{(e-h)}\right)^2 + \Delta E_{hom}^2}. \tag{2.8}$$

Here, $\hbar\omega$ is the emitted photon energy, $E_G^{(e-h)}$ is the energy separation between the electron (in this case on the ground level) and the hole state, and the homogeneous broadening width is $\Delta E_{hom} = \hbar/T_2$. The value of T_2 or $\Delta E_{hom} = \hbar/T_2$ is less easy to estimate from experiments than homogeneous broadening, but there is no reason to believe it to be fundamentally different from the value used for bulk and QW structures. In the literature, a figure of 1–10 meV, in other words, several times to one order of magnitude smaller than the inhomogeneous broadening, is usually taken. Therefore, a generic model for spectrally dependent gain should, in principle, include both homogeneous and inhomogeneous broadening:

$$g_G = A_G \varrho_G N_D \int \left(f_G^{(e)}\left(E_G^{(e)}\right) + f_G^{(h)}\left(E_G^{(h)}\right) - 1\right) D_{hom}\left(\hbar\omega, E_G^{(e-h)}\right) D_{inh}\left(E_G^{(e-h)}\right) dE_G^{(e-h)}. \tag{2.9}$$

Here, the homogeneous and inhomogeneous broadening functions are given by the expressions above. $E_G^{(e)}$ and $E_G^{(h)}$ are the energies of the electron and hole states involved in the transition, respectively, $\Delta E_G^{(e-h)}$ is the energy separation between them, and the constant A_G contains the transition matrix element and has the same physical meaning (though a different dimensionality) as the constant σ_G used in (2.3). In terms of the dipole matrix element μ of the optical transition, $\sigma_G = \mu^2 \omega T_2 / \hbar c \eta \varepsilon_0$ [50], where η is the refractive index and ε_0 is the dielectric permittivity of vacuum; the relation between A_G and the spontaneous recombination time will be discussed below.

An approximate form of (2.9) is obtained assuming $\Delta E_{\text{hom}} \ll \Delta E_{\text{inh}}$, which is not a particularly accurate approximation given the fact that ΔE_{hom} is in reality only a few times smaller than ΔE_{inh} but leads to a useful simple estimate. In this case, the integral in (2.9) disappears and we obtain

$$g_G = g_{G0} = A_G \varrho_G N_D \left(f_G^{(e)}\left(E_G^{(e)}\right) + f_G^{(h)}\left(E_G^{(h)}\right) - 1 \right) D_{\text{inh}}\left(E_G^{(e)}\right), \tag{2.10}$$

where the electron and hole energies are determined from the condition that the separation energy between them is exactly resonant with the amplified photon: $\Delta E_G^{(e-h)} = \hbar\omega$. Thus, the highest gain theoretically possible in ground-state transitions is achieved at the peak of the inhomogeneous broadening function (notional electron–hole energy level separation), with *full inversion* ($f_G^{(e)} = f_G^{(h)} = 1$) and weak homogeneous broadening, and equals

$$g_{G0}^{(\max)} \approx \frac{A_G \varrho_G N_D}{\sqrt{\pi} \Delta E_{\text{inh}}^{(G)}}. \tag{2.11}$$

In more accurate simulations, the peak gain value is always smaller as full inversion is not achieved and the homogeneous broadening is not negligible. Expression (2.11), incidentally, identifies the relation between the gain constants in the energy-resolved and simplified gain models in the case of $\Delta E_{\text{hom}} \ll \Delta E_{\text{inh}}$:

$$A_G \approx \sqrt{\pi} \Delta E_{\text{inh}}^{(G)} \sigma_G = \frac{\sqrt{\pi} \Delta E_{\text{inh}}^{(G)} \mu^2 \omega}{c \eta \varepsilon_0 \Delta E_{\text{hom}}^{(G)}}. \tag{2.12}$$

In order to apply either the simplified formula (2.3) or (2.4) or the energy-resolved formula (2.9) or (2.10) in practice, several more questions need to be addressed. In particular, in the energy-resolved formula, it is not immediately obvious how, given the electron–hole-level separation $E_G^{(e-h)}$ that we use as the free parameter of integration in (2.9), to find how it is shared between the electron and hole levels $E_G^{(e)}$ and $E_G^{(h)}$ for a given dot size. In bulk and quantum well materials, where both electrons and holes have continuous bands of energy, this is done from energy and momentum conservation considerations (*k*-selection rule), which also result in the notion of a joint density of states. In QDs, where the energy levels are discrete, the electron and hole energies can, in principle, be calculated from first principles by solving the Schroedinger equation for the electron and the hole in a dot of each size and shape involved; it can be said that the *k*-selection rule of QWs is substituted by the *size selection rule* in QDs [51]. In some recent papers, for example, in Ref. [44], indeed, the inhomogeneous broadening was taken into account by solving the Schroedinger equation for representative dots of a number of sizes and then self-consistently calculating not only the level energies but also the matrix elements for all allowed transitions, as well as the characteristic times of dot kinetics (see Chapter 3). In a simplified model, relying on $E_G^{(e-h)}$ extracted from the experiment rather than calculated rigorously, some form of a simpler heuristic procedure has to be adopted. One possibility is to assume that the energy spread is mainly due to electrons, whereas the heavier holes all have the energy approximately equal to the average value

$E_G^{(h)} = E_G^{(h0)}$. A more accurate way [52, 53] is to note that in quantum mechanical analysis of a potential well of any size and shape, so long as it is sufficiently deep, the energy level position, counted from the bottom of the well, is to a good accuracy inversely proportional to the mass of the particle. Then, given the effective masses of the electron and hole in the dot material, it is natural to take

$$E_G^{(h)} - E_G^{(h0)} = \frac{m_e^*}{m_h^*} E_G^{(e)} - E_G^{(e0)}, \qquad (2.13)$$

which appears to be in reasonable agreement with the full quantum mechanical calculations [33]. For the purpose of determining the energies involved in transitions, this is sufficient, as the energy separation $\Delta E_G^{(e-h,0)}$ between the notional positions of the electron and hole levels can be estimated from experiment. For determining the kinetics of particle population, the absolute position of the notional levels $E_G^{(e0)}$ and $E_G^{(h0)}$ with respect to, say, the wetting layer conduction and valence bands is necessary. These may be evaluated using different procedures. One may, for example, assume that the positions of the notional electron–hole levels themselves satisfy an equation similar to (2.13) – if the energies are measured from the effective bulk band edges of the dot material (Figure 2.1), this means simply $E_G^{(h0)} = (m_e^*/m_h^*) E_G^{(e0)}$, and so

$$E_G^{(h0)} = \frac{m_e^*}{m_h^* + m_e^*} (\Delta E_G^{(e-h,0)} - E_g^{(D)}), \quad E_G^{(h0)} = \frac{m_h^*}{m_h^* + m_e^*} (\Delta E_G^{(e-h,0)} - E_g^{(D)}), \qquad (2.14)$$

where $E_g^{(D)}$ is the "bulk" bandgap of the QD material that is assumed known along with $\Delta E_G^{(e-h,0)}$, and the energies are counted (upward for electrons and downward for holes) from the "bulk" band edges of the QD material. Similar relations may be taken for excited levels, and finally the band edges of the dot material and the wetting layer are aligned given the measurable wetting layer bandgap, the WL thickness and composition, and the relative bandgap difference between the conduction and valence bands in the material system used (say, 0.65 conduction band and 0.35 valence band in GaAs-based devices). An alternative procedure has been proposed by Berg and Mork [53], involving fitting the densities of states of both electrons and holes with an effective bulk density of states.

A formula identical to (2.9) may be written for excited-state transitions, with the superscript G substituted by E to characterize excited-state parameters.

It has to be mentioned, however, that evaluating the gain spectra, even for photon energies that fall in the ground-state band, using (2.9) has some limitations that may be of some importance for modeling fast optical processes.

First, expression (2.9) describes only the contribution to the gain made by transitions between electrons and holes that are both in the ground state (grown to ground-state transitions). In reality, a certain contribution from the *excited- to excited-state*(s) transitions will be present in the ground-state gain, and vice versa, due to the tails of the inhomogeneous broadening distributions of both levels, and also to the relatively slow decay of the Lorentzian homogeneous broadening function with frequency detuning. Moreover, as shown in Ref. [47], for the gain in the excited-state band, a contribution of ground- to excited-state transitions may be of some importance. Thus, a more accurate, and more general, formula for the gain may need to be expressed as

$$g(\hbar\omega) = \sum_{\mu,\nu=G,E,...} \varrho_\nu N_D \int A_{\mu\nu}\left(E_\nu^{(e-h)}\right)\left(f_\nu^{(e)}(E_G^{(e)}) + f_\nu^{(h)}(E_\nu^{(h)}) - 1\right)$$
$$D_{\text{hom}}\left(\hbar\omega, E_\nu^{(e-h)}\right) D_{\text{inh}}\left(E_\nu^{(e-h)}\right) dE_\nu^{(e-h)}. \tag{2.15}$$

Here, the gain is not explicitly associated with either ground- or excited-level transitions (though of course for photon energies resonant with the center of either ground or excited transition band, the contributions from the corresponding levels will dominate); the constant $A_{\mu\nu}$ includes the matrix element of transition between holes in the state μ and the electrons in the state ν with the summation over all the relevant states (multiple excited states may be taken into account, particularly for holes).

The second limitation is due to the fact that with the simple Lorentzian expression (2.8) for the homogeneous broadening function, formulae (2.9) or (2.15) may not be accurate in the area of small gain; in particular, they do not always predict correctly the transparency point in the gain spectrum, which from the thermodynamic considerations should correspond to the energy separation between electron and hole Fermi quasi-levels. This computational artifact is well known for QW gain calculations and is also present in the calculated QD gain. The accuracy of the transparency point can be improved, in frequency domain calculations, for example, using a more complex non-Lorentzian homogeneous broadening function representing non-Markoffian relaxation of polarization. However, while the Lorentzian expression (2.8) can be relatively easily implemented in time domain, which may be needed in simulations of ultrafast processes (see Sections 2.4 and 5.4), the known more complex versions of $D_{\text{hom}}(\hbar\omega, E_\nu^{(e-h)})$ cannot.

Neither of the above limitations is too critical, however, if the calculations are restricted to the vicinity of the center of the gain and/or excited-state transition band, which is the most important optical frequency (wavelength) for most amplifier and laser operating regimes.

Figure 2.3a shows a typical gain spectrum calculated using (2.15) and also illustrates the roles of homogeneous and inhomogeneous broadening and carrier statistics in determining the gain spectrum. Only ground-to-ground- and excited-to-excited-state transitions have been taken into account, and the matrix elements of the two transitions have been assumed identical and independent of transition energy. The shapes of homogeneous and inhomogeneous broadening are shown as dashed and dash-dotted lines. The occupancies of electron and hole levels have been assumed to satisfy separate Fermi distributions for electrons and heavier holes, and the statistical factor featuring in the gain is also plotted for illustration. The total numbers of electrons and holes (summed over the dots and QW layers) have been assumed to be equal. The gain is normalized to $g_G^{(\text{max})}$ as introduced in (2.11); note though that the homogeneous broadening energy value $\Delta E_{\text{hom}}^{(G)} = 10$ meV used in the calculations is a sizeable fraction of both the temperature $k_B T = 26$ meV and the inhomogeneous broadening $\Delta E_{\text{inh}}^{(G)} = 18$ meV, making (2.10) and (2.11) rather crude estimates. In fact, it will be seen below that the calculated ground-state gain saturates at a value significantly (by a factor of about 0.7) smaller than $g_G^{(\max)}$. As can be expected, with the difference between the ground- and the excited-state band (taken as 60 meV) exceeding $\Delta E_{\text{inh}}^{(G)} = 18$ meV and $\Delta E_{\text{inh}}^{(E)} = 21$ meV, respectively, the gain in the ground- and excited-state bands is virtually entirely formed by the corresponding transitions. Contributions from both states are significant simultaneously only in the intermediate spectral area, where the gain is relatively weak. Thus, as the current/total carrier density is increased, the ground- and excited-state transition bands form two distinct peaks in the gain spectrum, with the excited-state peak overtaking the ground-state peak at a certain pumping level (Figure 2.3b). At this point, the photon energy or wavelength corresponding to the global gain maximum sees an abrupt jump from the ground- to the excited-state value. Thus, in a laser construction where the threshold level can be varied externally (e.g., by introducing controlled absorption in the laser cavity), an abrupt switching from the ground- to excited-state lasing will be seen at such conditions. This is what is indeed observed in a GaAs-based QD mode-locked laser operating near 1.2 µm, as will be discussed later (Chapter 5). By contrast, in a QD system where $\Delta E_{\text{inh}}^{(G,E)}$ are comparable to the interlevel separation, the ground- and excited-state bands merge to form a single, gradually evolving broad band; as the pumping is increased, the position of the global gain maximum moves smoothly, rather than abruptly, from the GS to the ES band through all the intermediate values. This tends to be the case for QD lasers on InP substrates, operating at 1.55 µm [54].

2.3
Kinetic Theory of Quantum Dots

In order to calculate the gain in quantum dots, the occupancies of the electron and hole states in the levels involved in transitions need to be known; in the description of dynamic behavior of a QD device, the time evolution of these populations on various timescales needs to be analyzed.

The approach for the *electron* dynamics in QDs, which is largely believed to be slower than the hole dynamics and thus the main process limiting the device behavior, is reasonably well established. In the general form, the dynamic model includes equations for the electron populations of the wetting layer, the excited level, and the ground level of the dots – the latter being the one involved in optical transitions in most devices of practical interest. In terms of level occupancies $f_G^{(e)}(E_G)$, $f_E^{(e)}(E_E)$, and the (three-dimensional) carrier density in the wetting layer N_{wl}, the dynamic equations are expressed as

$$\frac{df_G^{(e)}(E_G)}{dt} = R_G^{(\text{rel})}(E_G) - R_G^{(\text{esc})}(E_G) - R_G^{(\text{spont})}(E_G) - R_G^{(\text{stim})}(E_G), \qquad (2.16)$$

2.3 Kinetic Theory of Quantum Dots

$$\frac{df_E^{(e)}(E_E)}{dt} = R_E^{(cap)}(E_E) - R_E^{(esc)}(E_E) - \frac{\varrho_G}{\varrho_E}\left(R_G^{(rel)}(E_G) - R_G^{(esc)}(E_G)\right) - R_E^{(spont)}(E_G),$$

(2.17)

$$\frac{dN_{wl}}{dt} = R^{(pump)} - \frac{\varrho_E V_D N_D}{V_{wl}} \int dE_E D_{inh}^{(E)}(E_E)\left(R_E^{(cap)}(E_E) - R_E^{(esc)}(E_E)\right) - R_{wl}^{(spont)}.$$

(2.18)

In Equations 2.16 and 2.17, the electron level energies E_G and E_E (omitting the superscript "e" for brevity) are assumed to belong to the same dot and so are related to each other, for example, via (2.13) and (2.14).

The processes described by the dynamic equations (2.16)–(2.18) and shown schematically in Figure 2.1 are as follows.

Carriers are pumped into the wetting layer by the current flow, as in any semiconductor laser:

$$R^{(pump)} = \frac{\eta I}{e V_{wl}},$$

(2.19)

where $I = JWL$ (W and L being the width and length of the sample, respectively, assuming a rectangular shape as in waveguide devices) is the current, η is the internal quantum efficiency, and V_{wl} is the wetting layer volume.

From the wetting layer, electrons are captured into the dots. The process is generally believed to consist of two stages. First, the electrons are captured from the wetting layer into the excited level in the dot. An opposite process of escape from the dot into the wetting layer is also open, although it requires thermal activation as discussed below. For a group of dots with a certain excited-state energy E_E, the rates of the capture and escape processes are written as

$$R_E^{(cap)}(E_E) = \frac{N_{wl} V_{wl}}{\varrho_E V_D N_D}\frac{1-f_E(E_E)}{\tau_{cap}^{(E)}}, \quad R_E^{(esc)}(E_E) = \frac{f_E(E_E)}{\tau_{esc}^{(E)}}f'_{wl}.$$

(2.20)

Here, V_D is the effective volume occupied by dots and introduced for determining the 3D density N_D. We chose it to equal the volume of the entire active layer including the dots, the wetting layers, and the spacer layers between them: $V_D = d_{act}WL$, d_{act} being the thickness of the active layer as introduced above. The alternative definition used by some authors would be the total volume of the actual dots – so far as the expressions (2.18) and (2.20) are concerned, this is a matter of definition as it is the total number of dots $V_D N_D$ that features in the formulas. $\tau_{cap}^{(E)}$ and $\tau_{esc}^{(E)}$ are the characteristic times of capture and escape processes, respectively, and f'_{wl} is the characteristic function determining the probability of finding an empty state in the WL into which the escape is possible; it will be discussed in more detail below.

From the excited state, the carriers then relax into the ground state with a fraction of them, again, escaping in the opposite direction in a thermally activated process; the rates of the corresponding processes are

$$R_G^{(rel)}(E_G) = \frac{\varrho_E}{\varrho_G}\frac{f_E(E_E)(1-f_G(E_G))}{\tau_{rel}^{(G)}}, \quad R_G^{(esc)}(E_G) = \frac{f_G(E_G)(1-f_E(E_E))}{\tau_{esc}^{(G)}},$$

(2.21)

where again phenomenological characteristic times of relaxation $\tau_{\text{rel}}^{(G)}$ and escape $\tau_{\text{esc}}^{(G)}$ have been introduced. The physical nature of capture and escape processes and their effect on the associated time constants are discussed in more detail below.

Note that if we formally introduce the effective distribution function and effective degeneracy for the wetting layer as

$$f_{\text{wl}} = \frac{N_{\text{wl}}}{N_c^{(\text{wl})}}, \tag{2.22a}$$

$$\varrho_{\text{wl}} = \frac{N_c^{(\text{wl})} V_{\text{wl}}}{V_D N_D}, \tag{2.22b}$$

where

$$N_c^{(\text{wl})} = n_l \frac{m_e k_B T}{d_{\text{wl}} \pi \hbar^2} \tag{2.23}$$

is the three-dimensional effective density of states in the QW wetting layer (n_l being the number of layers in the stacked structure and d_{wl} the *total* thickness of the wetting layers associated with all the stacked dot layers – the value used in the definition of the volume V_{wl}), then (2.20) and (2.21) can be written in a symmetric form as

$$R_E^{(\text{cap})}(E_E) = \frac{\varrho_{\text{wl}} f_{\text{wl}} f'_E(E_E)}{\varrho_E} \frac{1}{\tau_{\text{cap}}^{(E)}}, \quad R_E^{(\text{esc})}(E_E) = \frac{f_E(E_E) f'_{\text{wl}}}{\tau_{\text{esc}}^{(E)}},$$

$$R_G^{(\text{rel})}(E_G) = \frac{\varrho_E f_E(E_E) f'_G(E_G)}{\varrho_G} \frac{1}{\tau_{\text{rel}}^{(G)}}, \quad R_G^{(\text{esc})}(E_G) = \frac{f_G(E_G) f'_E(E_E)}{\tau_{\text{esc}}^{(G)}}, \tag{2.24}$$

where the probabilities of finding an empty ground or excited level state, the counterparts of f'_{wl} in (2.20), are $f'_G = 1 - f_G(E_G)$ and $f'_E = 1 - f_E(E_E)$. Note that $f_{\text{wl}} = N_{\text{wl}}/N_c^{(\text{wl})}$ is only an *effective* distribution function; unlike a genuine distribution function, it can take values >1 at large enough N_{wl}.

The expressions of the type of (2.20) and (2.21) (or (2.24)) are believed to describe well the dynamics of carriers in dots grown on GaAs substrates and operating at the wavelength range of 1.2–1.3 μm. In 1.55 μm operating dots grown on InP substrate, possibly because of the closer spacing between ground and excited levels, *direct* capture of carriers from the wetting layer into the ground level, alongside the two-stage process described by (2.20) and (2.21), is believed to be efficient [54]. In this case, the downward relaxation and upward escape rate in (2.21) need to be substituted by

$$R_G^{(\text{rel,total})}(E_G) = R_G^{(\text{rel})}(E_G) + R_G^{(\text{cap,direct})}(E_G), \quad R_G^{(\text{cap,direct})}(E_G) = \frac{\varrho_{\text{wl}} f_{\text{wl}}(1 - f_E(E_E))}{\varrho_G} \frac{1}{\tau_{\text{cap}}^{(G,\text{direct})}},$$

$$R_G^{(\text{esc,total})}(E_G) = R_G^{(\text{esc})}(E_G) + R_G^{(\text{esc,direct})}(E_G), \quad R_G^{(\text{esc,direct})}(E_G) = \frac{f_E(E_E)}{\tau_{\text{esc}}^{(G,\text{direct})}} f'_{\text{wl}}.$$

$$\tag{2.25}$$

Equation 2.18 for the wetting layer carrier density needs to be modified accordingly, to include direct interchange (capture and escape) between the wetting layer and the ground state:

$$\frac{dN_{wl}}{dt} = R^{(pump)} - \frac{\varrho_E V_D N_D}{V_{wl}} \int dE_E D_{inh}^{(E)}(E_E)\left(R_E^{(cap)}(E_E) - R_E^{(esc)}(E_E)\right)$$
$$- \frac{\varrho_G V_D N_D}{V_{wl}} \int dE_G D_{inh}^{(G)}(E_G)\left(R_G^{(cap,direct)}(E_E) - R_G^{(esc,direct)}(E_G)\right) - R_{wl}^{(spont)}.$$

(2.26)

At the time of writing, dots on GaAs substrates are more widely used in ultrafast applications.

The carriers recombine through spontaneous and stimulated channels. It is assumed in the above model that the optical signal generated, amplified, or absorbed in the device is resonant with the ground-state transitions as is usually the intention when designing a QD optical device. Therefore, the net stimulated recombination rate $R_{stim}^{net}(E_G) = R_{stim}(E_G) - R_{stim}^{abs}(E_G)$, taking into account stimulated emission and absorption of light, is included only for ground-state dots, whereas spontaneous recombination terms are present for all the dot states and the WL. Under high pumping conditions, stimulated transitions involving the excited level and possibly the wetting layer may become important, in which case the model can be modified to include stimulated emission at higher levels. The exact form of the stimulated recombination term and its effects on the kinetics will be considered in the next chapter; for the time being, we shall concentrate on the situation when it is not significant (e.g., a laser amplifier under small-signal conditions or a laser below threshold). The spontaneous recombination rate is expressed in the simplest form as

$$R_{G,E}^{(spont)}(E_{G,E}) = \frac{f_{G,E}(E_{G,E})}{\tau_{sp}^{(G,E)}}.$$

(2.27)

In the first approximation, the spontaneous recombination time $\tau_{sp}^{(G,E,wl)} = \tau_{sp} \sim 1$ ns, by the order of magnitude, is constant and can be assumed approximately the same for all states. Due to Einstein relations between the spontaneous and stimulated emission probabilities, this time is related to the squared matrix element of the transition and hence to the gain coefficients σ_G and A_G in (2.3), (2.4), and (2.9); specifically [55]

$$A_G = \frac{\hbar \lambda^2}{4\eta^2} \frac{1}{\tau_{sp}}$$

(2.28)

with η the refractive index and λ the wavelength. A more accurate form of (2.27), involving the bimolecular nature of spontaneous recombination, is $R_{G,E}^{(spont)}(E_{G,E}^{(e)}) = f_{G,E}^{(e)}(E_{G,E}^{(e)}) \times f_{G,E}^{(h)}(E_{G,E}^{(h)})/\tau_{G,E}^{(spont)}$, where $E_{G,E}^{(h)}$ is the hole energy level in the dot of the size that has an electron energy level $E_{G,E}^{(e)}$.

In the quantum well wetting layer, bimolecular recombination $R_{\text{wl}}^{(\text{spont})} = B N_{\text{wl}}^2$ with a bimolecular recombination coefficient B is assumed most often as normal for QW materials. The value of the coefficient B is on the order of $10^{10}\,\text{cm}^3\,\text{s}^{-1}$. More accurately, the nondegenerate statistics at high carrier densities lead to a saturated law, heuristically expressed as

$$R_{\text{wl}}^{(\text{spont})} = \frac{B N_{\text{wl}}^2}{1 + B_2 N_{\text{wl}}}, \qquad (2.29)$$

so that at high carrier densities, the WL may be assigned a constant spontaneous recombination time. This is often done in simple models, as discussed in Chapters 3 and 5.

Since there is no direct exchange of populations between different groups of dots but only indirect exchange through escape and recapture, Equations 2.16–2.18 can be used either in a model not considering the inhomogeneous broadening, together with Equation 2.3 or 2.4 for gain, or in the spectrally resolved model, in conjunction with Equation 2.9. In the former case, the size distribution in (2.18) is effectively $D_{\text{inh}}^{(E)}(EE_{(e)}) = \delta(E_E - E_E^{(e)})$ and so the integration in the capture and escape term disappears, giving this term as just

$$-\frac{\varrho_E V_D N_D}{V_{\text{wl}}} \left(R_E^{(\text{cap})}\left(E_E^{(e)}\right) - R_E^{(\text{esc})}\left(E_E^{(e)}\right) \right). \qquad (2.30)$$

The capture, relaxation, and escape times introduced in (2.20) and (2.21) and the physical processes they describe are very important in determining the QD device performance. In the early days of QD research, it was sometimes assumed that the main process leading to capture of carriers from the WL into the dots and their subsequent relaxation into the ground energy state would be phonon emission. This led to pessimistic expectations of relatively slow capture, with characteristic times on the order of units to tens of picoseconds, as opposed to subpicosecond capture in QW laser devices, hampering the device performance (the so-called *phonon bottleneck*). The modern understanding is that there are *two* major processes that contribute to the capture and relaxation. The first is indeed the phonon emission, which dominates at low pumping levels. The second process has been described as the Auger-type process [48], or nonlocal Coulomb interactions, whereby the energy released by the captured or relaxing electron is not emitted as a phonon but transferred either to an electron or to a hole in the wetting layer near the dot. Naturally, the rate of this process may be expected to speed up with an increase in the wetting layer carrier density N_{wl}. A heuristic expression first proposed by Berg et al. [48] and widely used in the literature is

$$\frac{1}{\tau_{\text{rel}}^{(G)}} = A_{\text{rel}}^{(G)} + C_{\text{rel}}^{(G)} N_{\text{wl}}, \qquad \frac{1}{\tau_E^{(\text{cap})}} = A_E^{(\text{cap})} + C_E^{(\text{cap})} N_{\text{wl}}. \qquad (2.31)$$

The parameters $A_E^{(\text{cap})}$ and $A_{\text{rel}}^{(G)}$ are characteristic rates of phonon-assisted capture and interlevel relaxation processes, respectively, and the coefficients $C_{\text{rel}}^{(G)}$ and $C_E^{(\text{cap})}$ describe the Auger-assisted contributions to the corresponding processes. The values of these coefficients at room temperature have been determined by Berg et al. [48] by fitting calculated pump–probe response of a InGaAs/GaAs QD SOA to the experiments, as $1/A_E^{(cap)} = 1$ ps, $1/A_{rel}^{(G)} = 10$ ps, $C_E^{(cap)} = 10^{-14}$ m^3 s^{-1}, and $C_{rel}^{(G)} = 7 \times 10^{-12}$ m^3 s^{-1}. Thus, in QD components with a low N_{wl} ($N_{wl} < \sim 10^{11}$ m^{-3}), both capture and relaxation processes are predominantly phonon assisted, with relaxation process being the slower of the two. However, as the pumping of a QD laser or amplifier is increased, the wetting layer density increases, first slowly, then in a more pronounced way. The reason for that is that initially, as the pumping is increased from zero, most of the carriers get captured into the QDs. However, the major difference between the QDs and other semiconductor media is that the number of electron states in a QD active layer is limited: due to the Pauli exclusion principle, only $\varrho_G V_D N_D$ states are available to electrons in the ground state and $\varrho_E V_D N_D$ states in the excited state. Therefore, as first the ground state and then the excited state are filled with electrons, the capture and relaxation into these states saturate, which is represented by the factor $(1-f_E(E_E))$ in expression (2.20) for $R_E^{(cap)}(E_E)$ and the factor $(1-f_G(E_G))$ in expression (2.21) for $R_G^{(rel)}(E_G)$. Thus, the drain of carriers from the wetting layer into the dots (Figure 2.1) slows down, causing the buildup of a *reservoir* of wetting layer carriers. This is illustrated in Figure 2.4, calculated by assuming near-equilibrium distribution of carriers – this is to a good accuracy equivalent to solving the equations above in steady state and with no stimulated recombination term. The carriers in the reservoir increase the Auger-assisted capture speed as indicated by Equation 2.31, so that at $N_{wl} \sim 10^{24}$ m^{-3}, the capture into the excited state as predicted by (2.31) with the parameters suggested in Ref. [48] is likely to be still mainly phonon assisted, with a characteristic time $\tau_E^{(cap)} \approx 1/A_E^{(cap)} = 1$ ps, but the excited- to ground-state relaxation is mainly Auger

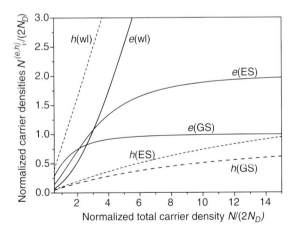

Figure 2.4 Steady-state total carrier density dependence of different populations in a QD laser amplifier (no optical signal present), illustrating the saturation of QD-bound level population and the emergence of WL carrier reservoir. Note that the hole-state populations are relevant only for a model without the excitonic approximation made.

assisted, characterized with a time of $\tau_{\text{rel}}^{(G)} \approx (C_{\text{rel}}^{(G)} N_{\text{wl}})^{-1} \sim 100$ fs. The longer of the two times involved in the downward relaxation is thus phonon assisted; this is consistent with the results of a recent paper [56] in which it was shown that the experimentally observed dominant process in the fast relaxation of a QD amplifier is better described by a constant relaxation time than by the one inversely proportional to the carrier density (excited-state dynamics was not explicitly considered in Ref. [56], so the capture and relaxation times studied could be interpreted as those of equivalent processes including both capture of carriers into the dots and subsequent relaxation to the ground state; see discussion below and Equation 2.41).

The carrier escape, being a thermally activated process, can be neglected at very low temperatures, resulting in only the capture terms being present; this is known as the *trickle-down approximation* and leads to a strongly nonequilibrium distribution of carriers. At room temperatures, however, the escape is very important and needs to be taken care of in the model.

In a semiphenomenological approach, the characteristic escape times can be evaluated, given the capture and relaxation times, from the principle of detailed (quasi)-equilibrium. This states that if all other processes are much slower than capture, relaxation, and escape, then the electron system should evolve to a Fermi distribution in energy, with a common Fermi quasi-level for both dot energy states and the wetting layer. For the ground- to excited-level relaxation and escape, this means that in the case of $f_E(E_E)$ and $f_G(E_G)$ satisfying a Fermi distribution $f_{E,G}(E_{E,G}) = f_F(E_{E,G}) = 1/[\exp((E_{E,G} - E_F^{(e)})/k_B T) + 1]$ (where k_B is the Boltzmann constant and T is the absolute temperature) with the WL electron Fermi quasi-level $E_F^{(e)}$ also applying to the excited and ground states in the dots, we should have $R_G^{(\text{rel})}(E_G) - R_G^{(\text{esc})}(E_G) = 0$. To ensure this, the capture and escape times must satisfy a relation

$$\tau_{\text{esc}}^{(G)} = \tau_{\text{rel}}^{(G)} \frac{\varrho_G}{\varrho_E} \exp\left(\frac{E_E - E_G}{k_B T}\right). \tag{2.32}$$

The wetting layer to excited-level capture and escape expressions should, in turn, ensure that $R_E^{(\text{cap})}(E_E) - R_E^{(\text{esc})}(E_E)$ when a Fermi distribution is substituted for the occupancies. The relation between characteristic capture and escape times in (2.20) is conveniently and naturally expressed [57] in a form similar to (2.32):

$$\tau_{\text{esc}}^{(E)} = \tau_{\text{cap}}^{(E)} \frac{\varrho_E}{\varrho_{\text{wl}}} \exp\left(\frac{E_{\text{wl}} - E_E}{k_B T}\right), \tag{2.33}$$

with E_{wl} being the edge of the conduction band in the WL (Figure 2.1). Relation (2.33) still leaves undetermined the factor f'_{wl} in (2.20), the effective probability of finding an empty wetting layer state into which the escape is possible. Describing the occupancy of the WL with this single parameter is not quite straightforward because the WL, being a QW, has a continuum of energy states in the conduction band. One approach is to treat this continuum as one single effective energy level, with the position E_{wl} and the degeneracy, including spin, of $\varrho_{\text{wl}} = N_c^{(\text{wl})} V_{\text{wl}}/V_D N_D$, as implied by (2.33). Then, we get

$$f'_{\text{wl}} \approx 1 - f_F(E_{\text{wl}}) = \frac{\exp\left((E_{\text{wl}} - E_F^{(e)})/k_B T\right)}{\exp\left((E_{\text{wl}} - E_F^{(e)})/k_B T\right) + 1}. \tag{2.34}$$

This approach gives the correct asymptotic behavior both at low populations of the WL [48, 57] and at very high populations. An alternative is to consider a continuum of wetting layer electron states $\varepsilon_e^{(\text{wl})}$ and assume that, as the carrier escape is a thermally activated process, the escape rate can be obtained by weighted averaging over all the states:

$$R_E^{(\text{esc})}(E_E) = \frac{f_E(E_E)}{\tau_{\text{esc}}^{(E)}} \int_{E_{\text{wl}}}^{\infty} \frac{d\varepsilon_{\text{wl}}^{(e)}}{k_B T} \exp\left(-\frac{\varepsilon_{\text{wl}}^{(e)} - E_{\text{wl}}}{k_B T}\right)\left(1 - f_F(\varepsilon_{\text{wl}}^{(e)})\right), \tag{2.35}$$

assuming quasi-Fermi distribution in the WL and a single electron subband with a flat density of states as in an ideal deep QW. Performing the integration and substituting the result into (2.20) and (2.33), we obtain, from the detailed quasi-equilibrium requirement, an expression for f'_{wl} in the form

$$f'_{\text{wl}} = \exp\left((E_{\text{wl}}^{(e)} - E_F^{(e)})/k_B T\right) \times \ln\left[1 + \exp\left((E_F^{(e)} - E_{\text{wl}}^{(e)})/k_B T\right)\right]. \tag{2.36}$$

This has the same asymptotic behavior at both low and high N_{wl} as (2.34) but predicts somewhat more free states, hence faster escape, at intermediate carrier densities (Figure 2.5); since both (2.34) and (2.36) are heuristic estimates, the difference is not too significant.

The process of relaxation from excited to ground state at room temperature is believed to be much faster than capture into dots. Therefore, it is in some cases – particularly if the details of subpicosecond dynamics are not important – an acceptable approximation to short-circuit (or adiabatically exclude) the excited-level dynamics and describe the dots by the dynamics of the ground state only. In this

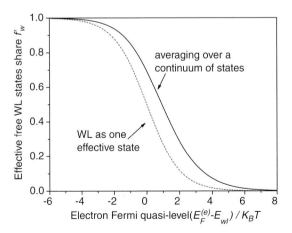

Figure 2.5 Effective complementary occupation factors f'_w in the wetting layer calculated using different approximations.

approximation, the dynamic equations reduce to a simplified form.

$$\frac{df_G^{(e)}(E_G)}{dt} \approx R_G^{(\text{cap})}(E_G) - R_G^{(\text{esc})}(E_G) - R_{\text{spont}}(E_G) - R_{\text{stim}}^{\text{net}}(E_G), \tag{2.37}$$

$$\frac{dN_{\text{wl}}}{dt} \approx R^{(\text{pump})} - \frac{\varrho_G V_D N_D}{V_{\text{wl}}} \int dE_G D_{\text{inh}}^{(G)}(E_G)(R_G^{(\text{cap})}(E_G) - R_G^{(\text{esc})}(E_G)) - R_{\text{spont}}^{(\text{wl})}, \tag{2.38}$$

where the effective capture and escape rates are described by phenomenological expressions similar to (2.20) but with the sub/superscript E for excited state (redundant in such a simplified model) substituted by G for ground state:

$$R_G^{(\text{cap})}(E_G) = \frac{\varrho_{\text{wl}} f_{\text{wl}}(1 - f_G(E_G))}{\varrho_G} \frac{1}{\tau_{\text{cap}}^{(G)}}, \quad R_G^{(\text{esc})}(E_G) = \frac{f_G(E_G)}{\tau_{\text{esc}}^{(G)}} f'_{\text{wl}}. \tag{2.39}$$

This description phenomenologically includes both the two-stage processes, involving carrier capture to the excited state and subsequent relaxation into the ground state, and the opposite escape process, and, if present (e.g., in long-wavelength InAs/InP dots), direct capture and escape from the wetting layer to the ground state. The effective capture and escape times thus depend on the relative (thermalized) populations of the ground and excited states and can be expressed as functions of the corresponding characteristic times. Expressions (2.36)–(2.38) can thus be seen as a special case of (2.16)–(2.18) (or (2.26)) under the condition of $\tau_{\text{rel}}^{(G)} \ll \tau_{\text{cap}}^{(E)}$. In this case, fast intralevel relaxation ensures that for each dot size/shape group i,

$$f_E(E_{Ei}) = \frac{t_i f_G(E_{Gi})}{1 - (1 - t_i) f_G(E_{Gi})}, \quad t_i = \exp\left(-\frac{E_{Ei} - E_{Gi}}{k_B T}\right) < 1. \tag{2.40}$$

By substituting this term into the rate equation for the excited-level population, we can establish the relations between the phenomenological capture and escape times that provide a direct correspondence between (2.36)–(2.38) and (2.16)–(2.18) (or (2.26)). In the simple model of the characteristic capture time $\tau_{\text{cap}}^{(E)}$ (and $\tau_{\text{cap}}^{(G,\text{direct})}$, if direct capture is present) being independent of the level populations, these take the form

$$\frac{1}{\tau_{\text{cap}}^{(G,i)}} \approx \frac{\varrho_G}{\varrho_E} \frac{1 - (1 - t_i) f_G(E_{Gi})}{t_i} \frac{1}{\tau_{\text{cap}}^{(E)}} + \frac{1}{\tau_{\text{cap}}^{(G,\text{direct})}},$$

$$\frac{1}{\tau_{\text{esc}}^{(G,i)}} \approx [1 - (1 - t_i) f_G(E_{Gi})] \frac{1}{\tau_{\text{esc}}^{(E,i)}} + \frac{1}{\tau_{\text{esc}}^{(Gi,\text{direct})}}. \tag{2.41}$$

In the case of InAs/GaAs dots described by (2.16)–(2.18) and at present more relevant for ultrafast applications, the direct capture/relaxation terms may be neglected. Notice that expressions (2.41) preserve the relation

$$\tau_{\text{esc}}^{(Gi)} = \tau_{\text{cap}}^{(Gi)} \frac{\varrho_G}{\varrho_{\text{wl}}} \exp\left(\frac{E_{\text{wl}} - E_{Gi}}{k_B T}\right), \tag{2.42}$$

which is expected from the detailed equilibrium principle. Note also that both capture and escape times in such an approach become dependent on the dot energy levels (size group) and on the dot population, even in the simple model when the capture time used in (2.16)–(2.18) is constant.

Equations 2.20–2.36 and 2.39 represent a semiphenomenological approach to dot kinetics, which requires the introduction of phenomenological capture and relaxation times, whether including the excited-state dynamics or not.

A potentially more accurate and powerful, but also much more complicated, approach is to calculate both the capture/relaxation and escape rates fully microscopically. A number of such investigations have been recently reported [44, 58, 59] and will be discussed below.

One of the advantages of such a fully microscopic approach is that it can be used not only for electron but also for *hole* capture and escape.

The dynamics of hole populations in QDs appears to be less well understood compared to that of electrons, and it is more difficult to extract useful parameters from experiments. Consequently, at least three significantly different approaches have been proposed in the literature.

The first one is to consider the electron–hole pair localized in a dot as an indivisible *exciton*, and so assume, at least implicitly, that for hole levels $E_G^{(h)}$ and $E_E^{(h)}$ belonging to the same dot group as the electron levels $E_G^{(e)}$ and $E_E^{(e)}$ and so calculated from $E_G^{(e)}$ as described above, the occupancy is the same as that of the corresponding electron level: $f_G^{(h)}(E_G^{(h)}) = f_G^{(e)}(E_G^{(e)})$ and $f_E^{(h)}(E_E^{(h)}) = f_E^{(e)}(E_E^{(e)})$. This leads to a gain model in the form of Equation 2.3 instead of the more general (2.4), if the inhomogeneous broadening is not taken into account, and a corresponding simplification to (2.9) in the model with inhomogeneous broadening. This model can be used to analyze the ultrafast optical processes because it takes into account both the slow processes of recombination and the fast processes of carrier capture, relaxation, and escape, and so is likely to predict correct trends if not numbers. A number of authors have indeed used this approach to analyze the short-pulse amplification (see, for example, Ref. [48]) and generation. It has to be admitted, however, that the excitonic assumption has little justification, either experimental or theoretical. In fact, it has been theoretically shown [60] that even the entire QD active layer is, strictly speaking, *not* neutral, let alone individual dots. There is also a certain ambiguity in the excitonic approach as to the physical meaning of the energy intervals involved in the kinetic analysis and, in particular, in the definition of the distribution functions. Some authors (e.g., the authors of Ref. [55]) explicitly treat these energy intervals, such as the ground-level depth $E_{wl} - E_G$, as pertaining to *electron* levels (by extension, the same may be assumed of the level separation $E_E - E_G$, though only the ground level was considered in Ref. [55]). Alternatively, the distribution function $f(E_G)$ can be treated as the distribution function of the *entire exciton*, or the electron–hole pair, with the energy intervals involved taken as the *sums* of the electron and hole contributions, in which case $E_E - E_G$ can be directly identified with the experimentally observed separation between the excited- and the ground-state luminescence levels. Such an approach is the easiest one because no explicit information at all is needed about hole states in the QDs.

The second possible approach to treating hole kinetics is, in a way, opposite to the first one, with the dynamics of electrons and holes completely decoupled. In this approach, one still solves the detailed dynamic rate equations for electron occupancies only. As regards the holes, the assumption is made that their capture and relaxation are much faster than those of electrons. Then, the hole occupancies at any moment in time, for any levels involved, can be taken to satisfy a Fermi distribution, with a common hole Fermi quasi-level for both the holes in the wetting layer and those in QDs. Then, one needs only one dynamic equation for the entire hole population, including both the WL and the dot states, with only pumping and recombination terms and no need to treat the internal kinetics (see below). This approach is more generic than the previous one as it does not require assuming quasi-neutrality either in the dots or in the wetting layer. It is also arguably more justified as there is some indication that the hole relaxation is indeed faster than that of the electrons – although not all experimental and theoretical evidence supports this, as discussed below. Relating the total dot concentration to the quasi-levels, however, requires knowledge of the hole density of states, including the wetting layer and all the localized states in the dots – of which there may be more than just two (the ground and the "excited"), and which are not very well known. In the paper where this approach was proposed and used for analyzing steady-state behavior of QD amplifiers [61], an effective bulk-like density of states was introduced for the dots. A similar, slightly modified approach has also been applied to fast dynamics of QD amplifiers (not lasers) in analyzing fast nonlinear effects such as four-wave mixing for signal processing [62]. While the dynamics of electron populations for various levels were described using multistage capture and escape processes as discussed above, dynamics of holes in all states (assigned an effective bulk-like density of states) were characterized using a single, subpicosecond ($\sim$100 fs) relaxation time.

The third, and the most rigorous, approach to treating the hole dynamics is to calculate the capture and escape rates fully microscopically, for both electrons and holes, as has been reported recently in Refs [58, 59, 63] and, independently, in Ref. [44]. The authors of Refs [58, 59, 63] made use of the fact that the relaxation rates inside the dots, for either electrons or holes, were faster than the capture rates, and so used the approach of the type of Equations 2.37 and 2.38, without separate equations for the excited-state carriers. The inhomogeneous broadening was taken into account approximately, by selecting a lasing group of dots and treating its population separately. Auger contributions to capture and escape have been calculated by integrating the rates of in-scattering (capture) and out-scattering (escape) processes over all states in the wetting layer. The physical mechanism of capture and escape was screened using Coulomb interaction between carriers (nonlocal Auger process). For each set of parameter values, integration was required over the *initial* states of the carrier captured into the dot and of the WL carrier to which the energy was transferred. The energy of the *final* state in the WL layer was determined from the requirement of energy conservation, and adherence to the principle of detailed equilibrium was ensured by including the population factors for all the relevant states. Both single-carrier (electron–electron in the case of electron capture and hole–hole in the case of hole capture, with the energy of, say, a captured electron transferred to another electron) and mixed electron–hole and hole–electron processes were taken into account.

The results obtained were presented in terms of the in-scattering $S_e^{(in)}$ and $S_h^{(in)}$ and out-scattering $S_e^{(out)}$ and $S_h^{(out)}$ rates for both electrons and holes. The values of the electron rates could be interpreted in terms of phenomenological capture and escape times used in (2.37)–(2.39) as

$$S_e^{(in)} = \frac{N_{wl} V_{wl}}{\varrho_G V_D N_D} \frac{1}{\tau_{cap}^{(G)}} = \frac{\varrho_{wl}}{\varrho_G} \frac{f_{wl}}{\tau_{cap}^{(G)}}, \quad S_e^{(out)} = \frac{f'_{wl}}{\tau_{esc}^{(G)}}. \quad (2.43)$$

The definitions for hole rates are similar. The results are shown in Figure 2.6 as functions of 2D electron and hole densities in the WL: $w_e = N_{wl} V_{wl}/A_s$ and $w_h = N_{wl}^{(h)} V_{wl}/A_s$, $A_s = WL$ is the surface area of the sample investigated. In the latest study [58, 59] these densities were assumed to be different from each other, but their ratio $g_c = w_h/w_e$, deduced from steady-state analysis, was kept constant. While the absolute values of the parameters obtained are specific to the structure modeled and were noted to depend strongly on the structure and operating conditions, several important, and probably generic, trends have been established.

First, for small carrier densities, the in-scattering rates $S_e^{(in)}$ and $S_h^{(in)}$ increase superlinearly with the WL carrier density, so that the heuristic law (2.31), which would imply $S_e^{(in)} \propto w_e^2$ (note that the phonon capture was not included in the microscopic analysis) would appear to be a reasonably good first approximation. However, at high values of w_h and w_e, this approximation breaks down

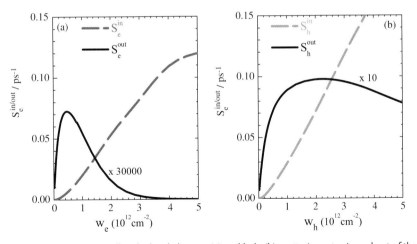

Figure 2.6 Microscopically calculated electron (a) and hole (b) scattering rates in and out of the dot as functions of the wetting layer carrier densities. $g_c = w_h/w_e = 1.5$. Reprinted with permission from Ref. [59].

dramatically, with both the electron and the hole in-scattering rates saturating, and then *decreasing* with WL carrier density, which the authors attribute to the filling of the final WL states. The in-scattering rates thus have a maximum at a certain value of $w_e \sim (2-6) \times 10^{12}\,\mathrm{cm}^{-2}$. The absolute value of the scattering rate is lower than that implied by the phenomenologically deduced values used in Ref. [48], which may be partly due to the fact that the microscopic calculations considered only Auger-type capture whereas the authors of Ref. [48] believed the capture in their structure to be mainly phonon assisted and partly due to different structure parameters.

Second, the out-scattering rates, after a very sharp initial increase with w_h and w_e, decrease almost exponentially with the wetting layer carrier density. This decrease starts at much more modest w_h and w_e than in the case of in-scattering, and is also due to the filling-in of the possible final states of transitions (Pauli blocking). In the simple phenomenological model, this is taken into account by the factor f'_w – the microscopic calculations did not explicitly introduce this effective parameter, performing rigorous summation over initial and final WL states instead. The general form of the presented curve thus appears to validate the phenomenological approach, though some quantitative deviations may occur – mainly at very high w_h and w_e, however, where the out-scattering is strongly suppressed, so these deviations will not be too important in the general dot kinetics.

The authors of Refs [58, 59, 63] applied their theory to a switch-on transient of a QD laser and concluded that an accurate account for both electron and hole dynamics was important in predicting the damping of the relaxation oscillations. In Ref. [44], the accurate calculations of all the kinetic processes were used to analyze the shaping and propagation of a single, intense subpicosecond pulse.

To conclude the discussion of various approaches to QD kinetics, we note that describing the kinetics of quantum dots in terms of the occupation functions f_E and f_G averaged over an entire ensemble of dots may, in principle, not be entirely accurate, as each dot is independent of the rest and so the ensemble averaging is, strictly speaking, not always meaningful. A different approach was presented in Ref. [64] in the form of the *master equations model* (MEM) for the occupancies of all the "microstates" of dots. The dynamic variables describing the electron populations in a dot in this model are the probabilities of finding a dot with a given number of carriers in the excited and ground state, W_{ij}, where $i = 0, \ldots, \varrho_G/2$ (neglecting spin dynamics) is the number of electrons in the ground level in the dot and $j = 0, \ldots, \varrho_E/2$ is the number of electrons in the excited level. Then, with $\varrho_G/2 = 1$ and $\varrho_E/2 = 2$, the ground-state occupancies that feature in the gain model are calculated as $f_G = W_{10} + W_{11}$ and $f_E = (W_{01} + W_{11})/2 + W_{02} + W_{12}$. The interlevel kinetic behavior is described using capture, relaxation, and escape times and does not need an explicit account for Pauli blockage for interlevel transitions (though the factor f'_w is still desirable for describing the wetting layer). The model has been claimed to be, in principle, more accurate than the rate equations approach. However, it is also considerably more complex than the rate equations: for $\varrho_G/2 = 1$ and $\varrho_E/2 = 2$, six equations are required for describing the electron populations in each size group of dots, instead of just two equations for f_G and f_E. Care should also be taken in

2.3 Kinetic Theory of Quantum Dots

introducing the escape processes in accordance with the principle of detailed equilibrium [65]. In principle, similar approach can be applied to holes, although the authors of the original master equation model were working in the "excitonic" approximation, assuming electrons and holes to have the same occupation function. The only mention of the application of MEM to ultrafast dynamics of optical QD devices is found in Ref. [48] where it was stated that the response of a QD amplifier to a set of short optical pulses was very similar in the rate equation model and MEM (though no explicit comparison was presented). Indeed, the main difference between the MEM and the rate equation model lies in the treatment, not of capture of carriers from the WL into dots but of the interlevel scattering between the excited and ground states. The latter is believed to be a fast, subpicosecond, process about an order of magnitude faster than the capture into the dots, so accuracy in its description is often less important. Therefore, for the overwhelming number of situations of practical interest, the "standard" approach including rate equations for the state occupancies appears to be sufficient. Virtually all the published work has thus been performed using this approach, which we also follow in this book.

In the steady state and in the absence of the stimulated recombination, the main result predicted by all the kinetic models is that at low temperatures, the distribution of carriers (both electrons and holes) in energy is strongly nonequilibrium, but at room temperature the distribution is very similar to a Fermi one. This is the result, first, of the capture and escape times satisfying the detailed equilibrium principle and, second, of the fact that at room temperature both the capture and escape processes are characterized by picosecond time constants and so both are much faster than the recombination, which in the absence of stimulated emission is characterized by $\tau_{spont} \sim 1$ ns (at low temperatures, the escape time becomes long, hence the nonequilibrium distribution). In the case of a Fermi level below the WL bandgap, when the carrier distribution in the WL is Boltzmann $f_{wl} = f_F(E_{wl}) = \exp(-(E_{wl} - E_F^{(e)})/k_B T) \ll 1$ (and so the factor f'_w in (2.39) can be approximately set to 1), an analytical expression for the electron distribution function can be obtained [55] by setting the time derivative in (2.37) to zero; in our notations,

$$f_G(E_{Gi}^{(e)}) = \frac{1}{1 + \exp\left((E_{Gi}^{(e)} - E_F^{(e)})/k_B T\right) + (\varrho_g/\varrho_{wl})(\tau_{cap}^{(G)}/\tau_{sp})\exp\left((E_{wl} - E_F^{(e)})/k_B T\right)}$$
$$= \frac{1}{1 + \left(1 + \tau_{esc}^{(G)}(E_{Gi}^{(e)})/\tau_{sp}\right)\exp\left((E_{Gi}^{(e)} - E_F^{(e)})/k_B T\right)},$$
(2.44)

where the formula has been written for the ground-state energy E_{Gi} and could, in principle, be applied to either the ground or the excited state of a group of dots (the authors of Ref. [55] did not consider the excited state directly, but the formula should hold for excited states due to the principle of detailed equilibrium). Due to the small value of the factor $\tau_{esc}^{(G)}/\tau_{spon}$, the correction to the Fermi distribution is indeed clearly small at room temperature (though can be significant at low temperatures [55]).

Then, if we consider a device in steady state, at high temperature, and under small signal conditions, then the gain calculation in quantum dots, using the gain

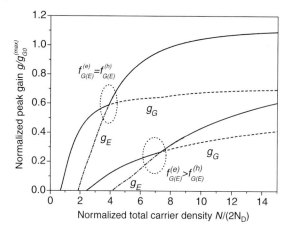

Figure 2.7 Calculated dependence of peak ground- and excited-state gain in a QD amplifier on the *total* carrier density N in the QD–wetting layer system, with and without the excitonic (quasi-neutrality) approximation. Equilibrium carrier distribution in energy is assumed; parameters and the rest of the model are the same as in Figure 2.3.

formula (2.3), (2.4), or (2.9) with the carrier distribution (2.44), is not that dissimilar to bulk and QW materials. This was the procedure used in producing the spectrum of Figure 2.3. We can also use the calculated distribution functions to plot the maximum gain in the QD active layer, both for the ground-state transitions and, if required, for the excited-state and wetting layer transitions as a function of the device current. The salient feature of such a curve (Figure 2.7) is the strong saturation of the dependence $g_G(J)$. This saturation, which is in agreement with experimental observations, is predicted in all models, with or without taking into account inhomogeneous broadening, with or without quasi-neutrality approximation, although if electron and hole statistics are treated as independent, the saturation of $g_G(J)$ happens at higher currents than those in the excitonic approximation and is thus noticeably less pronounced under the typical operating conditions of a QD laser or amplifier. Indeed, the saturation occurs when the dot population approaches *total inversion* of the dot energy levels ($f_G^{(e)}(E_G^{(e)}) \to 1$, $f_G^{(h)}(E_G^{(h)}) \to 1$), meaning that with the increase in current, no further increase in occupancy of the states in the ground-level band can happen, and hence no further contribution to gain. In principle, saturation of gain (at a given size quantization level) with current due to total inversion is also possible in quantum wells (indeed, the gain–current relation in QWs is sublinear, as in QDs). However, in QD devices, the saturation is more pronounced and occurs at much lower/more realistic currents, due to the density of states in quantized QD levels being smaller than that in continuous QW bands. Indeed, the phenomenological expression for the gain–current relation in QD devices is [51, 66]

$$G_{QD}^{(peak)}(J) = G_{QD}^{(0)}\left(1 - \exp\left(-\frac{\gamma(J - J_0)}{J_0}\right)\right). \tag{2.45}$$

Here, $G_{QD}^{(0)}$ is the gain at total inversion, $J = I/WL$ is the current density, J_0 is the transparency current density, and γ is a fitting parameter dependent on inhomogeneous broadening of various levels. The measured gain–current curves are indeed well approximated by this fit. This has to be compared with the sublinear biparametric fit $G_{QW}^{(peak)}(J) = G_{QW}^{(0)} \ln(J/J_0)$ used in QW lasers (the alternative three-parameter fit is even more weakly saturating with J). As discussed above, the other effect of the (near)total inversion of the ground and excited states is the gradual buildup of a reservoir of carriers in the wetting layer (and possibly the bulk barrier). Both of these features significantly affect the interaction of light and matter in QDs, as explained below.

At higher currents, similar saturation is seen in the excited-level band, at a higher gain value than that seen for the ground state.

2.4
Light–Matter Interactions in Quantum Dots

In the previous sections, we considered the small-signal optical properties of dots with given occupation functions of levels and, separately, the dot kinetics in the absence of light. Here, we shall consider the kinetics of dots in the presence of strong light signal in a laser or amplifier and its effect on the optical properties of the QD sample, and thus on the optical signal.

Presenting a generic, yet manageable model of light interaction with quantum dots, suitable for treating all ultrafast optoelectronic devices, including in-plane and vertical cavity lasers, amplifiers, and saturable absorber mirrors, both in steady state and in dynamics, is hardly practical. Specific models of quantum dot devices will be considered in the following chapters; here, we shall discuss some general considerations concerning light amplification and absorption in quantum dot media. We can start from the general form of the wave equation for light propagation:

$$\frac{\partial^2}{\partial \underline{r}^2} \Upsilon - \frac{1}{c^2} \frac{\partial^2}{\partial t^2} [\hat{\eta}^2 \Upsilon] = \frac{1}{c^2} \frac{\partial^2}{\partial t^2} P. \tag{2.46}$$

Here, $\Upsilon(\underline{r}, t)$ is the electric field of the light wave at the point $\underline{r}$ at time t, $\eta(\underline{r})$ is the background refractive index that describes the *nonresonant* (built-in) dielectric properties, and finally, P is the *resonant* component of the dielectric polarization associated with the active material – in our case, the QDs. The symbol ˆ over a parameter notation, in this case η^2, means that the corresponding parameter is, strictly speaking, dispersive (optical frequency dependent), so in time domain it is an operator

$$\hat{\eta} \Upsilon = FT^{-1}[\eta(\underline{r}, \omega) \cdot FT(\Upsilon)], \tag{2.47}$$

where FT and FT^{-1} are the direct and inverse Fourier transforms, respectively, between time (t) and optical frequency (ω) domains. In realistic devices, the main manifestation of the operator nature of the background refractive index η is the emergence of the group velocity of light, rather than phase velocity, in the slow-wave formulas below. Although the fastest of timescales we will be interested in may be subpicosecond, it can still be assumed comfortably long compared to the light oscillation period of $\sim 10^{-15}$ s. Therefore, the rotating wave and slow-amplitude approximations can be introduced, though the former requires specifying the geometry of the device under analysis. To begin with, we shall assume an *in-plane* waveguide device, such as a QD SOA treated in the next chapter, or an in-plane QD laser of Chapter 5. The natural formalism for devices in a *vertical* geometry, such as vertical cavity surface emitting lasers and semiconductor saturable absorber mirrors, is somewhat different though the fundamental principles are similar. Thus, we can take z as the propagation (longitudinal) direction and introduce the rotating wave notation

$$Y(z,t) = \exp(-i\omega_{\text{ref}}t)(Y_{\text{f}}(z,t)\exp(i\beta_{\text{ref}}z) + Y_{\text{b}}(z,t)\exp(-i\beta_{\text{ref}}z)) + \text{c.c.}, \quad (2.48)$$

where ω_{ref} is the reference optical frequency (for a laser, it is natural though not necessary to take the optical frequency of gain peak as the reference, and for an amplifier, the carrier frequency of the incident signal), $\beta_{\text{ref}} = \eta_0 \omega_{\text{ref}}/c = \eta_0 k_{\text{ref}}/c$ is the longitudinal wave vector determined by ω_{ref}, and Y_f and Y_r are the "slow amplitudes" ("slow" meaning

$$\left|\frac{\partial}{\partial t} Y_{f,r}\right| \ll |\omega Y_{f,r}|, \quad \left|\frac{\partial}{\partial z} Y_{f,r}\right| \ll |\beta Y_{f,r}|) \quad (2.49)$$

of the forward- and reverse-propagating fields, respectively. The reference frequency is chosen arbitrarily, but close to the operating wavelength of the device investigated (e.g., at the peak ground-state gain if the device operates via ground-state transitions). The z-dependent field $Y(z,t)$ is related to the full three-dimensional distribution as

$$\Upsilon(\mathbf{r},t) = Y(z,t)\psi(x,y), \quad (2.50)$$

where $\psi(x,y)$ is the modal distribution of light in the waveguide in the transverse (x) and lateral (y) directions. It is assumed to satisfy the waveguide equation for light $(d^2\psi/dx^2) + (d^2\psi/dy^2) + [\eta^2(x,y,\omega) - \eta_{\text{mod}}^2(\omega)](\omega^2/c^2)\psi = 0$ at the reference frequency $\omega = \omega_{\text{ref}}$, with $\eta_0 = \eta_{\text{mod}}(\omega_{\text{ref}})$ being the effective modal refractive index, and normalized to unity ($\int_{-\infty}^{\infty}\int_{-\infty}^{\infty} |\psi|^2 dx, dy = 1$). Note that the approach (2.48) with assumption (2.49), which is usually comfortably valid in QW and bulk semiconductor lasers and amplifiers, may get strained in more broadband QD devices, particularly when operation at both ground and excited levels is involved; the model would need to be modified in this case (e.g., introducing slow-wave approximations separately for light resonant with ground- and excited-state transitions, with separate reference frequencies and wave vectors).

With approximation (2.48), we arrive at the often-used slow-wave equations for the amplitudes of left- and right-propagating waves:

$$\pm \frac{\partial Y_{f,b}}{\partial z} + \frac{1}{v_g}\frac{\partial Y_{f,b}}{\partial t} = \left(\frac{1}{2}(\hat{g}_{\text{mod}} - a_i) + ik_{\text{ref}}\widehat{\Delta\eta}_{\text{mod}}\right) Y_{f,b} + F_{\text{spont}}(z,t). \quad (2.51)$$

Here, $v_g = c/\eta_g$ is the group velocity of light, $\eta_g = \eta_o + \omega(d\eta_{mod}/d\omega)|_{\omega_{ref}}$ is the waveguide group refractive index, and $\hat{g}_{mod}$ and $\widehat{\Delta\eta}_{mod}$ are, respectively, the modal gain and the resonant (active layer) contributions to the modal refractive index that arise from the resonant polarization P in (2.46). The latter term describes effects of self-phase modulation and, in principle, group velocity dispersion. In general, both terms are dispersive and so in time domain may need to be described as operators, depending on the timescale used. The term a_i is the internal (dissipative and scattering) loss in the waveguide and F_{spont} is a random term describing spontaneous emission.

In *QW and bulk* lasers and amplifiers, the modal gain is calculated (in optical frequency, or photon energy, domain) as

$$g_{mod}(\omega) = \Gamma_{xy}(\omega) g(\hbar\omega), \tag{2.52}$$

where $\Gamma_{xy} \approx \iint_{active} |\psi|^2 dx\, dy$ is the optical confinement factor, with the integration over the cross section of the active layer. The resonant contribution $\widehat{\Delta\eta}_{mod}$ to the refractive index, in principle, consists of two parts: the interband contribution that can be calculated from $g(\hbar\omega)$ using Kramers–Kronig relations and the free-carrier contribution that comes from plasma effects associated with a large number of free carriers. The former is believed to be stronger. In most situations of practical interest, $\widehat{\Delta\eta}_{mod}$ can be linearized as

$$\Delta\eta_{mod} \approx -\frac{\alpha_H}{k_{ref}} \left(g_{mod} - g_{mod}^{ref}\right), \tag{2.53}$$

where the coefficient α_H is known as the Henry linewidth enhancement factor and g_{mod}^{ref} is the reference value of gain, usually the threshold value in laser analysis and the steady-state value when analyzing amplifier dynamics. Neglecting the waveguide dispersion, the linewidth enhancement factor is determined as

$$\alpha_H = -(\partial \text{Re}\chi/dN)/(\partial \text{Im}\chi/dN) \approx -\frac{1}{2\lambda}\frac{dn}{dN}\left(\frac{dg}{dN}\right)^{-1}, \tag{2.54}$$

where χ is the complex dielectric susceptibility of the active layer material, the imaginary part of which is proportional to gain or loss. In practice, the dispersion and waveguide nature of semiconductor lasers and amplifiers make the situation more complicated, and the parameter α_H was shown to depend not only on the material but also on the specific device and even operating conditions (temperature and even current). For bulk materials at laser and amplifier carrier densities, it is in the range of 2–7 and for QW materials in the range of 1–3.

The situation in quantum dot devices is somewhat more complex. The modal gain is still determined by (2.52) as in the bulk and QW cases (note that if we choose the volume V_D to be only a part of the active layer V_{act}, then we would have to introduce an extra $V_{QD}/V_{act} < 1$: $g_{mod}(\omega) = \Gamma_{xy}(\omega)(V_D/V_{act})g(\hbar\omega)$.

If the transverse dimensions of the waveguide are much thinner than the lateral width W, as is usually the case, then the active layer confinement factor is mainly due to the transverse rather than the lateral extent of the mode: $\Gamma_{xy} \approx \Gamma_x \approx \int_{-\infty}^{\infty} |\psi|^2 dx$. If we then introduce the effective transverse width of the waveguide mode

$$w_{mod}(\omega) = d_{act}/\Gamma_x(\omega), \tag{2.55}$$

then the modal gain for the ground-state transitions can be expressed, in the simplest approximation not considering the inhomogeneous broadening, in the form

$$g_{mod} = \sigma_G \varrho_G \frac{n_l N_D^{(2d)}}{w_{mod}} \left(f_G^{(e)} + f_G^{(h)} - 1 \right) = \frac{\sigma_G}{w_{mod}} \left(N_G^{(e,2d)} + N_G^{(h,2d)} - \varrho_g n_l N_D^{(2d)} \right), \tag{2.56}$$

where we have introduced, alongside the total two-dimensional density of dots $n_l N_D^{(2d)}$ ($N_D^{(2d)}$ being the value per layer in a stacked structure), the two-dimensional densities of electrons and holes $N_G^{(e,2d)}$ and $N_G^{(h,2d)}$. The excited-state gain is written in a similar way. The stimulated recombination term in (2.16) in this case is simply

$$R_G^{(stim)} \left(E_G^{(e-h)}, z, t \right) = v_g \sigma_G \left(f_G^{(e)} \left(E_G^{(e)}, z, t \right) + f_G^{(h)} \left(E_G^{(h)}, z, t \right) - 1 \right)$$

$$\times \left(|Y_f^2(z,t)| + |Y_b^2(z,t)| \right), \tag{2.57}$$

if the units of the slow amplitudes $Y_{f,b}$ are chosen so that $S(z,t) = |Y_f^2(z,t)| + |Y_b^2(z,t)|$ means the *local photon density*. It is related to the local light signal intensity (power per unit cross-section area)

$$I_{sign} = v_g \hbar \omega_{ref} S(z,t). \tag{2.58}$$

A similar stimulated recombination term can be introduced into (2.17) for the excited state if an optical signal resonant with the excited state is present in the device (or to take into account amplified spontaneous emission in an amplifier).

As regards the *refractive index* term in the model without inhomogeneous broadening, some authors take into account the self-phase modulation in QD lasers in the same simple form (2.53) as that used in bulk and QW devices. However, such a formula is likely to be an oversimplification for QDs, and a more accurate approach may be preferable. To illustrate this, and also to analyze the details of the gain properties of the QD devices, it is illustrative to consider the full picture of optical properties of a waveguide QD device, with spectral phenomena and inhomogeneous broadening taken into account.

In this case, using expression (2.9) for the material gain $g(\hbar\omega)$, we can obtain the modal gain as

$$g_{mod}^{(G)} = A_G \varrho_G \frac{n_l N_D^{(2d)}}{w_{mod}} \int \left(f_G^{(e)} \left(E_G^{(e)} \right) + f_G^{(h)} \left(E_G^{(h)} \right) - 1 \right)$$

$$D_{hom} \left(\hbar\omega, \Delta E_G^{(e-h)} \right) D_{inh} \left(E_G^{(e-h)} \right) dE_G^{(e-h)}. \tag{2.59}$$

A similar expression, with the subscript G substituted by E for excited state, can be used for excited-state transitions. The use of the Lorentzian homogeneous broadening function (2.8) means that the *same approach* can be used for calculating the resonant refractive index contribution without recourse to Kramers–Kronig transformation. For example, complementing (2.9), the contribution from the ground state to the (material) refractive index becomes

$$\Delta\eta_G(\hbar\omega) = \frac{A_G \varrho_G N_D}{k_{ref}} \int \left(f_G^{(e)}\left(E_G^{(e)}\right) + f_G^{(h)}\left(E_G^{(h)}\right) - 1\right)$$
$$D_{hom}^{\prime\prime(G)}\left(\hbar\omega, \Delta E_G^{(e-h)}\right) D_{inh}\left(E_G^{(e-h)}\right) dE_G^{(e-h)}. \qquad (2.60)$$

Here, most of the notations are the same as in (2.9) and (2.59), and we have introduced the complementary homogeneous broadening function in the form

$$D_{hom}^{\prime\prime(G)}\left(\hbar\omega, \Delta E_G^{(e-h)}\right) = \frac{1}{\pi} \frac{\hbar\omega - \Delta E_G^{(e-h)}}{\left(\hbar\omega - \Delta E_G^{(e-h)}\right)^2 + \Delta E_{hom}^2}. \qquad (2.61)$$

Again, a similar formula can be written for the excited-state contribution, only with the sub/superscript G substituted by E for excited level. Note that $D_{hom}^{\prime\prime(G)}$ and $D_{hom}^{(G)}$ are the real and imaginary parts of the same complex function with a single pole, thus ensuring automatic adherence of (2.60) and (2.9) to the principle of causality and to Kramers–Kronig relations, as in the standard theory of optical properties of two-level systems. Also similarly to two-level systems, in the case of $\Delta E_{hom} \gg \Delta E_{inh}$, formula (2.60) predicts $\Delta\eta_G|_{\hbar\omega=\Delta E_G^{(e-h)}} = 0$, which gave grounds to the early expectations of a zero linewidth enhancement factor in QD lasers. Unfortunately, in QD laser structures fabricated so far, the opposite relation holds, $\Delta E_{hom} \ll \Delta E_{inh}$, meaning that these predictions are far from guaranteed. Moreover, $D_{hom}^{\prime\prime(G)}$ is a function only slowly decaying with its argument, the difference $\hbar\omega - \Delta E_G^{(e-h)}$ between the photon energy and the electron–hole pair energy separation – in fact, the integration in (2.60) converges only because of the sharp decay of the wings of the inhomogeneous broadening function. This means that, with the homogeneous broadening in both the ground and excited states of the order of $\Delta E_{hom} \sim 5$–10 meV ($T_2 \sim 100$–200 fs) and the separation between them of <100 meV, the *excited-state* transitions affect the refractive index in the band of photon energies primarily resonant with the ground-state transitions, and vice versa. Another contribution is made by the gain or loss in the wetting layer, so that in general, for any photon energy,

$$\Delta\eta_{mod}(\hbar\omega) = \Gamma_{xy}(\Delta\eta_G(\hbar\omega) + \Delta\eta_E(\hbar\omega)) + \Gamma_{wl}\Delta\eta_{wl}(\hbar\omega) \qquad (2.62)$$

with $\Gamma_{wl}(\omega) \approx (V_{wl}/V_{act})\Gamma_{xy}(\omega)$ being the WL confinement factor, or the overlap of the waveguide mode with the wetting layer. The gain or loss spectrum in the WL may, in principle, be calculated using the standard theory of QW materials, and $\Delta\eta_{wl}(\hbar\omega)$ then evaluated using Kramers–Kronig transformation. All three contributions in (2.62) depend on both the photon energy and the relevant populations – the first two on the various distribution functions in the ground and excited states, respectively, and the third one on N_{wl}. As follows from the dynamic equations (2.16)–(2.18), these populations do not necessarily vary in phase with each other during dynamic laser operation; a time delay may be expected due to the finiteness of the capture and delay times that, in addition, may be different for different levels. Therefore, strictly speaking, the introduction of a single linewidth enhancement factor

$\alpha_H = -(\partial \text{Re}\chi/dN)/(\partial \text{Im}\chi/dN)$, with N the total number of carriers in the active layer (dots and WL), is not entirely justified, and attempts at experimentally determining such a parameter may be expected to result in spurious values being measured. Indeed, depending on the method of measurement and operating conditions, values of α_H from almost zero to almost a hundred have been reported in quantum dot lasers. This caveat may become particularly important when analyzing or modeling short-pulse operation regimes such as mode locking.

When the optical signal present in the device is slow enough to assume

$$\frac{dY(z,t)}{dt} \ll \frac{\Delta E_{\text{hom}}}{\hbar} Y = \frac{Y}{\tau_h} \tag{2.63}$$

(which is the case, for example, in a laser with the operating spectrum considerably narrower than $\sim (\Delta E_{\text{hom}}/\hbar) = (1/T_2) \sim 10^{-13}\,\text{s}^{-1}$ in terms of optical frequency, or in an amplifier with a single-wavelength incident signal modulated at a rate much slower than $\sim (\Delta E_{\text{hom}}/\hbar) = (1/T_2)$), the operator nature of expressions (2.60) and (2.62) may be ignored, and time and photon (or electron) energy taken as independent variables. The gain and refractive index in this case may be calculated using expressions (2.9), (2.60), and (2.62) at the photon energy $\hbar\omega_{\text{ref}}$, and their time dependence is determined by the time evolution of the electron and hole occupation functions in the ground and excited states of the QDs and in the wetting layer. These are given either by Equations (2.16)–(2.18) if the dynamics of the excited state are taken into account separately from the ground state or by (2.37) and (2.38) if the excited state is adiabatically excluded.

The rate of stimulated recombination of an electron in (2.16) is given in the spectrally resolved model by

$$R_G^{(\text{stim})}(E_G, z, t) = v_g A_G \left(f_G^{(e)}(E_G^{(e)}, z, t) + f_G^{(h)}(E_G^{(h)}, z, t) - 1 \right) \\ \times D_{\text{hom}}\left(\hbar\omega_{\text{ref}}, \Delta E_G^{(e-h)}\right) \left(|Y_f^2(z,t)| + |Y_b^2(z,t)| \right). \tag{2.64}$$

The hole recombination term would be identical, and a similar term for the excited-state electron or hole can be introduced into (2.17).

If we can use the excitonic dot neutrality approximation $f_G^{(e)}(E_G^{(e)}, z, t) = f_G^{(h)}(E_G^{(h)}, z, t)$, then, from (2.64), the effective stimulated recombination time of an electron in a QD ground state with the energy $E_G^{(e)}$ is

$$\tau_G^{(\text{stim})}(E_G) = \left(2 v_g A_G D_{\text{hom}}(\hbar\omega_{\text{ref}}, \Delta E_G^{(e-h)}) \left(|Y_f^2(z,t)| + |Y_b^2(z,t)|\right)\right)^{-1}. \tag{2.65}$$

As the stimulated recombination in the presence of light signal is an alternative channel of carrier recombination in (2.16) that is added to spontaneous recombination, the approximate solution (2.44) of Equation 2.16 for a ground-state electron distribution function can be modified in the presence of a light signal as

$$f_{Gi}^{(e)}\left(E_{Gi}^{(e)}\right) = f_{Gi}^{(h)}\left(E_{Gi}^{(h)}\right)$$

$$\approx \frac{1 + \left(\tau_{\text{esc}}^{(G)}\left(E_{Gi}^{(e)}\right)/2\tau_G^{(\text{stim})}\left(E_{Gi}^{(e)}\right)\right)\exp\left(\left(E_{Gi}^{(e)} - E_F^{(e)}\right)/k_B T\right)}{1 + \left(1 + \left(1/\tau_{\text{sp}} + 1/\tau_G^{(\text{stim})}\left(E_{Gi}^{(e)}\right)\right)\tau_{\text{esc}}^{(G)}\left(E_{Gi}^{(e)}\right)\right)\exp\left(\left(E_{Gi}^{(e)} - E_F^{(e)}\right)/k_B T\right)}. \tag{2.66}$$

2.4 Light–Matter Interactions in Quantum Dots

With the assumption opposite to the excitonic one, when the distribution of holes is assumed fully equilibrium, the formula is rewritten as

$$f_G^{(e)}(E_{Gi}^{(e)})$$

$$\approx \frac{1 + \left(\tau_{esc}^{(G)}(E_{Gi}^{(e)})/2\tau_G^{(stim)}(E_{Gi}^{(e)})\right)\left(1 - f_G^{(h)}(E_{Gi}^{(h)})\right)\exp\left((E_{Gi}^{(e)} - E_F^{(e)})/k_B T\right)}{1 + \left(1 + \left(1/\tau_{sp} + 1/2\tau_G^{(stim)}(E_{Gi}^{(e)})\right)\tau_{esc}^{(G)}(E_{Gi}^{(e)})\right)\exp\left((E_{Gi}^{(e)} - E_F^{(e)})/k_B T\right)}$$

(2.67)

with $f_G^{(h)}(E_{Gi}^{(h)})$ being the quasi-equilibrium value determined by the hole Fermi quasi-level.

Expression (2.66) is an approximation, like (2.44). However, it is useful in that it demonstrates the nature of *spectral hole burning* in a QD laser or amplifier. Indeed, the stimulated recombination time $\tau_G^{(stim)}$ (2.65) is shortest for a dot with the electron–hole separation resonant with the light: $\Delta E_G^{(e-h)} = \hbar\omega_{ref}$. This leads to a local minimum in $f_G(E_G^{(e)})$ given by (2.66), with a position at the resonant condition $\Delta E_G^{(e-h)} = \hbar\omega_{ref}$ and a width on the order of ΔE_{hom} in terms of electron energy. By virtue of (2.9) or (2.55), this also leads to a depression in gain at photon energies around $\hbar\omega_{ref}$ – the *spectral hole*. It has to be mentioned though that with $\Delta E_{hom} \sim 10$ meV and realistic light intensities, the "hole" is rather smeared and not very pronounced, as illustrated in Figure 2.8. This is calculated for the same parameter values as Figure 2.3, with the excitonic approximation made and the optical signal that causes hole burning incident at the photon energy coinciding with the notional ground-state position $\hbar\omega_{ref} = \Delta E_G^{(e-h)}$ (indicated by the arrow in the figure). The strength of the signal is characterized by $\tau_{esc}^{(G)}(E_{G0}^{(e)})/\tau_{stim}^{(G)}(E_{G0}^{(e)}) = 0.3$ for the exactly resonant dot group, which corresponds to a rather substantial optical power. The hole in the *distribution function* is clearly visible. The hole in the corresponding *gain* spectrum is, however, smeared out by the integration; its magnitude would be reduced further if the excitonic approximation were not made.

Note that due to the fast relaxation time for interlevel transitions, an echo hole also appears in the energy distribution for the *excited* level and the excited level gain (though this was not considered in the calculation of Figure 2.8).

It is worth noting also that although formula (2.66) appears to be a closed-form solution of the rate equations for the dot occupancies, the expression for the stimulated recombination time contains the local light intensity that is determined by gain, whether in a laser or in an amplifier, and the gain itself is a function of the distribution functions.

The evolution of $Y_{f,b}$ in the case of light slow enough to satisfy (2.63) is governed by Equation 2.51 in which the right-hand side can, in this case, be treated as a relatively simple algebraic equation:

$$(\hat{g}_{mod} - a_i + ik_{ref}\widehat{\Delta\eta}_{mod})Y_{f,b} \rightarrow (g_{mod}(\omega_{ref}, z, t) - a_i + ik_{ref}\Delta\eta_{mod}(\omega_{ref}, z, t))Y_{f,b},$$

(2.68)

where, however, the gain and refractive index contributions need, in general, to be calculated by integration over all energy states as in (2.59) and (2.62).

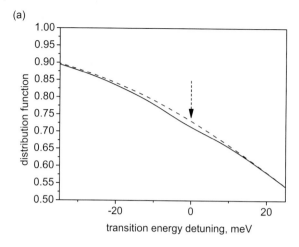

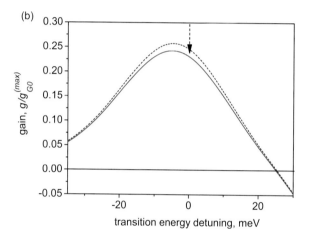

Figure 2.8 Evolution of a distribution function of electrons in a quantum dot (a) and the corresponding gain profile (b) with an increase in optical power: the emergence of a spectral hole. Dashed curve: no optical signal; solid curve: strong optical signal at the center of the ground-state band (indicated by an arrow). The excitonic approximation is made in the calculations; total carrier density $N = 2.7 N_D$.

A still more complex approach is necessary if the timescale on which the optical signal varies is comparable to the dephasing time T_2 or, in spectral terms, the spectral width $\Delta\omega_{\text{sign}}$ of the signal generated or amplified in the device is comparable to, or greater than, homogeneous broadening $\Delta\omega_{\text{sign}} \sim \Delta E_{\text{hom}}/\hbar = 1/T_2$. In this case, the complex operator describing the gain and the contribution to self-phase modulation due to the ground-state transition would become

$$(\hat{g}_{\text{mod}} + ik_{\text{ref}}\widehat{\Delta\eta}_{\text{mod}})Y_{f,b} = \hat{g}^{(G)}_{\text{mod}} = \varrho_G \frac{n_1 N_D^{(2d)}}{w_{\text{mod}}} \int dE_G^{(e\text{-}h)} D_{\text{inh}}(E_G^{(e-h)}) \hat{\sigma}_{\text{mod}}(E_G^{(e-h)}) Y_{f,b}.$$

(2.69)

Here, the complex gain cross section operator includes the Lorentzian homogeneous transition broadening implemented in time domain:

$$\hat{\sigma}_{mod}(E_G^{(e-h)})Y_{f,b}(z,t) = \frac{\sigma_G}{T_2}\int_0^\infty d\tau \exp\left[\left(-\frac{1}{T_2}-i\frac{\Delta E_G^{(e-h)}-\hbar\omega_{ref}}{\hbar}\right)\tau\right]$$
$$\times (f_G^{(e)}(E_G^{(e)},z,t-\tau)+f_G^{(h)}(E_G^{(h)},z,t-\tau)-1)Y_{f,b}(z,t-\tau). \quad (2.70)$$

The stimulated recombination rate becomes

$$R_G^{(stim)}(E_G^{(e-h)},z,t) = v_g \text{Re}[Y_f^*(z,t)\hat{\sigma}_{mod}(E_G^{(e-h)})Y_f(z,t)$$
$$+ Y_b^*(z,t)\hat{\sigma}_{mod}(E_G^{(e-h)})Y_b(z,t)]. \quad (2.71)$$

Contributions of other states (excited state and wetting layer) to ground-level transitions will be considered in Section 5.4.2.

An immediate conclusion we can draw from (2.70) and (2.71) is that with very short pulses, with a duration $\Delta\tau_p \ll T_2$, the interaction of light with dots is considerably weakened. In particular, with short and intense pulses, coherent phenomena such as self-induced transparency become possible. This is not unique for quantum dots, but may be facilitated in QD media, particularly at low temperatures, because the time T_2 is somewhat longer in QDs than in other materials.

An alternative to the formalism of (2.70) and (2.71) is known as *semiconductor Bloch equations* and consists of introducing, for each pair of electron and hole states i and j (where both can be either in the excited or in the ground state and must satisfy the size selection rule, that is, belong to the same group of dots), a separate complex variable, the transition polarization [44].

As regards the choice of a QD model for a particular theoretical and numerical study, the following hierarchy of approaches may thus be envisaged:

1) In the most generic case, when it is not clear whether or not the optical signal involved satisfies condition (2.63) for the adiabatic elimination of polarization (such as when modeling generation or amplification of subpicosecond pulses with a duration comparable to, or smaller than, T_2), the full formalism of either (2.70) and (2.71) or the Maxwell–Bloch equations needs to be implemented.

    In a numerical simulation, the temporal integrals in (2.70) and (2.71) may be implemented using digital filters/recursive formulas at a relatively modest extra computational cost compared to the numerical realization of (2.59) and (2.64); alternative numerical approaches are also available for implementing the dispersive operators (see Section 5.2.3). However, the need to take very small temporal steps in calculations ($\Delta t \ll 2\pi/\Delta\omega_{sign}$, $\Delta t \ll T_2$), combined with the need to resolve the variables in carrier energy as well as time and space, makes this approach a rather daunting computational task, given the present state of numerical facilities, and limits the timescales that can be usefully simulated. Another difficulty is that although the gain caused by, and the self-phase modulation contribution from, transitions closely resonant with the light – for example, ground-state transitions in the case of light resonant with ground-state emissions – can be quite rigorously taken into account by (2.70) and (2.71), it is more computationally taxing to do so for nonresonant light, for example, for excited state and the WL contribution to the self-phase modulation of light resonant with ground-state transitions. Mathematically, it is straightforward to extend the integral over energies in (2.70) to include (with an appropriate modification for the degeneracy factor) carriers in the excited state, which has indeed been performed in Ref. [44]. However, numerical implementation of such an approach is quite taxing as it requires numerical steps on the order of $\Delta t \ll 2\pi\hbar/(E_E - E_G) \sim 10$ fs. The same holds for the Maxwell–Bloch equations, where fast pulsations of the polarization need to be taken into account accurately. Including the effects of the wetting layer in a numerical implementation of this rigorous approach is even more difficult. An approach using Maxwell–Bloch equations has been reported for the analysis of ultrashort-pulse propagation in QD amplifiers [44], including the effects of lateral diffraction of light in a broad waveguide. Propagation without significant distortion was predicted in a broad range of parameters, with the coherent effects likely to play some role. However, the extremely high computational resources required impose a limit on the application of such an approach; in Ref. [44], only a single picosecond pulse propagation was reported.

2) In a large number of cases of practical interest in applications such as communications, the optical signal consists of pulses of picosecond, rather than subpicosecond, duration, and with the central frequency known in advance. Then, condition (2.63) holds and so the polarization can be eliminated adiabatically and an approach of the type of (2.59)–(2.62) used. Such approach has been implemented in a number of papers analyzing short-pulse amplification [67–69]. This saves some computational effort compared to either the operator implementation of the homogeneous and inhomogeneous broadening (2.70) and (2.71) or the semiconductor Bloch equations. Indeed, with homogeneous broadening functions introduced as instantaneous, the time steps used in calculations need not satisfy the condition $\Delta t \ll T_2$, saving the computation time. However, the dynamics of all sizes of dots is still taken into account independently, using energy-resolved rate equations (2.17)–(2.19) (or (2.37) and (2.38) if the excited-state dynamics is not considered separately from ground state) for electrons and possibly for holes. This is the only option at low temperatures (or very high light intensities), when it is possible that the distribution of carriers is significantly nonequilibrium under pulse amplification, and ensures accurate calculation at room temperature. However, it is still quite computationally intensive, as the electron and hole populations have to be resolved in space, time, and energy, and the instantaneous gain coefficient and the refractive index determined from (2.59)–(2.62) at each moment in time and space. Besides, as with all comprehensive numerical models, an accurate account for all the factors present can make it difficult to distinguish the dominant contributions and establish the main tendencies. Thus, when possible, a simpler approach is useful.

3) Such an extra simplification is indeed possible if, in addition to the condition (2.63), it is known that the temperature is high enough, and the optical intensities moderate enough, to ensure that the carrier escape time is much smaller than the combined spontaneous and stimulated recombination time:

$$\left(\frac{1}{\tau_{\text{spon}}} + \frac{1}{\tau_G^{(\text{stim})}(E_G^{(e)})}\right)\tau_{\text{esc}}^{(G)} \ll 1. \tag{2.72}$$

Then, as implied by (2.66), the energy distribution of carriers in the dots and the WL will be near equilibrium, not only in steady state but also in dynamics. Therefore, it is possible to use the approach common in analyzing bulk and QW devices, in which the distribution functions for all electron and hole states and at each moment in time and space will be mainly determined by the *total* carrier (electron and hole) densities N_{tot} in the active layer (including the carriers in the QD and the WL), with a small correction caused by nonequilibrium effects such as dynamic hole burning. For example, for electrons, we get

$$f_G^{(e)}(E_G^{(e)}) = f_{QE}^{(e)}(E_G^{(e)}) + \Delta f^{(e)}(E_G^{(e)}), \tag{2.73}$$

where $f_{QE}^{(e)}(E_G^{(e)}) = 1/1 + \exp((E_G^{(e)} - E_F^{(e)}(N_{\text{tot}}))/k_B T)$ is the quasi-equilibrium, thermal, Fermi distribution function, and $\Delta f^{(e)}(E_G^{(e)})$ ($|\Delta f^{(e)}(E_G^{(e)})| \ll 1$) describes the hole burning. A similar expression can be written for holes, if required, and the wetting layer population is described as

$$N_{\text{wl}}(E_G^{(e)}) = N_{\text{wl}}^{(QE)} + \Delta N_{\text{wl}} \quad \text{or} \quad f_{\text{wl}}(E_G^{(e)}) = f_{\text{wl}}^{(QE)} + \Delta f_{\text{wl}}, \tag{2.74}$$

where, again, $N_{\text{wl}}^{(QE)} = N_c^{(\text{wl})} f_{\text{wl}}^{(QE)}$ is a function of the total carrier density only, and the second term describes the fast varying correction due to the finite capture and escape time. This can greatly simplify the numerical analysis of QD devices, and in some cases, may even allow to obtain analytical solutions, as shown in Chapter 3. In this case, the gain coefficient due to the ground-state transitions can be expressed in a form similar to (2.73):

$$g_G(\hbar\omega) = g_G^{(QE)}(\hbar\omega) + \Delta g_G(\hbar\omega), \tag{2.75}$$

where the first term is the quasi-equilibrium gain describing the evolution of the optical properties of the QD laser or amplifier on a timescale much longer than the capture and escape of carriers and is related to the quasi-equilibrium distribution function. The second term is the nonlinear correction to the gain, taking into account the fast dynamics and related to the nonequilibrium correction:

$$g_G^{(QE)}(\hbar\omega) = A_G \varrho_G N_D \int \left(f_{QE}^{(e)}(E_G^{(e)}) + f_{QE}^{(h)}(E_G^{(h)}) - 1\right)$$

$$D_{\text{hom}}(\hbar\omega, E_G^{(e-h)}) D_{\text{inh}}(E_G^{(e-h)}) dE_G^{(e-h)} \tag{2.76}$$

and

$$\Delta g_G = \Delta g_G^{(e)} + \Delta g_G^{(h)}, \qquad (2.77a)$$

$$\Delta g_G^{(e,h)} = A_G \varrho_G N_D \int \Delta f_G^{(e,h)}\left(E_G^{(e,h)}\right) D_{\text{hom}}\left(\hbar\omega, E_G^{(e-h)}\right) D_{\text{inh}}\left(E_G^{(e-h)}\right) dE_G^{(e-h)} < 0. \qquad (2.77b)$$

By integrating the energy-resolved rate equations, we can obtain the rate equations for the total carrier density similar to those commonly used for bulk and QW materials:

$$\frac{dN_{\text{tot}}}{dt} = R_{\text{pump}} - \frac{N_{\text{tot}}}{\tau_{\text{sp}}(N_{\text{tot}})} - v_g (Y_f^* \hat{g}_G Y_f + Y_r^* \hat{g}_G Y_r). \qquad (2.78)$$

Here, the spontaneous recombination rate $N_{\text{tot}}/\tau_{\text{sp}}(N_{\text{tot}})$ can be calculated assuming equilibrium distribution over dot energy levels and the wetting layer, with the recombination rates of electrons in the dots and the WL being, in general, different, as described above (e.g., using Equation 2.26 for the QD electrons and a bimolecular recombination law (2.29) for WL electrons); the gain $\hat{g}_G$ contains both the quasi-linear and the nonlinear contribution. The operator nature of gain represents gain dispersion. It is important in the operation of *laser* devices, if the spectral properties of the laser are to be accurately represented, and for *amplifiers* if the incident signal is significantly broadband. When considering amplification of a signal narrowband enough to satisfy (2.63), the operator nature of the second term in (2.78) can be neglected, leaving simply

$$\frac{dN_{\text{tot}}(z, t)}{dt} \approx R_{\text{pump}} - \frac{N_{\text{tot}}}{\tau_{\text{sp}}(N_{\text{tot}})} - v_g g_G (|Y_f(z, t)|^2 + |Y_r(z, t)|^2), \qquad (2.79)$$

where, in the case of amplifiers, the light propagation is usually unidirectional to a good accuracy so that $|Y_r(z, t)|^2$ in the last term disappears leaving only $|Y_f(z, t)|^2$.

The nonequilibrium corrections are calculated from the kinetic equations within the first-order perturbative approach. To implement this, we assume in kinetic equations that the capture and escape terms for the equilibrium contributions $f_{\text{QE}}^{(e)}(E_G^{(e)})$ vanish. We shall follow Refs [44, 48], in assuming that under laser/amplifier operating conditions, relaxation between the ground and the excited level is faster than capture into dots, and so will use the kinetic equations in the form of (2.37)–(2.39). Then, for the quasi-equilibrium part of the electron distribution function, we get

$$\frac{\varrho_{\text{wl}} f_{\text{wl}}^{(\text{QE})} \left(1 - f_{\text{QEi}}^{(e)}\right)}{\varrho_G \tau_{\text{cap}}^{(G)}} - \frac{f_{\text{QEi}}^{(e)}}{\tau_{\text{esc}}^{(Gi)}} f_{\text{wl}}^{\prime(\text{QE})} - \frac{f_{\text{QEi}}^{(e)}}{\tau_{\text{sp}}^{(G)}} = 0, \qquad (2.80)$$

where, for brevity, i indicates the dot size, determining the electron–hole level separation $E_{Gi}^{(e-h)}$ and the energies $E_{Gi}^{(e)}$ and $E_{Gi}^{(h)}$ – the former of these is the argument of the function $f_{\text{QE}}^{(e)}(E_G^{(e)})$ in (2.80). The effective fraction of empty WL states

$f''^{(QE)}_{wl} = f'_{wl}(N^{(QE)}_{wl})$ can be calculated from the quasi-equilibrium WL population using Equation 2.34 or 2.36.

A similar equation can be written for holes. Below, we shall assume that the characteristic electron and hole capture times $\tau^{(Ge)}_{cap}$ and $\tau^{(Gh)}_{cap}$ can be assumed to be independent of the QD-level populations, as implied in the analysis above – though a generalization should be possible if necessary. Moreover, as the capture is believed to be mainly phonon assisted, we shall neglect the dependence of the characteristic times on the WL density as well (the approach can be relatively straightforwardly generalized to include such a dependence, at least to the first order). Then, the linearization of (2.37)–(2.39), taking into account (2.80), gives for the nonequilibrium corrections $\Delta f^{(e)}(E^{(e)}_G) = \Delta f^{(e)}_i$ the following relation:

$$\frac{\partial \Delta f^{(e)}_i}{\partial t} = -\frac{\Delta f^{(e)}_i}{\tau^{(i)}_{rel}} + \Delta f_{wl}\left(\frac{\varrho_{wl}}{\varrho_G}\frac{1-f^{(e)}_{QEi}}{\tau^{(G)}_{cap}} + \left|\frac{\partial f_{wl}}{\partial f_{wl}}\right|\frac{f^{(e)}_{QEi}}{\tau^{(Gi)}_{esc}}\right) - R^{(stim)}_{QEi}. \quad (2.81)$$

Here, the effective relaxation rate for the QD electron is defined as

$$\frac{1}{\tau^{(i)}_{rel}} = \frac{\varrho_{wl}}{\varrho_G}\frac{f^{(QE)}_{wl}}{\tau^{(Gi,dyn)}_{cap}} + \frac{f'^{(QE)}_{wl}}{\tau^{(Gi,dyn)}_{esc}} + \frac{1}{\tau_{sp}} + \frac{1}{\tau^{(stim)}_G(E^{(ei)}_G)}, \quad (2.82)$$

with the "dynamic" capture and escape rates introduced as

$$\frac{1}{\tau^{(Gi,dyn)}_{cap}} = \frac{1}{\tau^{(Gi)}_{cap}} - (1-f^{(QE)}_{Gi})\frac{\partial}{\partial f_{Gi}}\left(\frac{1}{\tau^{(Gi)}_{cap}}\right), \quad \frac{1}{\tau^{(Gi,dyn)}_{esc}} = \frac{1}{\tau^{(G,i)}_{esc}} + f^{(QE)}_{Gi}\frac{\partial}{\partial f_{Gi}}\left(\frac{1}{\tau^{(Gi)}_{esc}}\right). \quad (2.83)$$

In practice, the first two terms (capture and escape) dominate in (2.82), as we need the conduction (2.72) to hold, or, in other words, the processes of capture and escape to be much faster than those of recombination, in order to introduce the approximation (2.73). The quasi-equilibrium stimulated recombination rate for the dot group i in (2.81) is given by an equation of the type of (2.64) with the quasi-equilibrium distribution functions:

$$R^{(stim)}_{QEi} = v_g A_G \left(f^{(e,QE)}_{Gi} + f^{(h,QE)}_{Gi} - 1\right) D_{hom}\left(\hbar\omega_{ref}, \Delta E^{(e-h)}_{Gi}\right)\left(|Y_f|^2 + |Y_b|^2\right). \quad (2.84)$$

We can derive the equation for the *total* nonequilibrium ground-state carrier density (which can be said to describe the effect known as QD–wetting layer hole burning, similar to well-barrier hole burning in QW lasers) by summing/integrating over all groups of dots (or integrating over transition energy):

$$\Delta N_G = \varrho_G N_D \sum_i \Delta f^{(e)}_i = \varrho_G N_D \int \Delta f^{(e)}\left(E^{(e-h)}_G\right) D_{inh}\left(E^{(e-h)}_G\right) dE^{(e-h)}_G. \quad (2.85)$$

Then, by summing Equation 2.81 for all groups of dots in the same way, we get

$$\frac{\partial \Delta N_G}{\partial t} \approx -\frac{\Delta N_G}{\tau^{tot}_{rel}} - v_g g_G\left(|Y_f(z,t)|^2 + |Y_r(z,t)|^2\right). \quad (2.86)$$

Here, we have used the relation

$$\Delta N_{wl} = -\frac{V_D}{V_{wl}}\Delta N_G \qquad (2.87)$$

and introduced the characteristic relaxation time for the QD–WL hole burning as

$$\frac{1}{\tau_{rel}^{(tot)}} \approx \frac{1}{\tau_{cap}^{(G,dyn)}}\left(\frac{f_{wl}^{(QE)}\varrho_{wl}}{\varrho_G} + 1 - \frac{N_G^{(QE)}}{\varrho_G N_D}\right)$$

$$+ \frac{1}{\tau_{esc}^{(G,dyn)}}\left(f_{wl}^{\prime(QE)} + \frac{V_D N_G^{(QE)}}{V_{wl}}\left|\frac{\partial f'_{wl}}{\partial N_{wl}}\right|\right) + \frac{1}{\tau_{sp}}. \qquad (2.88)$$

The expression includes an additional approximation of neglecting the transition energy (or i) dependence of the dynamic capture and relaxation times. Assuming the spectral width of the signal and the homogeneous broadening smaller than the inhomogeneous broadening and k_BT, the dependence is indeed not strong over the integration energy range, and the characteristic times can be estimated at the transition energy $E_{Gi}^{(e-h)} = \hbar\omega_{ref}$.

Note that Equation 2.88 does not have a stimulated recombination contribution; it is included implicitly in (2.86) since the last term in (2.86) includes the *total* gain, both linear and nonlinear. In practice, this is not too important since, as in (2.82), the first two terms dominate in expression (2.88).

The dynamic equation for the electron contribution to the nonlinear gain is then obtained from (2.77) and (2.81) in the following form:

$$\frac{\partial \Delta g^{(e)}}{\partial t} = -\frac{\Delta g^{(e)}}{\tau_{rel}^{(g,e)}} + \Delta N_{wl}\left(\frac{V_{wl}}{\varrho_G V_D N_D}\frac{g_{max}-g_{QE}^{(e,G)}}{\tau_{cap}^{(G,dyn)}} + \left|\frac{\partial f'_{wl}}{\partial N_{wl}}\right|\frac{g_{QE}^{(e,G)}}{\tau_{esc}^{(G,dyn)}}\right) - R_{NL}^{(stim)}. \qquad (2.89)$$

Here, the gain relaxation time is estimated as

$$\frac{1}{\tau_{rel}^{(g,e)}} \approx \frac{N_{wl}^{(QE)}V_{wl}}{\varrho_G V_D N_D}\frac{1}{\tau_{cap}^{(G,dyn)}} + \frac{f_{wl}^{\prime(QE)}}{\tau_{esc}^{(G,dyn)}} + \frac{1}{\tau_{sp}} + v_g\frac{A_G}{\Delta E_{hom}^{(e)}}|Y|^2, \qquad (2.90)$$

where, as in (2.88), the capture and escape times are evaluated in the first approximation at $E_{Gi}^{(e-h)} = \hbar\omega_{ref}$.

In the bracketed term in (2.89), the parameter g_{max} is the gain at the spectral point in question under full inversion:

$$g_{max}(\hbar\omega) = A_G \varrho_G N_D \int D_{hom}\left(\hbar\omega_{ref}, E_G^{(e-h)}\right) D_{inh}\left(E_G^{(e-h)}\right) dE_G^{(e-h)} \qquad (2.91)$$

and $g_{QE}^{(e,G)}$ is the electron contribution to the total quasi-equilibrium gain:

$$g_{QE}^{(e,G)}(\hbar\omega) = A_G \varrho_G N_D \int f_{QE}^{(e)}\left(E_G^{(e)}\right) D_{hom}\left(\hbar\omega, E_G^{(e-h)}\right) D_{inh}\left(E_G^{(e-h)}\right) dE_G^{(e-h)}. \qquad (2.92)$$

Note that with the hole contribution $g_{QE}^{(h,G)}$ defined in a similar way, the total quasi-equilibrium gain is given, from (2.76), by $g_G^{(QE)}(\hbar\omega) = g_{QE}^{(e,G)}(\hbar\omega) + g_{QE}^{(h,G)}(\hbar\omega) - g_{max}(\hbar\omega)$. In a dynamic simulation, both $g_{QE}^{(e,G)}$ and $g_{QE}^{(h,G)}$ may be precalculated and tabulated prior to the start of simulation to save computation time – or just calculated approximately since, as shown below, the term that contains them is not the most important one in the calculation of the nonlinear gain.

Finally, the nonlinear stimulated recombination term, describing the spectral hole burning in the QD active layer, is given by

$$R_{NL}^{(stim)} = A_G \varrho_G N_D \int R_{QE}^{(stim)}(E_G^{(e-h)}) D_{hom}(\hbar\omega, E_G^{(e-h)}) D_{inh}(E_G^{(e-h)}) dE_G^{(e-h)} \quad (2.93)$$

or, in shorthand notation, $R_{NL}^{(stim)} = A_G \varrho_G N_D \sum_i R_{QE,i}^{(stim)} D_{hom}(\hbar\omega, E_{G,i}^{(e-h)})$. Substituting (2.84),

$$R_{NL}^{(stim)} = F_{NL}^{(stim)}(N)(|Y_f|^2 + |Y_b|^2), \quad (2.94)$$

where

$$F_{NL}^{(stim)}(N) = v_g A_G^2 \varrho_G N_D \sum_i D_{hom}^2(\hbar\omega_{ref}, E_{G,i}^{(e-h)})(f_{Gi}^{(e,QE)} + f_{Gi}^{(h,QE)} - 1) \quad (2.95)$$

contains only quasi-equilibrium occupation probability values and can thus be precalculated before the start of dynamic simulations as a function of carrier density, as can $g_{QE}^{(e,G)}(\hbar\omega, N)$.

Equation 2.89 is a useful result, as it allows the nonlinear gain dynamics in a QD laser or amplifier to be described without the need to analyze the energy-resolved dynamics of dot populations. Some points need to be noted, however, when applying it. First, as discussed above, it relies substantially on the approximation of neglecting the energy dependence of dynamic capture and escape times. However, the fact that the experimentally studied dynamics of QD gain is well fitted by an exponential with a picosecond time constant (plus a subpicosecond one that may be attributed to interlevel relaxation and a much slower recombination one) [39] implies that describing the gain relaxation by a single time is indeed a reasonable approximation. Second, the nonlinear gain described by (2.89) is calculated at the spectral point corresponding to the optical signal frequency/wavelength, in other words, at the bottom of the spectral hole. For analyzing moderately broadband signals, (2.89) may be modified to include a digital filter simulating the spectral shape of the hole, as will be discussed in chapter 5. Third, of course, (2.89) is a small-signal approximation based on the assumption $|\Delta f^{(e)}(E_G^{(e)})| \ll 1$, which restricts its applicability to moderate light intensities.

2.5
The Nonlinearity Coefficient

In the case when the optical signal is slow enough to treat (2.86) and (2.89) as instantaneous, we get from (2.89) and (2.86)

$$\Delta g^{(e)}(\hbar\omega) \approx -\varepsilon^{(e)}\left(|Y_f^2| + |Y_b^2|\right)g_G^{(QE)}(\hbar\omega), \tag{2.96}$$

where $\varepsilon^{(e)}$ is the *electron contribution* to the gain compression coefficient, determined by

$$\varepsilon^{(e)}(\hbar\omega, N) \approx v_g A_G \tau_{rel}^{(g,e)} \left\{ \frac{\sum_i D_{hom}^2(\hbar\omega_{ref}, E_{G,i}^{(e-h)})(f_{Gi}^{(e,QE)} + f_{Gi}^{(h,QE)} - 1)}{\sum_i D_{hom}(\hbar\omega_{ref}, E_{G,i}^{(e-h)})(f_{Gi}^{(e,QE)} + f_{Gi}^{(h,QE)} - 1)} \right.$$

$$\left. - \left(\frac{1}{\varrho_G V_D} \frac{\tau_{rel}^{(tot)}}{\tau_{cap}^{(G,dyn)}} (g_{max} - g_{QE}^{(e,G)}) + \frac{V_{QD}}{V_{wl}} \left| \frac{\partial f'_{wl}}{\partial N_{wl}} \right|_{g_{QE}^{(e,G)}} \frac{\tau_{rel}^{(tot)}}{\tau_{esc}^{(G,dyn)}} \right) \right\}. \tag{2.97}$$

With the hole contribution defined in a similar way, the *total* gain compression coefficient can be introduced as

$$\Delta g(\hbar\omega) \approx -\varepsilon(|Y_f^2| + |Y_b^2|)g_G^{(QE)}(\hbar\omega), \tag{2.98a}$$

$$\varepsilon(\hbar\omega, N) = \varepsilon^{(e)}(\hbar\omega, N) + \varepsilon^{(h)}(\hbar\omega, N). \tag{2.98b}$$

The notion of gain compression coefficient is well known in the analysis of nonlinear gain in semiconductor lasers and SOAs and is often introduced phenomenologically. Microscopically, in the context of bulk and quantum well active media, it includes well-barrier and spectral hole burning, as well as dynamic heating of carriers. The latter is not relevant for quantum dots, as the carrier–carrier interactions that establish a carrier temperature separate from the lattice temperature are not a feature of quantum dot materials, and is consequently absent in (2.97). For carriers in the wetting layer, the concept of a carrier temperature different from the lattice temperature may apply, and some authors have proposed to introduce it [70], which may somewhat modify the expression for ε. However, due to the relatively long times of determining the magnitude of spectral hole burning nonlinearity (the capture and escape times of ~1 ps, as opposed to carrier scattering times on the order of 0.1 ps in bulk and QW materials), it may be expected that the hole burning nonlinearity will dominate in QD lasers and SOAs.

A heuristic expression often used for bulk and QW materials can then be adopted for QD materials as well [50]:

$$g_G(\hbar\omega) \approx \frac{g_G^{(QE)}(\hbar\omega)}{1 + \varepsilon(|Y_f^2| + |Y_b^2|)}. \tag{2.99}$$

In the case of $\Delta g(\hbar\omega) \ll g_G^{(QE)}(\hbar\omega)$ (which, technically speaking, is the only case when both approximations (2.77a) and (2.99) are valid anyway), (2.99) is equivalent to (2.77a) and (2.98a); however, heuristically it is sometimes used in the case when the nonlinear gain is only a few times smaller than linear, and is believed to be somewhat more accurate than approximations (2.77a) and (2.98a).

2.5 The Nonlinearity Coefficient

An accurate evaluation of the nonlinearity coefficient value and its spectral dependence, obviously, requires numerical integration and relies on careful analysis of the capture and escape rates for the calculation of the dynamic capture and escape times. A simple estimate is possible [50] if, to start with, we assume that the negative correction represented by the bracketed term can be neglected (which is indeed the case at high wetting layer carrier densities, when $(g_{max}-g_{QE}^{(e,G)})$ and $|\partial f'_{wl}/\partial N_{wl}|$ are both small factors). To obtain a very rough estimate, we can approximate the Lorentzian homogeneous broadening function (2.8) by a "top hat" function:

$$D_{hom}\left(\hbar\omega, E_G^{(e-h)}\right) = \begin{cases} \dfrac{1}{\Delta E_{hom}}, & \text{if } \left|\hbar\omega - E_G^{(e-h)}\right| < \dfrac{\Delta E_{hom}}{2}, \\ 0, & \text{otherwise.} \end{cases} \qquad (2.100)$$

In this case, (2.97) can be evaluated analytically and gives a rough estimate for the electron contribution to the nonlinearity as

$$\varepsilon^{(e)}(\hbar\omega, N) \sim \frac{v_g A_G \tau_{rel}^{(g)}}{\Delta E_{hom}}. \qquad (2.101)$$

A similar estimate is possible for the hole contribution. The phenomenological constant A_G can be expressed through microscopic parameters or related to measured gain. The order of magnitude estimate [50] assuming a characteristic relaxation time of 1 ps gives a value of $\sim 10^{-23}$ m^3, which is a few times, possibly up to about one order of magnitude, greater than the values typically quoted in the literature (see, for example, Ref. [71]) for quantum well or bulk active layers in lasers or SOAs. This slightly, but not dramatically, larger nonlinearity coefficient is consistent with the fact that the relaxation time $\tau_{rel}^{(g)}$ of the nonlinear gain (to which it is proportional according to (2.97)) also appears to be a few times, but not several orders of magnitude, greater than the corresponding times in bulk or QW materials (namely, the times of electron–electron collisions for spectral hole burning and carrier cooling through phonon emission for dynamic carrier heating, usually evaluated as 100 and 300–500 fs, respectively). Thus, arguably the most important difference between the QD and the more traditional active media under the conditions discussed above is the very strongly sublinear (saturating) dependence of the quasi-equilibrium gain on the total carrier density in the active layer.

It is worthwhile to note that while the exact quantitative values of the carrier density-dependent quasi-equilibrium gain and the nonlinearity coefficient are functions of both homogeneous and inhomogeneous broadening as well as the specific values of the characteristic times involved, the most salient features of the above model – the representation of the gain in the form of (2.99) (or (2.77a) and (2.98a)), the strongly nonlinear dependence (saturation) of the quasi-linear gain with carrier density, the order-of-magnitude value of the nonlinearity coefficient, and the characteristic times of its relaxation – are readily reproduced by the simplest model neglecting the inhomogeneous broadening (with a gain in the form of (2.3) and (2.4)). This partly explains the success of such a model in explaining and predicting the behavior of QD optical devices in the cases when the spectral properties of linear and nonlinear gain are not too important.

At very high light intensity values, yet another nonlinear effect may become significant in reducing the gain in a QD (or any other semiconductor) amplifier or laser, namely, the two-photon absorption (TPA). To take TPA into account, the propagation equation (2.51) is modified, in the simplest approximation, in the form (see, for example, Ref. [72]):

$$\pm \frac{\partial Y_{f,b}}{\partial z} + \frac{1}{v_g} \frac{\partial Y_{f,b}}{\partial t} \approx \left(\frac{1}{2}(\hat{g}_{mod} - a_i) + ik_{ref}\widehat{\Delta\eta}_{mod} \right) Y_{f,b} \\ - \frac{1}{2} \kappa_{TPA}\left(1 - i\alpha_H^{(TPA)}\right) \frac{P_{f,b}}{A_X} Y_{f,b} + F_{spont}(z,t). \quad (2.102)$$

where, following Ref. [72], κ_{TPA} is the two-photon absorption coefficient of the laser waveguide, $\alpha_H^{(TPA)}$ is the effective Henry factor associated with two-photon absorption, similar to the phenomenological Henry linewidth enhancement factor α_H (Equation 2.53) in linear gain or absorption, $P_{f,b}$ is the power of light propagating in the forward/backward direction along the waveguide, and A_X is the effective modal cross section. Recalling that $P_{f,b} = v_g \hbar \omega A_X |Y_{f,b}|^2$, we note that, at least in the case of unilateral propagation (e.g., in an amplifier or in a mode-locked laser if the circulating pulses do not collide in the gain section), the effect of TPA at modest optical powers is essentially to introduce an extra effective contribution to the gain compression factor in (2.98a), so that, instead of (2.98b), we have

$$\varepsilon(\hbar\omega, N) \approx \varepsilon^{(e)}(\hbar\omega, N) + \varepsilon^{(h)}(\hbar\omega, N) + \varepsilon^{(TPA)}(\hbar\omega, N), \quad (2.103)$$

where

$$\varepsilon^{(TPA)}(\hbar\omega, N) = \frac{v_g \hbar \omega \kappa_{TPA}}{g_{mod}(\hbar\omega, N)}. \quad (2.104)$$

An estimate with $\kappa_{TPA} = 70$ cm GW^{-1} [72] gives the value of $\varepsilon^{(TPA)} = 1 \times 10^{-18}$ cm^3 = 1×10^{-24} m^3 for modal gain values of around 10 cm^{-1}, which is over an order of magnitude smaller than the spectral hole burning contribution estimated from (2.98a), (2.98b), and (2.101). However, it may become an important source of nonlinearity either for small gain values or in the case of very high power. In the latter case, the effect of the gain saturation due to hole burning (or any other effect to do with nonequilibrium carrier distribution, such as dynamic heating of carriers) will have to be described by the saturating curve (2.99), whereas the TPA, as follows from (2.102), will still follow the stronger dependence of the type given by (2.77a) and (2.98a) and may even result in negative net gain, which is impossible with gain saturation caused by hole burning.

3
Quantum Dots in Amplifiers of Ultrashort Pulses

3.1
Optical Amplifiers for High-Speed Applications: Requirements and Problems

Semiconductor optical amplifiers (SOAs) are devices of primary interest in telecommunications and data communications due to their small size, high gain (up to 40 dB), and ease of on-chip optoelectronic integration. SOAs broadly fall into two categories: traveling wave amplifiers and Fabry–Perot cavity amplifiers, in particular vertical cavity ones. The former are more broadband, while the latter can have higher gain. Here, we shall concentrate on traveling wave amplifiers (TWAs), since broadband devices are essential for high-speed applications, and bring out at least some of the advantages of QD materials most prominently.

The general theory of traveling wave optical amplifiers is covered in most textbooks on optical communications; here, we shall briefly recall the most necessary information, mainly following the logic of G.P. Agrawal's textbook [73]. In assessing the performance of any optical amplifier, the main numerical parameters usually assessed are as follows:

- Optical gain, determined from the ratio of the output to input power $G = P_{out}/P_{in}$, or $G = 10 \lg(P_{out}/P_{in})$ in dB units. In terms of the modal gain, the small-signal gain of an amplifier is $G_0 = \exp(G_0) = \exp(g_{mod}^{(0)} L)$, ($L$ being the amplifier length and $g_{mod}^{(0)}$ the small-signal modal gain) assuming ideally nonreflecting amplifier facets. With a finite reflectance, the formula needs to be modified, and the standard of suppression of facet reflections sets the ultimate gain limit if the amplifier is intended to be used as a traveling wave device.
- Saturation power P_{sat}^{out} – generally, the gain *in all types* of laser amplifiers decreases with power, due to the depletion of population inversion by stimulated recombination; P_{sat}^{out} is the (CW) output power at which the gain falls by $\sim$3 dB compared to the small-signal value. The simplest, if not entirely accurate, approximation is to assume a linear gain–carrier density dependence:

$$g_b(N) = \sigma_b(N - N_{tr}) \tag{3.1}$$

Ultrafast Lasers Based on Quantum Dot Structures: Physics and Devices.
Edik U. Rafailov, Maria Ana Cataluna, and Eugene A. Avrutin
Copyright © 2011 WILEY-VCH Verlag GmbH & Co. KGaA, Weinheim
ISBN: 978-3-527-40928-0

where σ_b is the (bulk) material gain cross section, N is the carrier density, and N_{tr} is its transparency value. This is believed to be quite accurate for a bulk amplifier within a broad range of carrier densities, and a reasonable first approximation for a QW and in some cases even QD (cf. the approximation (2.3)).

The dynamics of carrier density in a SOA is described by a local rate equation similar to the laser case

$$\frac{dN}{dt} = R_{pump} - \frac{N}{\tau_{sp}} - v_g g_b S, \qquad (3.2)$$

where all the parameters have a similar meaning to the QD case, $v_g = c/\eta_g$ is the group velocity of light, and S is the photon density (in the notation of equations and assuming forward propagating signal and no reflections, $S = |Y_f|^2$. Then, in steady state ($\partial/\partial t = 0$), the small-signal gain is $g_0 = g|_{S \approx 0} = \sigma_b(R_{pump}\tau_{sp} - N_{tr})$, and the saturated gain in the presence of light,

$$g = \frac{g_0}{1 + S/S_{sat}}, \qquad (3.3)$$

$$S_{sat} = (v_g \sigma_b \tau_{sp})^{-1}. \qquad (3.4)$$

From a steady-state form of Equation 2.51 with $S = |Y_f|^2$, neglecting for simplicity the internal loss (as well as dispersion) and using (3.3), the equation for the photon density is

$$\frac{dS}{dz} = \Gamma_{xy} g S = \frac{\Gamma_{xy} g_0 S}{1 + S/S_{sat}}. \qquad (3.5)$$

This, given the boundary conditions $S|_{z=0} = S_{in}$ and $S|_{z=0} = S_{out} = G S_{in}$, gives for the saturated steady-state total gain an implicit equation

$$G = G_0 \exp\left(-\frac{G-1}{G}\frac{S_{out}}{S_{sat}}\right), \qquad (3.6)$$

$$G_0 = \exp(\Gamma_{xy} g_0 L). \qquad (3.7)$$

This gives for the output photon density, at which the gain falls to $G = G_0/2$, a value

$$S_{sat}^{out} = \frac{G_0 \ln 2}{G_0 - 2} S_{sat}. \qquad (3.8)$$

Recalling the connection of optical power with photon density (cf. (2.58))

$$P = A_X I_{sign} = \hbar \omega A_X v_g S, \qquad (3.9)$$

where $A_X = W d_{act}/\Gamma_{xy}$ is the waveguide cross-section area, we get for the 3 dB saturation power

$$P_{\text{sat}}^{\text{out}} = \hbar\omega A_X \frac{G_0 \ln 2}{G_0 - 2} (A_b \tau_{\text{sp}})^{-1} \approx \hbar\omega A_X \ln 2 (A_b \tau_{\text{sp}})^{-1} \qquad (3.10)$$

since, in practice, usually $G_0 \gg 1$. The typical values in bulk and QW SOAs are about 5–10 mW.

- The gain bandwidth in terms of optical frequency, photon energy, of wavelength, say, at 3 dB value. This parameter is particularly important for multiwavelength applications, or if short subpicosecond pulses are to be amplified. In typical bulk and QW SOAs, the bandwidth is on the order of 30–50 meV, depending on the gain.
- Amplified spontaneous emission noise, usually quantified by the noise figure $F_n = \text{SNR}_{\text{in}}/\text{SNR}_{\text{out}}$ (the ratio of the values of optical signal power to noise at the input and the output of the amplifier); in bulk/QW SOAs, the value is estimated as $F_n \approx N/(N - N_{\text{tr}})$; the actual values are on the order of 5–10 dB.
- Polarization sensitivity (quantified as gain variance depending on light polarization).

An additional performance measure at high bit rates, which is less straightforward to quantify by a single parameter but as important as those measured above, is what is known as transparency to format. Essentially, this is the degree of *distortion* to the signal wavefront of each signal pulse introduced by the amplifier, as well as the degree of *patterning* (the dependence of pulse gain on the prehistory), when a digitally modulated stream is incident on the amplifier. In a small-signal regime, distortions can be caused by dispersion in the amplifier medium and by any resonator effects – intentional in the case of Fabry–Perot SOAs, residual in the case of TWA; when the amplifier operates in a partly saturated regime, the distortions and patterning are caused by dynamic, time-dependent saturation. Self-phase modulation can also be introduced alongside the amplitude distortion.

The main, though not the only, alternative to SOAs comes from rare-earth-doped solid-state amplifiers. In particular, at the wavelength range around 1.55 μm, SOAs face a very strong competition from erbium-doped fiber amplifiers (EDFAs). These devices, though much less compact and not suitable for optoelectronic integration, have a number of advantages of their own, including the ease of smooth integration into fiber links with reflections as low as 10^{-5} (hence high gain, up to 50 dB, possible), broad gain bandwidth (~40 meV), a lower noise figure (~3 dB as opposed to 5–10 dB in SOAs), polarization insensitivity, and, possibly the most important, the slow response (with a characteristic time on the order of milliseconds) meaning that the gain in EDFAs is saturated by the *average*, rather than *instantaneous*, power. This in turn leads to amplification with very little distortion and excellent transparency to format (EDFAs support RZ or NRZ signals at virtually any realistic bit rate). At the wavelength range of 1.3 μm, the technology of rare-earth fiber amplifiers is not fully developed so the competition to SOAs is less intense (though still exists, partly in the form of Raman optical amplifiers). Still, in order to be usable for future high bit rate metropolitan area networks at 1.3 μm, and in particular to compete with EDFAs for a share in the 1.55 μm applications, SOAs need to match their competitors in as many respects as possible.

Some inherent disadvantages of SOAs, such as polarization sensitivity, can be, to a degree, compensated by waveguide design or, when using the device in a fiber link, by schemes using the amplifier in a reflecting geometry with polarization rotation, or two amplifiers, with epitaxial planes at a right angle, in series. In an on-chip integrated scheme, the polarization is predetermined and so polarization sensitivity is not an issue. However, there are other limitations of traditional SOAs that may be more difficult to overcome, and the lack of transparency to format at high bit rates is one of them.

The theory of pulse amplification in a (bulk) SOA and the associated distortions was developed by Agrawal and Olsson [74] and has been widely used since, sometimes with relatively minor modifications. It starts with a simple form of the time-dependent traveling wave equation (2.51) for the forward traveling wave (assuming ideally traveling wave operating regime, with no reflected wave), with the internal loss and dispersion neglected, and the self-phase modulation described by a single Henry factor α_H:

$$\frac{\partial Y_f}{\partial z} + v_g \frac{\partial Y_f}{\partial t} = \frac{1}{2}(\Gamma_{xy} A_b (N - N_{tr})(1 - i\alpha_H)) Y_f. \tag{3.11}$$

For the carrier density, the rate equation (3.2) is used, in which for short enough pulses, the recombination term can be omitted allowing a closed form solution.

The analysis then switches to a traveling time frame $z, \tau = t - z/v_g$, which allows a semianalytical treatment of pulse gain. The resulting value of pulse gain at the local time τ is then obtained [74] in the form

$$G(\tau) = \left\{ 1 - \left(1 - \frac{1}{G_0}\right) \cdot \exp\left[-\frac{U_0(\tau)}{U_{sat}^0}\right] \right\}^{-1}, \tag{3.12}$$

where G_0 is the small-signal gain (3.7) and

$$U_0(\tau) = \int_{-\infty}^{\tau} P_{in}(\tau') d\tau' \tag{3.13}$$

is the energy contained in the part of the input pulse up to the local pulse time τ ($P_{in}(\tau) = v_g \hbar \omega A_X S_{in}(\tau) = v_g \hbar \omega A_X |Y(z = 0, \tau)|^2$ being the input pulse power profile), and

$$U_{sat}^0 = \frac{\hbar \omega A_X}{\sigma_b} = \frac{\hbar \omega A_X}{dg/dN} \tag{3.14}$$

is the pulse saturation energy (*not* power). The output pulse form is then given by

$$P_{out}(\tau) = P(z = L, t = L/v_g + \tau) = G(\tau) P_{in}(\tau). \tag{3.15}$$

Taking an example of a bulk amplifier, with a cross section of, say, $A_x = 5\,\mu m \times 1\,\mu m = 5 \times 10^{-12}\,m^2$, and $\sigma_b = 4 \times 10^{-20}\,m^2$ (a typical value from the literature), with $\hbar \omega \approx 1.5 \times 10^{-19}$ J (light at 1.3 µm), we may estimate $U_{sat}^0 \approx 20\,pJ$. The input light pulse energies in a realistic amplifier are likely to be much lower (e.g., with an input pulse of 0.1 mW peak power and 10 ps duration, the pulse energy is on the order of $U_p \sim 10^{-3}$ pJ). This does not, however, mean that the distortions described by (3.12) have to be weak. Indeed, with $U_p \ll U_{sat}^0$ (and given the obvious fact that $U_0(\tau) < U_p = U_0(\tau \to \infty)$), we can expand $\exp[-U_0(\tau)/U_{sat}] \approx 1 - U_0(\tau)/U_{sat}$ and so rewrite (3.12) in a simplified, very instructive form

$$G(\tau) \approx \frac{G_0}{1 + G_0 U_0(\tau)/U_{sat}^0}. \tag{3.16}$$

From (3.16), it is clearly seen that significant distortions can be expected in the pulse shape, particularly at the trailing edge, at the input pulse energies of the order, *not* of U_{sat}^0, but of a significantly smaller magnitude of U_{sat}^0/G_0. In the example above, with $U_p \sim 10^{-3}$ pJ, the distortions will become very significant at $G_0 = 30$ dB, a realistic value for an SOA.

Due to the presence of the linewidth enhancement factor in (3.11), the distortions are also seen in the phase profile of the pulse, in the form of the saturation-induced self-phase modulation. If the incident pulse is unchirped, the chirp (the correction to the instantaneous optical frequency) introduced by the propagation takes the form

$$\Delta\omega = \frac{\partial}{\partial \tau} \arg(Y_f(\tau)) = -\frac{\Gamma_{xy} \alpha_H P_{in}(\tau)}{2 U_{sat}^0}(G(\tau) - 1). \tag{3.17}$$

With the same numerical example as above, and assuming $\Gamma_{xy}\alpha_H \approx 1$, a reasonable estimate in bulk and QW amplifiers, we may estimate $\Delta \nu = \Delta\omega/2\pi \sim 1$ GHz – not a significant chirp value for most applications, but may be a case for concern in some schemes, particularly those involving cascading amplifiers, when the phase distortions may accumulate and contribute to the limitations to the transparency to format.

The semianalytical SOA theory can also be used to evaluate, not only the distortions of the pulse but also the patterning effects, when amplifying a pseudorandom digital stream, and the maximum bit rate at which the amplifier can avoid patterning. To quantify these, we can compare the two limiting cases that occur when a digital stream passes through the amplifier. The first is a long stream of zeros, equivalent to single-pulse amplification as discussed above, and the second, a long stream of ones, equivalent to the amplification of a continuous stream of pulses with a repetition period of $T = 1/B$, B being the bit rate. In the former case, the gain at the start of each pulse is G_0; in the latter case, the starting gain $G_- < G_0$ is affected by pulse saturation during the pulse and the incomplete (in general) recovery in between pulses. The gain G_- may be evaluated by noting that the gain at the end of each pulse in a stream is $G_+ \approx G_-/(1 + G_- U_p/U_{sat}^0)$ (from (3.16) and with G_0 substituted by G_-). In between the pulses, assuming the pulse duration $\tau_p \ll T$, we can neglect the stimulated recombination term in (3.2), so that with the approximation of a constant recombination time T, the gain recovery is approximately exponential. Then, for an estimate, the gain at the start of the next pulse is approximately evaluated as $G'_- \approx G_+ + (G_0 - G_+)(1 - \exp(-T/\tau_{sp}))$ (more accurately, this would be written for $G = \ln G$). By observing that in a steady pulse stream, we must have $G'_- = G_-$, we obtain an equation for G_-. Then, by setting an acceptable margin ΔG of pulse patterning and requiring that

$$G_0 - G_- < \Delta G, \qquad (3.18)$$

in order to ensure that the relative variation in the output pulse amplitude is $\Delta P_{out}/P_{out} < \Delta G/G_0$, it is possible to evaluate the highest value of B for the patterning to remain within the specified limit. The result is

$$B_{max} = \left(\tau_{sp} \ln \frac{G_0(G_0-1)u_p + (1-G_0 u_p)\Delta G}{\Delta G(1 + u_p(G_0 - 1 - \Delta G))} \right)^{-1}, \qquad (3.19)$$

where $u_p = U_p/U_{sat}^0$. The dependence of B_{max} on the input pulse energy for several gain and gain margin values is shown in Figure 3.1. As expected, under low input powers ($G_0 U_p/U_{sat}^0 \ll 1$), the amplifier is linear and so the patterning negligible at any bit rate; however, with $G_0 U_p/U_{sat}^0 \sim 0.1$, the patterning-free bit rates fall to less than $1/1/\tau_{sp} \sim 1$ Gbit s^{-1}, hardly acceptable for modern optical communication applications. As will be shown below, this result is partly an artifact of the (rather approximate) linear gain–carrier density relationship (3.1). Still, it is indeed the case that patterning is a significant problem in QW and particularly bulk amplifiers – as will be shown later, this is the area in which QD amplifiers promise one of their main advantages.

In addition to the "slow" saturation due to carrier density depletion by light, the interband effects such as spectral hole burning and carrier heating are sometimes important in bulk/QW amplifier saturation. This type of saturation is dealt with by introducing a first-order gain saturation parameter ε, used in a similar way to (2.99):

$$g(N, S) = g(N, 0)/(1 + \varepsilon S) \approx g(N, 0)(1 - \varepsilon S), \qquad (3.20)$$

where, as discussed in Chapter 2, the microscopic processes described by ε are different in bulk, QW, and QD materials, but the phenomenological form is the same.

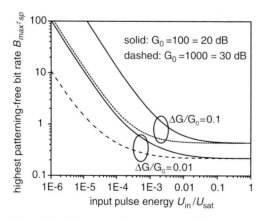

Figure 3.1 Estimated maximum patterning-free bit rate in a generic optical amplifier as a function of input pulse power for several values of the acceptable margin and unsaturated gain.

Values of $1-3 \times 10^{-23}$ m^3 are a typical order of the strength of gain saturation used in literature (see, for example, Ref. [71]). Equation 3.20 has the same functional form as (3.3), but is very different in the sense that (3.3) is valid only at steady state; however, (3.20) holds in dynamics too, so long as the pulses in question are longer than about 0.1 ps (which is the relaxation time of spectral hole burning and carrier heating nonlinearities that contribute to the saturation parameter ε in bulk and QW amplifiers. As will be shown later, the notion of the gain compression factor may be useful in QD amplifiers as well, though will require some care). A simple estimate shows that with typical values for most bulk and QW semiconductors ($\sigma \sim 10^{-16}$ cm^2, $\tau \sim 1$ ns), $\varepsilon S_{sat} \sim 10^{-2} \ll 1$, meaning that the "slow" gain saturation is much more important than the fast saturation in the CW regime. However, when working with short pulses, the importance of the fast saturation increases as the pulses become shorter and shorter. Mork and Mecozzi [75] and [76] have proposed a formula for the critical pulse duration at which the fast saturation is as important as slow saturation:

$$\tau_{crit} \approx \xi \frac{\varepsilon}{v_g (dg/dN)}, \qquad (3.21)$$

where ξ is a numerical constant of the order of one determined as

$$\xi = \tau_p \int_{-\infty}^{\infty} P_{in}^2(\tau') d\tau' \bigg/ \left(\int_{-\infty}^{\infty} P_{in}(\tau') d\tau' \right)^2 \qquad (3.22)$$

and thus depending on the pulse shape (e.g., $\xi = 1$ for a rectangular pulse and $\xi = 0.664$ for a Gaussian). With typical semiconductor parameters, the critical time is on the order of 1–10 ps. For much longer pulses ($\tau_{pulse} \gg \tau_{crit}$), fast processes may be neglected, and for much shorter pulses ($\tau_{pulse} \ll \tau_{crit}$), they completely dominate (although in that case, it may not be possible to treat Equation 3.20 as instantaneous; instead, the recovery time of the fast nonlinearity itself, on the order of 0.1–1 ps as mentioned above, may need to be taken into account).

The pulses of interest for current and next-generation optical communications are of the duration $\sim$1–10 ps and thus in the intermediate range, $\tau_{pulse} \sim \tau_{crit}$, so in general, both types of saturation need to be taken into account. The papers by Mecozzi and Mork show that, in the first approximation, the effect of the pulse nonlinearity is to modify the value of the saturation energy of a pulse as

$$U_{sat}^{\varepsilon} = \frac{U_{sat}^0}{1 + \tau_{crit}/\tau_p}, \qquad (3.23)$$

where U_{sat}^0 is the value given by Agrawal and Olsson's theory as discussed above, τ_{crit} by (3.21), and τ_p is the incident pulse duration. One can then substitute $U_p/U_{sat}^{\varepsilon}$ instead of U_p/U_{sat}^0 in all the formulas above. This formula does not, however, fully account for any modifications to the pulse shape due to the presence of fast nonlinearities, so a more rigorous approach is needed for detailed simulations.

Note that although a linear gain/carrier density dependence (3.1) has been used in deriving the analytical results outlined above, they can be applied, within the first approximation, to an arbitrary g(N) if the carrier density during the pulse changes only weakly around its average value $\bar{N}$ (for low bit rates, $\bar{N} = N_J = R_{pump}\tau_{sp}$ is the unsaturated value given by the pump and determining G_0; for high bit rates, it is influenced by gain saturation). Then, g(N) can be linearized around $\bar{N}$, and the analytical results above hold approximately, with $G_0 = G(\bar{N})$ and $A_{QD} = dg/dN|_{N=\bar{N}}$. As we shall see below, this allows us to apply some of the formulas to explain the advantages of quantum dot SOAs. However, there are some differences between the behavior of a QD SOA and SOAs of other types, leading to the need to modify the theory and predicting new exciting possibilities.

3.2
Quantum Dot Optical Amplifiers: Short-Pulse Operating Regime

QD semiconductors are particularly promising materials for fabricating SOAs for the amplification of subpicosecond and femtosecond pulses because one of the key parameters that influence the emission spectrum and the optical gain in QD devices is the spectral broadening associated with the distribution of dot sizes. With this in mind, a number of groups have investigated QD devices as amplifiers for both CW and ultrashort pulse signals. Impressively, a high-gain amplification (exceeding 18 dB) was demonstrated for CW operation [77] and for sub-200 fs pulses passing through a QD SOA [78]. It is important to note that this latter work also demonstrated that QD devices can amplify ultrashort pulses across a relatively broad spectral range that extends to more than 100 nm, as exemplified in Figure 3.2. This is a noticeable

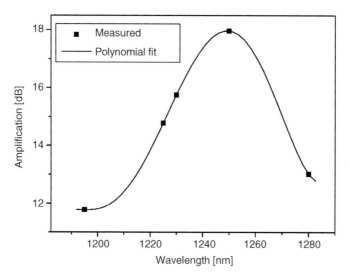

Figure 3.2 Spectral bandwidth of amplified femtosecond pulses, using a quantum dot-based semiconductor optical amplifier.

improvement over the more standard bulk and QW amplifiers, and a direct result of significant inhomogeneous broadening in QD SOAs.

Similar amplifiers based on QD materials have been used to boost the optical power of ultrashort pulses generated from semiconductor mode-locked lasers, also based on QD media. Choi et al. [79] used a 3.2 mm amplifier containing 10 layers of InAs QDs, in tandem with a QD mode-locked laser that was fabricated from the same wafer. The SOA exhibited an output saturation power of 16 dBm, while the maximum small signal gain was 19 dB (including coupling losses). The amplified pulses had durations between 3.5 and 38 ps. As a result of amplification and compression, it was possible to generate pulses with a peak power of 1.22 W (at a repetition rate of 5 GHz). In the future, we believe that it should be possible to raise the saturation power in this type of a device by increasing the number of dot layers from the typical 3–10 layer design, to perhaps 20–50. Further optimization could be achieved by increasing the density of QDs per layer.

3.3
Quantum Dot Optical Amplifiers at High Bit Rates: Low Distortions and Patterning-Free Operation

A step further in the assessment of QD-based SOAs is to evaluate the behavior of these devices at much higher repetition rates. This is especially relevant to communication applications and in particular to optical time domain multiplexing schemes [80]. Recently, a study has been performed [81], in which ultrafast and distortion-free amplification were reported at pulse repetition rates of 20, 40, and 80 GHz, with pulse durations of 710, 1.9, and 2.2 ps, respectively. The source of these ultrashort pulses was also a mode-locked QD laser, with the 4 mm long SOA being fabricated from the same wafer structure (15 layers of InGaAs QDs, emitting around 1.3 µm). The net gain of the SOA was 10 dB. At 20 GHz, there was no change in the pulse shape after being amplified, and only at 40 GHz and 80GHz was observed a modest increase of only 4% and 14%, respectively, in pulse duration. The authors attributed this increase to the biexponential gain recovery of the QD SOA, which may have more detrimental effects as the frequency of gain modulation increases. In this study, it was also reported that the SOA did not affect the jitter of the amplified pulses, which is of fundamental significance in telecommunication and data communication applications.

The conditions for low distortion propagation of ultrashort pulses in QD amplifiers have been analyzed within the framework of a complex theoretical approach involving Maxwell–Bloch equations for individual energy levels in an inhomogeneously broadened QD active layer [82]. In Ref. [50], a simplified semianalytical treatment of this operating regime was proposed, which could also be used to predict the potential of QD SOAs for patterning-free amplification. One possible way of analyzing the limits of such operation is to use Equation 3.1 with $G_0 = G(\bar{N})$ and σ_b substituted by $\sigma_{QD} = dg/dN|_{N=\bar{N}}$. As shown in the previous sections, due to the finite, and relatively small, density of states (DoS) in the QD, the dependence $g(N)$

is strongly sublinear at high carrier densities, meaning that the parameter $\sigma_{QD} = dg/dN|_{N=\bar{N}}$ tends to be zero at large values of N. Thus, the relative pulse energy u_p is small even for considerable absolute pulse energies, so the amplification remains, in effect, linear, with little patterning, below the bit rate given by (3.1).

A somewhat different estimate for the patterning-free bit rate is given by an approach more closely tailored for the performance of QD SOAs operating within the ground-state transition band, as proposed by Uskov et al. [50]. Analytical approximations could be obtained for the maximum patterning-free bit rates in the two limiting cases. The first one, arguably more relevant in practice, is the case of optical pulses of a duration τ_p long enough ($\tau_p \gg \tau_{rel}^{(g,e)}$, $\tau_{rel}^{(h,e)}$, the latter values being the (sub) picosecond relaxation times for the electron and hole contributions to the fast nonlinear gain) for the notion of instantaneous gain compression coefficient ε to be introduced. Then, the main effect that contributes to the patterning is the variation in the total carrier density N and thus the gain $g(N)$, as in bulk and QW SOAs. Keeping for the time being a generic dependence $g(N)$, we can determine the carrier density $N_{min}(\Delta g)$ from the condition

$$g(N_{min}) = g_0 - \Delta g, \qquad (3.24)$$

where g_0 is the unsaturated gain and Δg is the acceptable gain variation margin. The authors set the requirement for a bit rate and pulse energy small enough that the amplification is free not only from patterning but also from distortion – meaning that the parameter that needs to remain within a specified tolerance from the unsaturated gain (G_0 in our notations) is not the *prepulse* gain G_- but the *postpulse* gain G_+. Then, in order to keep the gain variation within the acceptable margin, the postpulse carrier density must not fall below $N_{min}(\Delta g)$:

$$N_+ > N_{min}(\Delta g). \qquad (3.25)$$

Evaluating the carrier density variation is particularly convenient in the situation when the amplifier is strongly pumped and so the gain during the pulse, instead of decreasing with carrier density, remains approximately constant at the maximum value: $g_0 = g_{max}$ and so $G \approx G_0 = \exp(\Gamma g_{max} L)$. Then, the dynamics of the total carrier density N (dots and reservoir combined) is described by a version of (2.79) (assuming the linear recombination as in Section 3.2, and substituting $g = g_{max}$):

$$\frac{dN}{dt} = \frac{N - N_J}{\tau_{sp}} - v_g g_{max}(1 - \varepsilon S) S. \qquad (3.26)$$

Here, $N_J(J)$ is the unsaturated carrier density in the amplifier (from (2.79) and assuming a constant recombination time $\tau_{sp} = \text{const}$, $N_J = R_{pump}\tau_{sp} = eJ\tau_{sp}/d_{act}$). During the short pulse, the recombination can be neglected, giving

$$N_- - N_+ \approx \frac{g_{max} U_p}{A_X \hbar \omega}\left(1 - \frac{U_p}{U_{nl}}\right). \qquad (3.27)$$

3.3 Quantum Dot Optical Amplifiers at High Bit Rates: Low Distortions and Patterning-Free Operation

Here, as in Section 3.2, $U_p = \int_{t \ll -\tau_p}^{t \gg \tau_p} P(t) dt = v_g A_x \hbar\omega \int_{t \ll -\tau_p}^{t \gg \tau_p} S(t) dt$ is the pulse energy, A_x is the waveguide mode cross-section, and $\hbar\omega$ is the signal photon energy. The energy U_{nl} is associated with the fast nonlinearity and is given by

$$U_{nl} = \frac{v_g A_x \hbar\omega \tau_p}{\varepsilon \xi}, \tag{3.28}$$

with τ_p being the pulse duration as above, the dimensionless pulse form factor ξ being determined by Equation 3.22, and the nonlinearity coefficient ε determined as in Section 2.4.

In between the pulses, stimulated recombination can be neglected, so, similar to the derivation of (3.19), we have

$$N_- = N_J - (N_J - N_+) \exp(-T/\tau_{sp}). \tag{3.29}$$

Combining (3.27) with (329) (3.25) and, we get for the "safe" operating rate an equation

$$B_{max}^{long} = \left(\tau_{sp} \ln \left(\frac{1}{1 - \tilde{U}/U_{sat}^{(long)}(\Delta N)} \right) \right)^{-1}. \tag{3.30}$$

Here,

$$\tilde{U} = U(1 - U/U_{nl}) \approx U/(1 + U/U_{nl}) \tag{3.31}$$

is introduced to take into account fast saturation of gain, and the expression

$$U_{sat}^{(long)}(\Delta g) = \frac{A_x \hbar\omega (N_J - N_{min}(\Delta g))}{g_{max}} \tag{3.32}$$

describes the saturation energy due to "slow" saturation. Obviously, the expression (3.30) is defined only in the case of $\tilde{U} < U_{sat}^{(l)}(\Delta g)$, with B_{max}^{long} falling to zero at $\tilde{U} = U_{sat}^{(l)}(\Delta g)$; in the case of $\tilde{U} > U_{sat}^{(long)}(\Delta g)$, by definition $B_{max}^{long} = 0$ because in this case, even a single pulse leads to a modulation of carrier density in excess of $\Delta N = N_J - N_{min}(\Delta g)$. This is because in Ref. [50], the condition (3.25) was used as the criterion for low patterning. If, as in deriving (3.19), we set a weaker condition

$$N_- > N_{min}(\Delta g) \tag{3.33}$$

as a cutoff for patterning-free (though not necessarily distortion-free) operation, then the estimate for the bit rate becomes

$$B_{max}^{long} = \left(\tau_{sp} \ln \left(1 + \tilde{U}/U_{sat}^{(long)}(\Delta N) \right) \right)^{-1}, \tag{3.34}$$

where all the parameters have the same notations as in (3.16). At high values of saturation, this expression, as (3.19) and unlike (3.30), still predicts a possibility of some patterning-free operation, albeit at very low bit rates ($B_{max}^{long} < \sim 1/\tau_{sp}$). At the range of parameters of most interest in practice (up to $\tilde{U}/U_{sat}^{(long)} \sim 0.1$, corresponding to B_{max}^{long} down to about $10/\tau_{sp}$), both (3.30) and (3.34) give the same results and

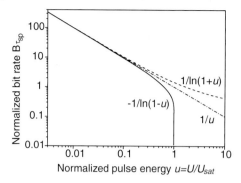

Figure 3.3 Estimated maximum patterning-free bit rate in a QD optical amplifier as a function of normalized input power using two possible approximations and their common asymptotic.

are well described by a common simple asymptotic (Figure 3.3).

$$B_{max}^{long} \approx \frac{U_{sat}^{(long)}(\Delta N)}{\tau_{sp} \tilde{U}}. \tag{3.35}$$

Expressions (3.32) and (3.30) (or (3.35)) clearly show that there is a degree of trade-off between patterning-free operation range and total gain in a QD SOAs, since the former increases and the latter decreases, as g_{max} is reduced, say, by reducing the number of QD layers or dot density.

In Ref. [50], the formula (3.30) (in somewhat different notations) was derived in the excitonic approximation ($f_{Gi}^{(e)} = f_{Gi}^{(h)} = f_i$), but the result is, in fact, quite generic; it does not rely on this approximation but only on the fact that under high pumping, the curve $g(N)$ reaches saturation. It has been pointed out in Ref. [50] that this saturation occurs, under high enough pumping, not only in QD amplifiers but also in QW and bulk ones, meaning that (3.30) is valid for those more traditional media too. Indeed, the very low patterning-free bit rates described by the formula (3.19) for high intensities are, to a significant degree, artifacts of the simple linear gain–carrier density dependence (3.1). This linear function is indeed a reasonable approximation for the carrier density dependence of the *peak* gain in bulk materials (and, more approximately, for QW materials at low to moderate N). However, the gain peak in semiconductors is not associated with a particular wavelength, but shifts to shorter wavelengths with increased N. Therefore, (3.1) is much less accurate, even in bulk materials, for describing the N dependence of the gain g at a *given spectral point* – for this purpose, it works only at relatively low N, whereas at carrier densities high enough to ensure that f_e and f_h for electron and hole states involved in transitions approach one, $g(N)$ does, in principle, saturate just as it does in QD materials. Then, the formulas (3.30)–(3.32) apply for those materials too, with g_{max} substituted by the saturated gain at a given spectral point, and the nonlinearity ε calculated for the corresponding active medium (and possibly negligible).

In practice, however, this is more difficult to achieve in QW and bulk materials than in QDs. Indeed, the condition $f_{e,h} \rightarrow 1$ in bulk/QWs requires both electron and hole statistics to become significantly degenerate, with the corresponding Fermi

3.3 Quantum Dot Optical Amplifiers at High Bit Rates: Low Distortions and Patterning-Free Operation

quasi-levels positioned inside the conduction and valence bands, respectively. As is known in semiconductor physics, this happens when the carrier density approaches and exceeds the reduced density of states for electrons and holes. In the case of bulk materials, where the density of states (per unit energy per volume) for a given carrier energy is proportional to the square root of the energy over the band edge, the reduced (DoS) is given by the formula

$$N_{c,v}^{(\text{bulk})} = \frac{1}{4}\left(\frac{2m_{e,h}k_B T}{\pi \hbar^2}\right)^{3/2}. \tag{3.36}$$

In the case of QWs, with the DoS for each energy level (subband) approximately constant for all carrier energies in that subband, the (three-dimensional) reduced DoS for the electrons (assuming the well is narrow enough for a single electron level to contain most of the electron population) is

$$N_c^{(\text{QW})} = \frac{m_e k_B T}{d_{\text{QW}} \pi \hbar^2}, \tag{3.37}$$

where d_{QW} is the well thickness (cf. Equation 3.37 for the reduced density of states in the wetting layer). For holes, at virtually all practical QW thicknesses, several levels are present, so the bulk value remains a reasonable approximation for the reduced DoS. For lighter electrons ($m_e \sim 0.04$–$0.1\, m_0$ in III–V compounds, m_0 being the free electron mass), the reduced DoS is evaluated as $N_c^{(\text{bulk,QW})} \sim 10^{23}\,\text{m}^{-3}$, and so the degeneracy condition is easily met under conditions of optical gain ($N \sim 10^{24}\,\text{m}^{-3}$). However, for heavier holes ($m_{hh} \sim 0.5\, m_0$) to become degenerate, carrier densities approaching $N \sim 10^{25}\,\text{m}^{-3}$ are required, which is not readily met even in lasers and amplifiers.

In QDs, electron/hole degeneracy involves the corresponding quasi-level moving into the ground-state energy range, and the role of energy-dependent densities of states in determining electron and hole Fermi quasi-levels is played by "DoS"$_G^{(e,h)} = \varrho_G N_D D_{\text{inh}}^{(e,h)}(E_G^{(e,h)})$, where the inhomogeneous broadening functions are given by (2.5). This expression, unlike the true DoS in bulk and QDs, has a finite integral over all energies, equal to the total spatial density of electron/hole states $\varrho_G N_D$ for both electrons and holes. Thus, the condition $f_{e,h} \to 1$ is met for *both types* of carriers at $N > \sim \varrho_G N_D$, and therefore the saturation $g(N) \to g_{\max}$ in QDs occurs at these lower values of N than in QWs and in bulk materials (this feature may be somewhat exaggerated in a theoretical calculation employing the excitonic approximation, as in Ref. [50], but also appears in more rigorous models, and is confirmed experimentally [51]). This carrier density at which $g(N) \to g_{\max}$ is essentially the parameter $N_{\min}(\Delta g)$ in (3.32). A small value of this parameter means that even with modest N_J, the saturation energy $U_{\text{sat}}^{(\text{long})}$ is high, making, in turn, for high values of the patterning-free bit rate $B_{\max}^{(\text{long})}$.

An additional factor that adds to the high saturation energy $U_{\text{sat}}^{(\text{long})}$ in QDs is the fact that the matrix element of radiative transitions, and hence the saturated gain $g_{\max}$ to which $U_{\text{sat}}^{(I)}(\Delta N)$ is inversely proportional (3.32), in QDs is reduced by the reduced overlap of the electron and hole wave functions compared to QWs [50].

Thus, the patterning-free operation in QD SOAs is achieved at much lower *current densities J* than in bulk or QW devices. (It has to be noted, however, that due to the higher material gain and confinement factor in typical bulk and QW SOAs, the length of the device required to achieve a certain absolute gain in dB is shorter than in QD SOAs. Thus, the conclusion regarding the *total current* required for patterning-free amplification with a given absolute gain is less straightforward).

It is also worth noting that, even if the gain, and hence intensity, patterning is low due to the small gain variation with N, this is not necessarily true for phase patterning due to self-phase modulation (SPM). Indeed, SPM, unlike gain, is generated not only by the dot population variation (which is small due to efficient capture of carriers from the reservoir) but also by the reservoir population variations, which are not necessarily negligible. This effect may thus require a further investigation. However, as mentioned above, SPM due to single-pass amplification is typically modest at realistic pulse energies.

Figures 3.4 and 3.5 ([50]) show the numerically evaluated dependence of gain in SOAs with different active layers on the output pulse energy and bit rate. The current in both cases is shown both in absolute units and in relative units of $x_J = N_J/\varrho_G N_D = eJ\tau_{sp}/d_{act}\varrho_G N_D$ (in the case of bulk and QW amplifiers, $\varrho_G N_D$ is substituted by the *joint reduced density of states* for bulk and QW, calculated using formulas similar

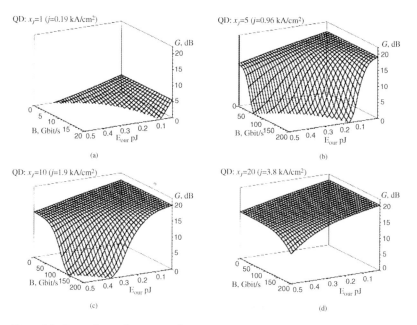

Figure 3.4 Dependence of gain G on the output pulse energy E_{out} (U_{out} in our notations) and bit rate B for a QD SOA 2.77 mm long, with three layers of dots, in the regime of long pulses (see text) for three different values of the applied current density. Note the different scale of B in (a). Reproduced with permission from Ref. [50].

3.3 Quantum Dot Optical Amplifiers at High Bit Rates: Low Distortions and Patterning-Free Operation

to (3.36) and (3.37) with the reduced effective mass $m_e m_h/(m_e + m_h)$ used as the carrier mass). The small-signal gain in all cases is set at 20 dB. In the case of QD SOA, the number of dot layers was set to $n_l = 3$ and the length to $L = 2.77$ mm to reach this value, with the other parameters (in our notations) set at $A_X = 0.4\,\mu m^2$ ($2 \times 0.2\,\mu m$), the areal density of dots $10^{15}\,m^{-2}$, $\Delta E_{hom} = 12$ meV ($T_2 = 100$ fs), $\Delta E_{inh} = 40$ meV. The results were calculated in the excitonic approximation, with the average exciton binding energy (the sum of electron and hole binding energies) of 100 meV and an effective carrier mass $m = 0.1\,m_0$ in both the QDs and the WL. For the dot gain cross section, $\sigma_G = A_G \times 1/\sqrt{\pi}\Delta E_{inh}^{(G)} = 1.66 \times 10^{-19}\,m^{-2}$ was taken. For bulk and QW SOAs, the excitonic approximation was of course not made, and the gain was estimated in a simple model with full energy and momentum conservation (a reasonable approximation for the spectral position used, $\hbar\omega - E_g^{bulk,QW} = 30$ meV), making the gain directly proportional to the population inversion $W = f_e + f_h - 1$ for the resonant transition. In determining Fermi quasi-levels for QWs, only one subband was taken for both conduction and valence bands. The latter implies that the carrier density at which hole saturation occurs in the QW may have been somewhat underestimated and so the predictions regarding g(N) saturation and hence the gain modulation in QW may need to be treated as best-case estimates. Despite this, the patterning-free amplification potential of QD SOAs was predicted to be distinctly superior to that of both bulk and QW SOAs. Figure 3.4a was calculated for values of current at which the gain g(N) was far from saturation (corresponding to the population inversion $g/g_{max} \approx W = f_e + f_h - 1 = 2f_e - 1 \approx 0.43$ for the transition resonant with the amplified light), in which case the analysis above is not applicable, and patterning-free operation is observed only in the linear regime (pulse energy $\to 0$). In Figure 3.4b–d, calculated for currents corresponding to the inversion $g/g_{max} \approx W = f_e + f_h - 1 = 2f_e - 1 \approx 0.95, 0.98, 0.99$, respectively. In these cases, the authors of Ref. [50] found a plateau of constant gain, corresponding to patterning-free operation, appearing at bit rates below a critical value that is in good agreement with the analytical theory presented above. For example, pulse trains at ≈ 150 Gbit s^{-1} (160 Gbit s^{-1} being widely expected to be used in the next generation of optoelectronic telecomunication and/or data communication systems) can be amplified to 0.2 and 0.4 pJ without patterning with the pump currents of ≈ 2 and ≈ 4 kA cm^{-2}, respectively. The plateaus of constant gain were found to have a weak slope along the pulse energy axis. The slope is related not to the total carrier density dynamics but to the fast gain saturation described by parameter U_{nl}(3.15). It therefore causes a degree of nonlinear amplification and associated pulse distortion (broadening), but does not contribute to patterning effects because in the long pulse approximation, the nonlinear gain is instantaneous and thus identical for all pulses of the same amplitude in the pulse stream, regardless of the bit rate.

In accordance with (3.32), an increase in current leads to an increase in the patterning-free bit rate, and this in the area of the plateau of constant gain in Figure 3.4 b–d.

In quantum wells under low pumping (Figure 3.5a), and particularly in bulk materials (Figure 3.5c), the plateau of patterning-free operation was found to be virtually absent, as in Figure 3.4a, for the same reason of low population inversion

QW: $x_J=5$ ($J=1.0$ kA/cm^2)

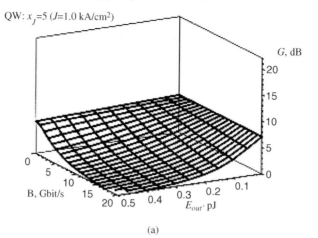

(a)

QW: $x_J=28$ ($J=5.8$ kA/cm^2)

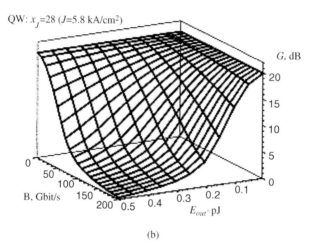

(b)

bulk: $x_J=8.2$ ($J=5.8$ kA/cm^2)

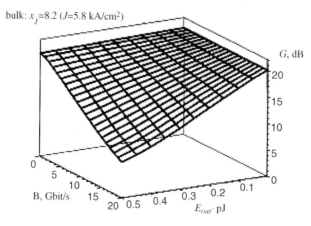

(c)

3.3 Quantum Dot Optical Amplifiers at High Bit Rates: Low Distortions and Patterning-Free Operation

($W < 0.5$ in all cases). In QWs under high pump (Figure 3.5b), a small plateau was predicted, but restricted to lower bit rates and energies than in QD amplifiers, partly due to a less pronounced sublinear (saturating) dependence of the inversion and hence gain on the carrier density and partly because of the higher unsaturated gain, meaning a lower saturation energy according to (3.32) (note that the QW and bulk amplifiers thus were significantly shorter than the QD SOA simulated to achieve the same absolute unsaturated gain of $G_0 = 20\,\text{dB}$). As in QD SOAs, the performance of QW amplifiers could be improved by increasing current, but to a smaller extent, for the same two reasons.

Thus, the authors of Ref. [50] could claim confidently that their calculations demonstrated the superiority of QD SOAs over QW and bulk counterparts as regards patterning-free operation.

Experimentally, indeed, reduced patterning at high currents was observed in QW amplifiers (and tentatively attributed by some authors, for example, to decrease in τ_{sp} due to Auger recombination, which may indeed play some part alongside the $g(N)$ saturation) [83]. However, arguably the most convincing study of patterning-free amplification was, as could be expected, reported with QD SOAs [84]. The authors analyzed a QD SOA with an active layer with 10 layers of InAs dots overgrown with $In_{0.17}Ga_{0.83}As$, in a GaAs waveguide layer with $Al_{0.3}Ga_{0.7}As$ claddings. The ground-state amplified spontaneous emission peak after overgrowth was around 1.3 μm; the experiments were conducted near this peak ($\lambda \approx 1.306\,\mu m$). The 25 mm long, 10 μm wide SOA had the peak gain of about 12 dB at the total current of 2.6 A (slightly below the 2.9 A threshold of laser operation due to residual facet reflections). The 3 dB saturation value of the output power was about 10 dBm (Figure 3.6, top). Experiments for 10 Gbit s^{-1} nonreturn to zero (NRZ) pulse streams showed no significant patterning (Figure 3.6, bottom) for powers beyond the 3 dB saturation, with only a very slight eye closing at the average power value corresponding to the gain saturation of about 5 dB. By contrast, in a bulk SOA with a similar gain, significant patterning, with two distinct "rails" in the eye diagram, was observed with a much lower (2 dB) gain saturation.

Both the theory presented above and the experiments in Ref. [84] relate to the case of *long pulses*, of a duration much longer than the relaxation times of QD states (evaluated in the experiments of [84] as less than 3 ps and taken to be 1 ps in Ref. [50]). At bit rates of hundreds of Gbit/s, the pulse durations used are likely to be on the order of 1 ps long, which is *comparable* to the relaxation time. This case is difficult to treat analytically, but the authors of Ref. [50] also considered the case of short pulses, of a duration much shorter than the relaxation times.

In this case, the wetting layer does not replenish the dot states during the pulse and so only the dots with energy states within a range of ΔE_{hom} around the resonant

Figure 3.5 Dependence of gain G on the output pulse energy E_{out} (U_{out} in our notations) and bit rate B for a QW and bulk SOAs, for a photon energy 50 meV above the bandgap. The amplifier lengths (0.59 mm for the QW SOA, 0.35 mm for the bulk one) chosen to give the same maximum unsaturated gain as the QD amplifier shown in Figure 3.4. Reproduced with permission from Ref. [50].

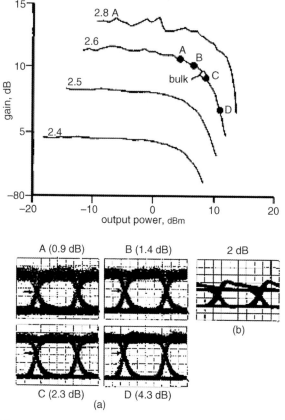

Figure 3.6 Experimental proof of patterning-free amplification at 10 Gbit s^{-1} in QD SOAs: Top: saturation curve; below: eye diagrams of a QD (a) and bulk (b) amplifier. Reproduced with permission from Ref. [84].

state – those dots involved in transitions – contribute to the depletion of carrier density. In the excitonic approximation ($f_{Gi}^{(e)} = f_{Gi}^{(h)} = f_i$) and neglecting all relaxation terms in Equation 3.38 for the ground-state occupancy of the dot i, we get the equation in the form

$$\frac{df_i}{dt} \approx -R_{\text{stim}(i)}^{\text{net}} = -A_G(2f_i - 1) D_{\text{hom},i} S, \tag{3.38}$$

where, unlike the analysis in the long-pulses case, the variation in f_i is not necessarily small. Assuming that before the pulse, the dots have time to revert to quasi-equilibrium (thermal) distribution, the integration of (3.38) gives

$$f_{i-} - f_{i+} \approx \frac{1}{2}(2f_i^{(\text{QE})} - 1)\left[1 - \exp\left(-2\frac{A_G D_{\text{hom},i} U}{\hbar \omega A_X}\right)\right]. \tag{3.39}$$

3.3 Quantum Dot Optical Amplifiers at High Bit Rates: Low Distortions and Patterning-Free Operation

The variation in the total density needs summing over all states:

$$N_{i-} - N_{+} = \varrho_G N_D \sum_i (f_{i-} - f_{i+}) \tag{3.40}$$

and strictly speaking can be evaluated only numerically.

Using the top-hat approximation (2.100) for $D_{\mathrm{hom},i}$, and the condition $\Delta E_{\mathrm{hom}} \ll \Delta E_{\mathrm{inh}}$, an analytical approximation for $N_{i-} - N_{+}$ may be obtained in the form

$$N_{i-} - N_{+} \approx \varrho_G N_D \Delta E_{\mathrm{hom}} D_{\mathrm{inh}}^{(\mathrm{res})} (2 f_{\mathrm{res}}^{(\mathrm{QE})} - 1)[1 - \exp(-U/U_s)], \tag{3.41}$$

where

$$\frac{1}{U_S} = 2 \frac{A_G D_{\mathrm{hom}}^{\mathrm{res}}}{\hbar \omega A_X} \tag{3.42}$$

determines the saturation pulse energy for this case, and $D_{\mathrm{inh}}^{(\mathrm{res})}$, $D_{\mathrm{hom}}^{(\mathrm{res})}$, and $f_{\mathrm{res}}^{(\mathrm{QE})}$ are all evaluated at the resonant energy $\Delta E_G^{(e-h)} = \hbar \omega$.

Applying the criterion (3.33) for the maximum patterning-free bit rate, assuming operation at or near the resonant point, and once again, as in the long-pulse case, setting the high inversion case $2 f_{\mathrm{res}}^{(\mathrm{QE})} - 1 \approx 1$ give the estimate of the patterning-free bit rate in the form

$$B_{\mathrm{max}}^{\mathrm{short}} = \left(\tau_{\mathrm{sp}} \ln \left(1 + \frac{\varrho_G N_D}{N_J - N_{\mathrm{min}}(\Delta g)} \times \frac{\sqrt{\pi} \Delta E_{\mathrm{hom}}}{\Delta E_{\mathrm{inh}}} (1 - \exp(-U/U_s)) \right) \right)^{-1}. \tag{3.43}$$

One clear result obtained from this formula is that, as in the case of long pulses, the patterning-free bit rate may be increased by increasing current and, obviously, suffers with the increase in pulse energy. Interestingly, at relatively low pulse energies, when $1 - \exp(-U/U_s) \approx U/U_s$, the expression (3.43) reproduces (3.34) (with a substitution $\tilde{U} \to U$), so the short-pulse limit gives similar results to the long-pulse limit. In general, expression (3.43) is less accurate and generic than its long-pulse limit counterpart (3.34) because its derivation relies on both the excitonic approximation for the level occupancies and the top-hat approximation for the homogeneous broadening function. Besides, with pulses short enough to neglect relaxation processes, coherent effects are likely to become important and a full model including polarization operators will become necessary for an accurate analysis. With this reservation, the general result from Equation 3.43 is that the patterning-free bit rate in the short-pulse approximation limit is *always higher than in the long-pulse limit* for the same pulse energy: $B_{\mathrm{max}}^{\mathrm{short}}(U) > B_{\mathrm{max}}^{\mathrm{long}}(U)$. A qualitative explanation for that, proposed by the authors of Ref. [50], is that in the short pulses regime, the dots are effectively isolated from the WL, so that the pulse can extract fewer carriers from the "dots-WL" system. In the long pulses regime, on the other hand, QDs are pumped by the WL during the pulse, thus depleting the WL. Therefore, the train of short pulses must have a bit rate higher than the train of long pulses with the same energy to exhaust the WL or, conversely, patterning-free amplification occurs at rates not higher for the regime of short pulses than for the regime of long pulses.

Thus, although the intermediate pulse regime, when the relation between the relaxation (capture and escape) time and the pulse duration is not known *a priori*, does not admit an easy analytical solution, it may be concluded that the simple, generic criterion for $B_{max}^{long}(U)$ gives a sensible, if somewhat conservative, estimate for this case as well.

Figure 3.7, also reproduced from Ref. [50], shows the simulated dependence of the gain on the bit rate and pulse energy for the short-pulse case, for the same parameters as in Figure 3.4. While a plateau of constant gains is still seen at low energies and bit rates at high pumping, the decrease in gain with pulse energy is stronger in this case than it is in the case of long pulses. This, too, has been attributed [50] to the absence of interaction between the dots and the WL during the pulse, meaning that the dots, which are not replenished by a supply of carriers from the WL, are depleted by pulses of lower energy.

One consideration that may affect the predictions of the analysis described above for both the long and the short pulse cases is that the analysis concentrated on ground to ground-state transitions. The role of excited states in the ground-state dynamics may be seen as treated through the expressions for the effective capture and escape times and thus the nonlinearity coefficients. However, strictly speaking, and particularly for InAs dots operating at 1.55 μm wavelength, where the separation between the ground and the excited state energies is comparable to the homogeneous (as well as inhomogeneous) broadening of either state, the transitions between electrons and holes in the excited states also make some *direct* contribution to amplification of light that notionally falls in the ground-state band. In addition, transitions between electrons in the *ground* state and holes in the *excited* states also make a contribution to gain [47]. These contributions slow down the saturation of the $g(N)$ curve and may somewhat decrease the bit-free error rates predicted by the analysis. However, the effects are relatively modest, particularly in GaAs-based dots working at 1.3 μm, where the levels are better separated.

Another effect on the degree of patterning is the two-photon absorption (TPA). As mentioned above, TPA can be seen as a source of additional gain compression in quantum dot (or indeed any other) semiconductor optical amplifier. However, in the context of QD SOAs under high bit rate operation, it has also been shown to lead to another, less straightforward, effect [72]. Namely, the TPA – which is not restricted to the dots, unlike linear gain or absorption – creates extra carriers in the layers surrounding the QDs. These carriers are then captured into the dots, effectively providing an extra source of synchronous pulsed optical pumping. It has been shown in Ref. [72] that this extra pumping can noticeably reduce patterning effects in the case of short, intense optical pulses – though the light intensities required for this were, arguably, in the extreme range, on the order of 10^{13} W m^{-2}, with pulse energies

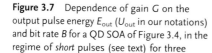

Figure 3.7 Dependence of gain G on the output pulse energy E_{out} (U_{out} in our notations) and bit rate B for a QD SOA of Figure 3.4, in the regime of *short* pulses (see text) for three different values of the applied current density. Note the different scale of B in (a). Reproduced with permission from Ref. [50].

3.3 Quantum Dot Optical Amplifiers at High Bit Rates: Low Distortions and Patterning-Free Operation

QD: $x_j=1$ ($J=0.19$ kA/cm^2)

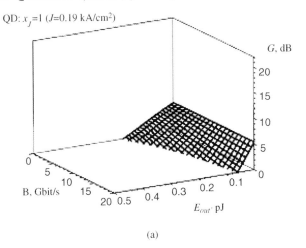

(a)

QD: $x_j=5$ ($J=0.96$ kA/cm^2)

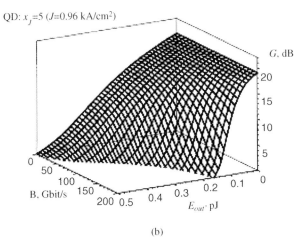

(b)

QD: $x_j=10$ ($J=1.9$ kA/cm^2)

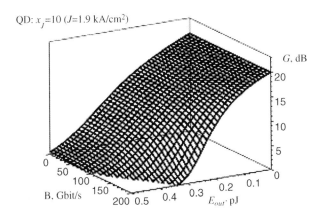

(c)

>1 pJ (that is, about an order of magnitude higher than those studied in Ref. [50]) contained in pulses only 0.1 ps long. With more moderate pulse parameters, the effect may be expected to be less significant.

3.4
Nonlinear Operation and Limiting Function Using QD Optical Amplifiers

The combination of high patterning-free bit rate and strong (and bit rate insensitive) gain saturation by short intense pulses, described in the previous section, paves the way for the use of QD SOAs as *limiting amplifiers* of short pulses. This, in turn, makes QD SOAs attractive for applications in all-optical signal processing.

In Ref. [85], it was proposed to use a limiting QD SOA as part of an all-optical clock recovery subsystem for operation with short data packets. The general idea was previously proposed and realized (see, for example, Ref. [86]) and involves using a passive Fabry–Perot resonator to extract a clock-like pulse sequence from the packet data stream. The clock-like sequence has still a strong residual modulation, which is subsequently reduced by a limiting element. Previously, QW SOAs in an interferometric construction and fiber elements were used as such limiting elements. In Ref. [85], it was shown that the QD SOA may be an attractive alternative for performing this function. Simulations using a simple model without inhomogeneous broadening taken into account showed that the AM modulation of the recovered clock sequence may be reduced from nearly 0 dB to about -10 dB using the ultrashort pulse saturation in the amplifier.

Later [69], the idea was extended to the use of the same QD amplifier for several channels at different wavelengths, exploiting the fact that in the short-pulse operating regime, different groups of dots do not interact and can be saturated independently.

It was also predicted by some authors [87] that several logical functions can be realized by QD amplifiers in an interferometric construction, utilizing fast nonlinear dynamics of refractive index that accompanies fast gain dynamics.

Finally, it has been proposed to use QD SOAs in microwave electronics for effecting frequency shift of the microwave envelope of the optical signal, by varying either the bias on the amplifier or the amplitude of a CW control light. The physical basis of the effect is either the shaping of the pulse by the nonlinear saturation of an amplifier (also known as coherent population pulsations) or cross-gain modulation in the presence of several frequency-detuned optical signals [62]. Neither effect is restricted to QDs, but the presence of capture and escape dynamics with picosecond characteristic times extends the operating frequencies of the proposed scheme to hundreds of GHz.

At the time of writing, experimental realization of most of such ideas remains pending, but the prospects appear realistic and attractive.

4
Quantum Dot Saturable Absorbers

4.1
Foundations of Saturable Absorber Operation

A saturable absorber (SA) is an optical material or component in which the absorption decreases with incident optical intensity. Thus, with increasing incident intensity, the absorption saturates, the loss decreases, and the transmittance increases. This feature can be implemented to suppress low-intensity signal discriminating in favor of short, intense optical pulses. The SA can thus be used as a free-standing element, or combined with an optical amplifier (see below), as an optical switch, for all-optical signal processing functions such as optical time-division demultiplexing, as well as reshaping and retiming of optical pulses in communication systems. However, the most widely used application of saturable absorbers is to facilitate or enhance the mode locking of a laser.

Both as a freestanding element and as part of a pulsating laser, the SA can have two main geometries.

The first one, used in in-plane semiconductor lasers and integrated optical circuits, is the *waveguide saturable absorber*, which is essentially a reverse biased section of a laser structure.

The second construction, obtained by combining a planar multistack DBR (*Distributed Bragg Reflector*) structure with a saturable absorber component, is the so-called SESAM (*SEmiconductor Saturable Absorber Mirror*) that is used as a saturable reflector, with a reflectivity increasing (and ideally approaching one) with power. SESAM technology has been used with considerable success in a variety of laser systems to reliably initiate and/or stabilize mode locking [88], and can also be used for Q-switching in other laser systems.

Several important parameters are to be considered when designing a saturable absorber, whether waveguide or SESAM type.

1) The first important parameter is the *modulation depth* (or modulation *strength*) that represents the difference between the saturated and the nonsaturated (linear) properties of the absorber. In the case of the waveguide absorber, a possible way of measuring this could be either the difference between the linear

Ultrafast Lasers Based on Quantum Dot Structures: Physics and Devices.
Edik U. Rafailov, Maria Ana Cataluna, and Eugene A. Avrutin
Copyright © 2011 WILEY-VCH Verlag GmbH & Co. KGaA, Weinheim
ISBN: 978-3-527-40928-0

4 Quantum Dot Saturable Absorbers

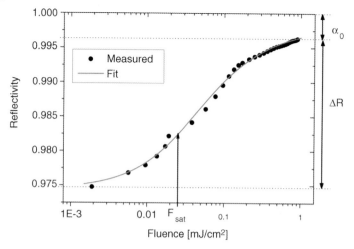

Figure 4.1 Reflectivity response of a semiconductor saturable absorber mirror. F_{sat} is the saturation fluence, ΔR is the modulation depth, and α_0 corresponds to the nonsaturable losses.

(unsaturated) and the fully saturated (residual) absorption coefficient, $\alpha_0 - \alpha_{res}$ or the associated extinction ratio E.R. $= \exp[\Gamma(\alpha_0 - \alpha_{ns})L]$, where Γ and L denote the confinement factor of the waveguide absorber (as in the laser) and the length of the SA segment, respectively. In the context of a SESAM geometry, the usually quoted parameter is the saturated minus unsaturated reflectivity difference ΔR, as depicted in Figure 4.1. From this figure, it is also possible to estimate the residual, or nonsaturable, loss α_{ns} that remains after saturation. In the design and growth of SESAM structures, the aim is to minimize such nonsaturable losses because they represent a source of CW loss that not only leads to a decrease in the intracavity power but also causes heating that may degrade the absorber properties.

2) The *absorber recovery time* is another crucial parameter. This is particularly true for high repetition operation because there is a requirement for the absorption to recover between pulses (in the context of a mode-locked laser, this means recovering within each resonator round-trip period, or a fraction thereof in case of harmonic operation). Thus, the characteristic time (or times) of the SA recovery must be significantly shorter than the *repetition period*. The relation between the absorber recovery time and the *pulse duration* is less definite. Depending on this relation, two limiting cases of the absorber operation are typically distinguished. The *slow saturable absorber* regime is the one when the recovery speed of the saturable absorber is longer than the optical pulse duration; the *fast saturable absorber* regime is the one where the opposite holds.

In practice, often neither of these limiting cases applies exactly to semiconductor SAs; instead, an intermediate case is often realized, although the slow absorber approximation is usually closer to describing realistic situations (typical pulse durations in semiconductor optoelectronics are in the range of 0.1–10 ps, with the measured absorber recovery time of 1–100 ps). Besides, as will be discussed below, it is often impossible to describe the recovery of SA by a single time constant, so for the same incident pulse, the saturable absorption of a semiconductor SA may (and often does) contain both a slow and a fast component.

3) Yet another SA parameter is needed to describe how "fast" (in the sense of light power or energy, not time) the absorber saturates with the light input. The appropriate parameter for describing this depends on the slow or fast absorber limit. In the slow SA case, in the absence of relaxation, the absorber "integrates" the power of light, and so the absorber saturation is approximated well by a formula

$$\alpha(F_{in}) \approx (\alpha_{uns} - \alpha_{ns})\exp\left(-\frac{F_{in}}{F_{sat}}\right) + \alpha_{ns}, \tag{4.1}$$

where α_{uns} is the unsaturated (linear) absorption, α_{ns} is the nonsaturable absorption residue,

$$F_{in}(t) = \frac{U_{in}}{A_X} = \frac{1}{A_X}\int_{-\infty}^{t} P_{in}(t')dt' \tag{4.2}$$

is the incident *fluence* of light, A_X is, as in Section 2.4, the cross section of the light beam, and P_{in} and $U_{in} = \int_{-\infty}^{t} P_{in}(t')dt'$, also as in Section 2.4, denote the power and the energy of the pulse up to the moment t.

The parameter F_{sat} is then known as the *saturation fluence* of the SA and is one of the most important absorber parameters. Technically speaking, this parameter applies only when the saturation of the SA is described well by (4.1); however, it is often quoted regardless of the exact form of $\alpha(F_{in})$, as the measure of how much optical energy density is needed to decrease the absorptance of the SESAM by a factor of $1/e \approx 0.37$, as in the exponential formula (4.1). The value of F_{sat} should be as low as possible to ensure that the SA is "sensitive" to fluctuations. In the context of a waveguide absorber, the cross section A_x (which in this case is the waveguide mode cross section $A_X = Wd_{act}/\Gamma_{xy}$) is fixed and so there is no difference between characterizing the absorber in terms of incident energy or fluence; however, in the case of SESAMs, where the area of the beam focused on the mirror is variable and tunable, it is the fluence, rather than the total energy, that has a clearer physical meaning.

One has to bear in mind that the saturation fluence can be a constant absorber parameter only within the slow SA approximation; as the pulse duration approaches the recovery speed of the SA, the value of F_{sat} begins to depend not only on the SA parameters but also on the pulse duration and shape.

In the extreme case of a *fast* absorber (e.g., for CW or quasi-CW operation), the SA responds to the *instant* incident power P_{in}. Then, neither (4.1) nor the notion of saturation fluence is applicable, and the appropriate parameter to characterize the absorber saturation is the input saturation *intensity* (power per unit area) I_{sat}:

$$\alpha(P_{in}) \approx \frac{\alpha_{uns} - \alpha_{ns}}{1 + I_{in}/I_{sat}} + \alpha_{ns}. \tag{4.3}$$

The spectral profile of the absorption and spectrally related dependence of all the parameters above is of great importance. Thanks to bandgap engineering, a SESAM can be designed and fabricated to accommodate the wavelengths of choice for the laser in question. Although particular lasers require specific saturable absorber properties, those based on quantum dots have added versatility allowing them to be designed to suit a range of laser requirements.

4) Finally, in applications such as mode locking of lasers, it is important to characterize any self-phase modulation/chirp the absorber may introduce to the pulse. In the case of QW and bulk saturable absorbers, this is often parameterized by assigning a Henry factor $\alpha_{H\alpha}$ to the slow part of the absorption, defined as the ratio of the carrier density derivatives of the real and imaginary parts of dielectric permittivities, similar to the Henry linewidth enhancement factor (2.53)–(2.54) associated with gain in SOAs. Assuming that the main physical principle of absorption is state filling (see (4.5)), the value of $\alpha_{H\alpha}$ is estimated as

$$\alpha_{H\alpha} = -(\partial \text{Re}\,\chi_\alpha/dN)/(\partial \text{Im}\,\chi_\alpha/dN) \approx -\frac{1}{2\lambda}\frac{dn_\alpha}{dN}\left(\frac{d\alpha}{dN}\right)^{-1}, \quad (4.4)$$

where all the variables are as in (2.54), with the subscript α for the absorber. The values of $\alpha_{H\alpha}$ vary widely in literature, with little direct experimental evidence available; most often, values of the order of $\alpha_{H\alpha} \sim 0\text{--}1$ are assumed. This is similar to the typical chirp parameter of an electroabsorption modulator (which is essentially the same device as a semiconductor saturable absorber, just operated in a somewhat different regime) and somewhat smaller than the typical linewidth enhancement factor of $\alpha_{Hg} \sim 2\text{--}3$ in QW optical amplifiers and lasers. The fast part of the absorption may also be associated with some nonlinear phase modulation; however, in the case of spectral hole burning effects, this nonlinearity is similar in nature to transitions in a two-level system, and so the index modulation at the transition frequency is believed to be weak. This may also expected to be the case in QD SAs, where the main fast contribution to absorber bleaching is indeed similar to spectral hole burning. This has indeed been confirmed by the recent experiments [89] that found that the measured phase shift of an optical pulse changed sign with wavelength variation, as in the case of a two-level system.

4.2
The General Physical Principles of Saturable Absorption in Semiconductors

4.2.1
Physical Processes in a Saturable Absorber

In principle, several types of processes contribute to a semiconductor saturable absorber operation:

1) The most straightforward mechanism of saturable absorption is caused by the dependence of the absorption coefficient on the carrier population in the SA material. In the context of slow saturable absorption (mainly in QWs and bulk

materials), this is usually quantified, in the first approximation, by the simple approximation for the saturable absorption coefficient $\alpha_{sat}(N) = \alpha(N) - \alpha_{ns}$:

$$\alpha_{sat}(N) \approx \alpha_{uns} - \sigma_\alpha N, \quad (4.5)$$

where N is the nonequilibrium (photo)carrier density and $\sigma_\alpha = (\partial \alpha / \partial N)|_{N=0}$ is the saturable absorption cross section (cf. Equation 3.1 for bulk and quantum well amplifiers). During the optical pulse, the dynamics of carrier density is given by an equation similar to (3.2), with the gain g substituted by $-\alpha_{sat}(N)$ and without the pumping and (given the slow SA assumption) relaxation terms:

$$\frac{dN}{dt} \approx v_g \alpha_{sat} \bar{S}, \quad (4.6)$$

where $\bar{S}$ is the average photon density *inside the SA*. Recalling that $\bar{S} = \bar{P}/(\hbar \omega A_X v_g)$, $\bar{P}$ being the average optical power inside the SA, and introducing the geometric ratio $X = \bar{P}/P_{in}$ of internal to incident optical power (which will be considered in more detail below), we obtain the saturation law $\alpha(F_{in})$ in the form of (4.1), with the saturation fluence given, in the case of a constant X, by the simple expression

$$F_{sat}^{all\text{-}opt} = \frac{\hbar \omega}{X \sigma_\alpha}. \quad (4.7)$$

The origin of the carrier dependence of the absorption coefficient, in itself, is determined by at least two physical mechanisms. First, and most obvious, it is due to the filling of electron and hole states by the excited carriers, represented by the statistical factor $1 - f_e(N) - f_h(N)$ that features one way or another in the expression for the saturable absorption in all semiconductor materials and in virtually all models. This is sometimes referred to as dynamic or nonequilibrium Moss–Burstein effect (as opposed to the static equilibrium effect, introduced by doping). Second, absorption near band edge in materials of all types is enhanced by Coulomb interaction between electrons and holes, which is screened by the photogenerated carriers, contributing to absorber saturation. Both of these are classified as *all-optical* mechanisms of absorber saturation (hence the superscript in (4.7)) and are present in all SAs.

2) An additional *electro-optical* SA saturation mechanism is relevant if the absorbing medium in the SA is placed in the electric field, either by applying an external bias voltage or by utilizing the built-in electric field of a p–n or p–i–n junction. Both are often implemented to facilitate the recovery of the absorption between pulses, sweeping the carriers out of the SA area by the field rather than relying on recombination to restore the absorber between pulses. The absorber, whether a waveguide one or a SESAM, may then operate as a version of a *self-electro-optic effect device* (SEED). The principle of this is based on the absorption in semiconductors, whether bulk, quantum well, or quantum dot, near the fundamental absorption edge depending on the electric field. This is due to, first, the field-dependent overlap between the electron and the hole wave functions (the

Franz–Keldysh effect or, in the case of QWs and QDs, the quantum confined Franz–Keldysh effect, QCFKE), and second, specifically in QWs and QDs, due to the shift of the carrier energy levels with the field (quantum confined Stark effect, QCSE). The net result of both effects is an increase in the absorption with the field magnitude, manifested, for example, in the increase in the threshold current of monolithic laser diodes with a contact split to form an SA section, with the reversed voltage applied to this section. This threshold increase is observed universally, whether in bulk, QW, or QD lasers.

Then, an additional (under some conditions, the main) effect of photocarrier generation in the absorber performance is caused by the fact that the photocurrent charges the barrier capacitance of the structure, creating the *screening potential* φ_s that partially screens the voltage applied to the absorbing layer. In the case of a p–i–n structure and assuming that the screening potential is much smaller than the total negative bias (built-in and applied externally: $\varphi_s \ll V_{ext} + V_{bi}$) so that the leakage current of the junction can be neglected, the connection between φ_s and the photocurrent i_{ph} is represented as

$$i_{ph}(\varphi_s) = \frac{\varphi_s}{R} + C\frac{d\varphi_s}{dt}, \qquad (4.8)$$

where R is the effective resistance representing the spreading of the photocarriers in the plane of the structure away from the SA and their exit through the external circuit, and C is the effective capacitance of the structure. From (4.8), between the pulses ($I_{ph} = 0$), the screening potential disappears (the capacitor is discharged) with the characteristic time RC. During the short pulse, if the pulse duration τ_p obeys $t_{tr} \ll \tau_p \ll RC$, t_{tr} being the transit time of carriers across the absorbing region, we can write

$$i_{ph} \approx C\frac{d\varphi_s}{dt}. \qquad (4.9)$$

The photocurrent, from (4.6), is given by

$$i_{ph} \approx X\alpha(\varphi)d\frac{eP_{in}}{\hbar\omega}, \qquad (4.10)$$

where the dependence of the absorption coefficient on φ_s is due to the fact that the screening potential modifies the electric field F_e applied to the absorbing layer by a value $\Delta F_e = \varphi_s/W_i$, W_i being the thickness of the i-region of a p–i–n structure into which the absorbing layer is assumed to be incorporated. Then, approximating

$$\alpha(\varphi) \approx \alpha_{uns} - \left|\frac{\partial\alpha}{\partial\varphi}\right|\varphi = \alpha_{uns} - \frac{1}{W_i}\left|\frac{\partial\alpha}{\partial F_e}\right|\varphi, \qquad (4.11)$$

with the derivative evaluated at $\varphi_s = 0$, $\alpha = \alpha_{uns}$, we can separate variables in (4.9) and obtain the saturation formula in the form of (4.1). By substituting for the capacitance C the standard parallel plate capacitor formula $C \approx \varepsilon_0\varepsilon_r A_X/W_i$, ε_r and ε_0 being, respectively, the relative dielectric permittivity of the semiconductor and the dielectric permittivity of vacuum (i.e., assuming homogeneous illumination of the entire

SESAM area), we can obtain the expression for the saturation fluence due to this electro-optical saturation in the form [90] (assuming X independent of the absorption coefficient:

$$F_{\text{sat}}^{\text{eo}} \approx \frac{\varepsilon_0 \varepsilon_r \hbar \omega}{eX |\partial(\alpha d)/\partial F_e|}. \tag{4.12}$$

In the general case, both electro-optical and all-optical saturation may be present, so

$$\frac{1}{F_{\text{sat}}} = \frac{1}{F_{\text{sat}}^{\text{all-opt}}} + \frac{1}{F_{\text{sat}}^{\text{eo}}}. \tag{4.13}$$

The relative importance of the two effects is determined by the material used and the wavelength (on which both A_α and $\partial(\alpha d)/\partial F_e$ are strongly dependent). Usually, the all-optical saturation is believed to be the dominant effect, but some authors have stated recently [91] that the electro-optical effects are likely to play an important part in quantum dot materials.

The relaxation time of the electro-optical saturable absorption, from (4.8), is the RC time of the circuit, which can be easily varied within the range of several picoseconds to very large values by changing the external resistance.

The relaxation time of the all-optical absorber is determined, in the case of QW absorbers under an applied electric field, by the carrier sweepout by the field. This is caused by a combination of two processes: first, thermionic escape of carriers (electrons and holes) with high enough in-plane kinetic energy due to the lowering of the QW barrier by the electric field and, second, tunneling through the barrier of carriers whose energies remain below the barrier. The former process dominates at weaker fields and the latter at stronger fields. In a simple model, the resultant escape time has been shown to decrease approximately exponentially with the applied electric field. A more accurate description takes into account the modification to the density of quasi-bound and unbound states in the QW by the field-dependent QW potential. Then, thermionic and tunneling escape are treated on the same footing, and the resulting dependence is somewhat different from a simple exponential, but is still described reasonably well by an exponential fit [92].

The strong dependence of carrier sweepout on field, together with the notion of the screening potential (4.8)–(4.10), means that the saturable absorber *recovery time* is sensitive to the properties of the external circuit, even at the spectral points where the absorption *coefficient* does not have a significant field dependence [93].

The operation of QD SAs has additional peculiar features and will be discussed in more detail below. Still, to a good approximation, SAs based on the principles discussed above, whether QD or QW, can be considered *slow saturable absorbers* when operated with pulses with a duration of 0.1–1 ps.

In addition to that, saturable absorption in bulk and QW semiconductors is known to contain what is classified as *fast* saturable absorption for pulses of the same range, with a recovery time $\tau_\alpha^{\text{fast}}$ in the range of tens to about 100 fs. This was verified experimentally in waveguide geometry SAs [94, 95] and is believed (see, for example,

Refs [96, 88]) to play a very significant part in the operation of QW SESAMs – in fact, for applications such as mode locking of solid-state lasers it may be dominant [96]. The physics of these fast processes has been attributed mainly to the spectral hole burning processes and also, specifically for QW absorbers operating near bandgap, to the dynamic ionization of bound excitons (calculations show that carrier heating, which is an important source of nonlinearities in *amplifiers*, plays a more modest role in *absorbers*, due to small densities of quasi-equilibrium carriers, and can in fact result in a small *negative* absorption suppression, which has not been observed experimentally, so this effect is unlikely to be dominant). τ_α^{fast} is then essentially the carrier–carrier scattering time in the absorber material. Similar to the gain nonlinearity coefficient introduced for semiconductor amplifiers, one can characterize this fast saturable absorption via "absorption suppression, or nonlinearity, coefficient" ε_α:

$$\alpha_{sat} = \alpha - \alpha_{ns} = \frac{\alpha_{sat}(N)}{1 + \varepsilon_\alpha S}. \tag{4.14}$$

In this case, the fast absorber saturation characteristic is described by (4.3), with the saturation intensity

$$I_{sat}^{all\text{-}opt} = \frac{v_g \hbar \omega}{X \varepsilon_\alpha}, \tag{4.15}$$

with the geometric factor X the same as above. The absorption coefficient can be represented in the form of $\varepsilon_\alpha = v_g \sigma_\alpha^{fast} \tau_\alpha^{fast}$, where σ_α^{fast} is the effective cross section of fast saturable absorption [96], which can be expected to be of the same order as the slow absorption cross section σ_α (though not equal to it, as it involves the same transition matrix element but a different interplay of carrier distribution, homogeneous, and inhomogeneous broadening; cf. Equation 2.97 for gain nonlinearity in QD amplifiers). Then,

$$I_{sat}^{all\text{-}opt} = \frac{\hbar \omega}{X \sigma_\alpha^{fast} \tau_\alpha^{fast}}. \tag{4.16}$$

4.2.2
Geometry of Saturable Absorber: SESAM versus Waveguide Absorber – The Cavity Enhancement of Saturable Absorption and the Standing Wave Factor in SESAMs

As mentioned above, the two main geometries of the SA are the waveguide and SESAM geometries. In the waveguide construction, the pulse propagation and attenuation are described by essentially the same equation as those describing the light amplification in a SOA or laser, with the gain g substituted by " $-\alpha$," α being the saturable absorption (the unsaturable contribution may be included in the background absorption a_i).

$$\pm \frac{\partial Y_{f,b}}{\partial z} + \frac{1}{v_g} \frac{\partial Y_{f,b}}{\partial t} = \left(-\frac{1}{2}(\Gamma_{xy}\hat{\alpha} - a_i) + ik_{ref}\widehat{\Delta \eta}_{mod}\right) Y_{f,b}. \tag{4.17}$$

The total absorption is then calculated by integrating over the length of the SA (in the traveling frame of reference, in case of a short pulse).

In the SESAM construction, the very short propagation distance means that the propagation approach is typically not required; instead, the complex amplitude Y_r of the reflected light is simply given as a function of the incident light amplitude Y_i as

$$Y_r = \hat{r}_{SESAM} Y_i, \tag{4.18}$$

where $\hat{r}_{SESAM}$ is the amplitude (saturable) reflectance, which, strictly speaking, needs to be represented as an operator in time domain because of a combination of the material dispersion of the absorption as in (4.17) and the resonator nature of the SESAM that also involves some dispersion. In the context of a semiconductor laser, with pulse durations of the order of picoseconds or slightly below, the effect of the dispersion is not too strong, however, and so the SESAM can be characterized by an intensity reflectance connecting the reflected power P_r to the incident power P_i

$$P_r(t) = R_{SESAM}(t) P_i(t); \quad R_{SESAM} = |r_{SESAM}|^2. \tag{4.19}$$

The time dependence of the reflectance is due to the saturation, caused by the light. As it is the *internal* light intensity that saturates the SESAM, as opposed to the incident intensity, it is necessary to introduce, besides the reflectance, also the geometric factor $X = |Y_{int}/Y_i|^2$, $|Y_{int}|^2$ characterizing the internal optical intensity inside the SESAM, or, more precisely, the intensity seen by the absorbing layer(s). It is this factor that features in Equations (4.7), (4.12), and (4.15).

In the simplest SESAM construction, in which the thin (of a thickness d satisfying a condition $\lambda \ll d \ll \alpha^{-1}$) absorbing layer rests on top of the mirror with a bottom reflectance R_b, the geometrical factor is simply due to the fact that the absorption is saturated by both the incident and the reflected wave and so

$$X \approx 1 + R_b \tag{4.20}$$

(in practice, usually $R_b \approx 1$ and so $X \approx 2$). In the more generic case of a *resonator* SESAM, with the top reflector (intensity reflectance R_t, phase ψ_t) and the bottom one (intensity reflectance R_b, phase ψ_b) forming a Fabry–Perot cavity containing the absorbing layer(s), the ratio X needs, in general, to take into account the Fabry–Perot enhancement factor. In the generic case of an arbitrary absorption and reflectances (but still neglecting absorption in all the layers except the absorbing layer), and of the absorbing layer thickness comparable to the wavelength, the formula for the resonant enhancement factor takes the form [97]

$$X(\alpha) = \frac{(1-R_t)(1+R_b \exp(-\alpha d))(1-\exp(-\alpha d))}{\alpha d (1 - 2\sqrt{R_t R_b} \exp(-\alpha d) \cos(2\beta L_{res} + \psi_t + \psi_b) + R_t R_b \exp(-2\alpha d))} f_{sw}, \tag{4.21}$$

where $\beta = 2\pi n/\lambda$ is the transverse wave vector in the SESAM, L_{res} is the thickness of the SESAM resonator, and f_{sw} is the standing wave factor, taking into account the fact that the absorption in a thin absorbing layer is affected by its position relative to the standing wave formed by light in the SESAM resonator – with the absorbing layer in the antinode of the standing wave, the interaction of the layer with the wave is maximized; with the layer in the node, it is all but eliminated. The factor f_{sw} can be taken to be 1 if the thickness d of the active layer in the SESAM layer is greater than the wavelength of light in the material λ/n, so that the standing wave distribution is averaged over the SA. In the arbitrary case, it is evaluated as

$$f_{sw} = 1 + \frac{2\sqrt{R_b}}{1+R_b}\frac{\sin(\beta d)}{\beta d}\cos(2\beta L_b + \beta d + \psi_b), \quad (4.22)$$

where L_b is the distance between the absorbing layer and the bottom reflector.

In the case of a SESAM designed to resonantly enhance the light (with L_{res} adjusted to minimize the denominator in (4.21), we have

$$X(\alpha) = \frac{(1-R_t)(1+R_b\exp(-\alpha d))(1-\exp(-\alpha d))}{\alpha d(1-\sqrt{R_t R_b}\exp(-\alpha d))^2} f_{sw}. \quad (4.23)$$

Note also that in the case of a thin SA layer, which is often a single quantum well (or layer of dots) under normal incidence, the absorption coefficient α, introduced for propagation distances greater than the wavelength, is not a good characteristic of light absorption – it is more meaningful to speak of the dimensionless normal transmittance of the layer $T_n = P_{transmitted}/P_{incident}$, or the normal absorptance $Q = 1 - T_n$. It is the latter quantity that can be calculated quantum mechanically, from first principles. The parameter $T_n = 1 - Q$ should then feature in (4.21) or (4.23) instead of $\exp(-\alpha d)$. Purely formally, the "traditional" absorption coefficient α can be introduced as $\alpha = Q/d \approx 1/d \ln(1/T_n)$ and then used in (4.21) or (4.23), but it becomes a function of the layer (well) thickness as well as of the material and wavelength, and does not necessarily equal the in-plane absorption coefficient.

Since X is, in this general form, itself a function of α, (4.7) needs to be generalized, for small to moderate αd, in the form

$$F_{sat}^{all-opt} \approx \frac{\hbar\omega}{\sigma_\alpha X'}, \quad X' = X + \alpha_{uns}\frac{\partial X}{\partial \alpha}\bigg|_{\alpha=\alpha_{uns}}. \quad (4.24)$$

Likewise, for the case of an electroabsorption mechanism, we can generalize (4.12) as

$$F_{sat}^{eo} \approx \frac{\varepsilon_0\varepsilon_r\hbar\omega}{eX'|\partial(\alpha d)/\partial F_e|}, \quad (4.25)$$

where, as discussed above, in the case of a subwavelength d, the product αd is understood to mean the normal dimensionless absorptance Q.

In the case of large $\alpha d > \sim 1$, the formula (4.1) is no longer a suitable approximation for the dependence of $\alpha(F_{in})$, and the definition of F_{sat} is not straightforward. In case of $d \ll \alpha^{-1}$ and if R_t is not too large (up to about $R_t \approx 0.3$, which is the reflectance of the uncoated facet), which is very often the case in practical SESAMs, we can simplify (4.23), in the case of a resonant enhancement of the SESAM, in the form [90]

$$X \approx \frac{(1-R_t)(1+R_b)}{(1-\sqrt{R_t R_b})^2} f_{sw}, \quad (4.26)$$

which is a constant value, so $X' = X$, hence allowing the use of (4.7) rather than the more complex (4.24).

Since the reflectance of the bottom mirror in SESAMs is usually nearly unity, the intensity reflectance of the entire SESAM chip and thus the saturation of this intensity are estimated as

$$R_{SESAM} \approx R_t + (1-R_t)(1-X\alpha d). \quad (4.27)$$

In the case of a *waveguide* absorber, the propagation of light is described by a propagation equation, so technically speaking, Equation 4.1 does not necessarily hold, and thus the interpretation of the saturation fluence sometimes quoted in some experimental papers (e.g., [98]) needs some caution. However, with small to moderate $\Gamma_{xy}\alpha L_a < 1$, L_a being the absorber section length, an expression of the type (4.23) may be used as a reasonable estimate, with $f_{sw} = 1$ as the standing wave profile in the long waveguide SA is certain to be averaged over, and the thickness d substituted by $\Gamma_{xy} L_a$.

Most often, the waveguide absorber will be used in a traveling-wave geometry (e.g., in a pulse-shaping amplifier–absorber module), in which case

$$X = X_{tw} \approx \frac{1-\exp(-\Gamma_{xy}\alpha L_a)}{\Gamma_{xy}\alpha L_a}, \quad (4.28)$$

which is approximated by $X = X_{tw} \approx 1$ in the case of a very short absorber ($\Gamma_{xy}\alpha L_a \ll 1$), or in a reflective geometry, backed with a mirror of a reflectance R_b (e.g., in a mode-locked laser), in which case

$$X = X_{refl} \approx \frac{(1+R_b\exp(-\alpha d))(1-\exp(-\alpha d))}{\alpha d}, \quad (4.29)$$

approximated, as in (4.20), by

$$X = X_{refl} \approx 1 + R_b \quad (4.30)$$

in the case of $\Gamma_{xy}\alpha L_a \ll 1$.

4.3
The Main Special Features of a Quantum Dot Saturable Absorber Operation

As all saturable absorbers, quantum dot SAs can be used both in a waveguide and in a SESAM configuration, and the broad principles of absorber saturation are the same in QDs as in QWs and bulk materials. There are, however, important differences as regards both the nature of saturation fluence and the recovery dynamics, some of which may be advantageous for a number of applications. Below, we shall consider these features and comment on potential advantages and caveats associated with them.

4.3.1
Bandwidth of QD SAs

The bandwidth of a saturable absorber component is defined by the characteristics of the absorber material that is responsible for the saturable absorption and, in the case of a SESAM, also to some extent by the characteristics of the SESAM resonator. The prime characteristic of the SA material is the spectral band within which the absorber can be efficiently saturated. Compared to the interband absorption in QWs and bulk materials, absorption due to transitions between localized states in QDs has a relatively narrow spectrum, particularly if only ground or only excited state absorption is involved in the SESAM operation (which is usually the case). Indeed, the absorption bandwidth due to QD transitions is *finite* and mainly determined by the inhomogeneous broadening $\Delta E_{\text{inh}} \sim$ 20–50 meV whereas in quantum wells and bulk materials, the absorption spectrum is effectively *semi-infinite*, with the light with any photon energy above the fundamental absorption edge (the effective bandgap) being absorbed. However, not all of this bandwidth can be efficiently saturated. Indeed, because of the relatively fast carrier–carrier scattering in bulk and QW materials, carriers are thermalized on a subpicosecond scale. Thus, for pulses in the picosecond and subpicosecond duration range, the slow saturation of absorption in QWs is seen only in a spectral area where changes in the quasi-equilibrium carrier distributions are significant, that is, in a relatively narrow spectral band ($\sim k_B T$) near the fundamental absorption cutoff. For transitions creating carriers with energies further in the band, only the (usually weaker) fast absorption saturation due to spectral hole burning is significant. In QDs, in contrast, carrier–carrier interactions are substituted by slower capture and escape times, so the carrier distribution during the pulse remains substantially nonequilibrium, and slow saturation of absorption is significant in the entire absorption band. Thus, QD-based SESAM structures may be seen as having a broader *saturable* absorption spectrum than their quantum well counterparts and hence offer more latitude for the generation of significantly shorter pulses. This feature not only facilitates their use in ultrafast broadband lasers but also allows extra tunability. Furthermore, the presence of excited states in the absorption spectrum of deep-level QDs opens up possibilities for the operation of saturable absorbers at very distinct spectral bands, notably in the ground and excited states. Thus, for some constructions of QD-SESAMs, the main limitation to the spectral bandwidth of operation might actually be the wavelength selectivity of the DBR rather than the gain medium.

4.3.2
Dynamics of Carrier Relaxation: Ultrafast Recovery of Absorption

As mentioned previously, quantum dot materials exhibit intrinsic ultrafast carrier dynamics. This feature is particularly useful for enabling saturable absorbers to be deployed in high repetition rate lasers, where the absorption recovery should occur within the round-trip period of the cavity. In this respect, several studies have been performed to evaluate the absorption recovery time of QD saturable absorbers, such as those described in Ref. [40]. This paper reported the first pump–probe measurement of a QD-based SESAM. This study showed two distinct time constants for the recovery in the absorption saturation. One was a fast recovery of around 1 ps followed by a slower recovery process that extended to 100 ps [40]. We have also measured directly by optical means the carrier escape time from the ground state of the quantum dots in a waveguide p–i–n structure [99]. The faster recovery time remained roughly constant with the reverse bias, while the "slow" absorption recovery time constant at room temperature decreased exponentially from 62 ps to about 700 fs by nearly two orders of magnitude, as the reverse bias applied to the structure varied from 0 to −10 V (Figure 4.2). This exponential dependence, similar to that observed in QW materials, suggests carrier sweepout by the field as the most likely mechanism of absorber relaxation. This initial observation was subsequently confirmed by a number of independent investigations (see, for example, Ref. [98, 100]).

A theoretical interpretation of the nature of the time constants involved was offered in Ref. [98]. The authors explained the dynamics of the absorber recovery by considering an initial population created by a short, intense optical pulse in the ground state (and thus determining the initial conditions of the problem), followed by free relaxation described by the standard kinetic equations (2.16) and (2.17) for the ground- and excited-state electron occupancies. As the sweepout processes to be described by the model are known to be much faster than spontaneous recombination, and no stimulated recombination is present after the pulse, the recombination processes can be omitted from the start, leaving only capture and escape.

$$\frac{df_G^{(e)}}{dt} = R_G^{(\mathrm{rel})} - R_G^{(\mathrm{esc})}, \qquad (4.31)$$

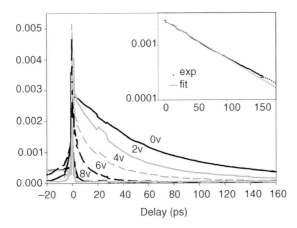

Figure 4.2 Kinetics of bleaching relaxation of the InAs p–i–n ridge waveguide QD modulator for increasing reverse bias. *Inset*: single exponential fit for 0 V. Reprinted from Ref. [99] with permission from AIP.

$$\frac{df_E^{(e)}}{dt} = R_E^{(cap)} - R_E^{(esc)} - \frac{1}{2}(R_G^{(rel)} - R_G^{(esc)}), \quad (4.32)$$

where $\varrho_G/\varrho_E = 1/2$ has been substituted. The capture and escape rates are described using the standard approach (2.21); in our notations (slightly different from those of Ref. [98]),

$$R_G^{(rel)}(E_G) = 2\frac{f_E(1-f_G)}{\tau_{rel}^{(G)}}; \quad R_G^{(esc)}(E_G) = \frac{f_G(1-f_E)}{\tau_{esc}^{(G)}}, \quad (4.33)$$

with all the variables having the same meaning as in Section 2.3. Contrary to the situation in the amplifiers, where the relaxation of carriers from the excited to ground state is subpicosecond (due to the Auger mechanisms facilitated by a high wetting layer carrier density), in absorbers, with the wetting layer carrier density absent or very low, the relaxation time is on the order of a few picoseconds, comparable with, or longer than, the capture from the wetting layer into the dots (which does not enter the model of Ref. [98]).

The difference from the forward biased situation is that under reverse bias, a new escape process from the excited level is postulated in the model, which substitutes the recombination and is believed to be responsible for the slower of the relaxation rates observed experimentally. The characteristic time of this process is taken to depend on the applied electric field (or, neglecting the screening potential, on the voltage):

$$R_E^{(esc)} = \frac{f_E}{\tau_w}; \quad \tau_w = \tau_0 \exp\left(-\frac{V}{V_0}\right), \quad (4.34)$$

where $\tau_0 \sim 10\,\mathrm{ps}$ (18 ps for the construction studied in Ref. [98]) and $V_0 \sim 1\,\mathrm{V}$ (2 V in the construction of Ref. [98]) are constants determined from the experiment. The authors of Ref. [98] interpreted the physical nature of this process as escape into the wetting layer, though the exponential dependence of the characteristic time on the transverse electric field (similar to that observed in QW SAs), as well as the absence of an opposite capture process, may be more consistent with escape into the barriers (possibly with the WL as an intermediate stage, as in the case of a stepped quantum well potential [101]).

Then, the population of the wetting layer (or at least the effect of this population on the absorption relaxation) remains negligible throughout the relaxation process, no equation of the type of (2.18) for the WL population is necessary, and we can assume, following [98], that $R_E^{(cap)} = 0$.

Equations 4.31 and 4.32 are, in general, nonlinear, so an analytical, exponential-like solution is possible only in limiting cases. These depend on the initial conditions. The easiest limiting case to solve is the *linear* one when the initial occupation of the ground state (assuming ground-state pumping) created by the pulse is small: $f_{G0} = f_G|_{t=0} \ll 1$. Then, throughout the relaxation process one can set $f_{G,E}(t) \ll 1$, omitting terms $1-f_{G,E}(t)$, and thus obtaining the relaxation as a sum of two exponential processes, fast and slow. The fast and slow times are estimated in the form

4.3 The Main Special Features of a Quantum Dot Saturable Absorber Operation

$$\frac{1}{\tau_{\text{fast,slow}}^{\text{linear}}} = \frac{1}{2}\left(\frac{1}{\tau_{\text{rel}}^{(G)}} + \frac{1}{\tau_{\text{esc}}^{(G)}} + \frac{1}{\tau_w} \pm \sqrt{\left(\frac{1}{\tau_{\text{rel}}^{(G)}} + \frac{1}{\tau_{\text{esc}}^{(G)}} + \frac{1}{\tau_w}\right)^2 - \frac{4}{\tau_{\text{esc}}^{(G)} \tau_w}}\right), \quad (4.35)$$

with the sign + corresponding to the fast time and − to the slow time. At low voltages, when $\tau_w \gg \tau_{\text{esc}}^{(G)} > \tau_{\text{rel}}^{(G)}$, this gives

$$\tau_{\text{slow}}^{(\text{low } V)} \approx \tau_w(V)\left(1 + \frac{\tau_{\text{esc}}^{(G)}}{\tau_{\text{rel}}^{(G)}}\right), \quad (4.36)$$

explaining the experimental voltage dependence of the slow relaxation. With the sweepout fast enough to ensure that $\tau_w < \tau_{\text{rel}}^{(G)} < \tau_{\text{esc}}^{(G)}$, (4.35) gives to

$$\tau_{\text{slow}}^{(\text{high } V)} \approx t_{\text{esc}}^{(G)}. \quad (4.37)$$

meaning that there is no more slow stage. Indeed, at high voltages, only one time constant is observed in the experiments and does not depend on voltage any more.

For the fast time constant, the linear approximation (4.35) would give, in the case of $\tau_w \gg \tau_{\text{esc}}^{(G)} > \tau_{\text{rel}}^{(G)}$, an estimate $1/\tau_{\text{fast}}^{\text{linear}} \approx 1/\tau_{\text{rel}}^{(G)} + 1/\tau_{\text{esc}}^{(G)}$, reminiscent of the fast relaxation time in QD amplifiers, determined by (2.82). However, in the linear case, the relative magnitudes of the slow and fast responses are independent of the initial conditions, which is not the case in the experimental observations in either [99] or [98]. Thus, the authors of Ref. [98] proposed an elegant *nonlinear* theory of the initial stage of relaxation. An analytical estimate in the case of $\tau_w \gg \tau_{\text{esc}}^{(G)} > \tau_{\text{rel}}^{(G)}$ is obtained by omitting terms proportional to τ_w^{-1} in (4.31) and (4.32)–(4.33), leading to conservation of the total populations of two levels: $f_G(t) + 2f_E(t) = f_{G0} = \text{const}$. Substituting $2f_E(t) = f_{G0} - f_G(t)$ into (4.31) and (4.32), a simplified equation is then obtained for the population $f_G(t)$ of the ground state, which admits an analytical solution in the form

$$f_G(t) = f_{G0} - 2\frac{f_{G+}[1 - \exp(-t/\tau_{\text{fast}})]}{1 - f_{G+}/f_{G-}\exp(-t/\tau_{\text{fast}})}, \quad (4.38)$$

with

$$\frac{1}{\tau_{\text{fast}}} = \frac{\varsigma}{\tau_{\text{rel}}^{(G)}};$$

$$\varsigma = \sqrt{\left[1 - f_{G0} + \frac{\tau_{\text{rel}}^{(G)}}{2\tau_{\text{esc}}^{(G)}}(2 + f_{G0})\right]^2 + 4f_{G0}\left(1 - \frac{\tau_{\text{rel}}^{(G)}}{2\tau_{\text{esc}}^{(G)}}\right)\frac{\tau_{\text{rel}}^{(G)}}{\tau_{\text{esc}}^{(G)}}} \quad (4.39)$$

and

$$f_{g\pm} = \left[-\left(1 - f_{G0} + \frac{\tau_{\text{rel}}^{(G)}}{2\tau_{\text{esc}}^{(G)}}(2 + f_{G0})\right) \pm \varsigma\right] / \left(4\left(1 - \frac{\tau_{\text{rel}}^{(G)}}{2\tau_{\text{esc}}^{(G)}}\right)\right). \quad (4.40)$$

Note that, somewhat counterintuitively, the fast time (4.39) is scaled by the downward relaxation time $\tau_{rel}^{(G)}$ rather than by the escape time $\tau_{esc}^{(G)}$.

The slow relaxation stage occurs at the end of the fast stage ("time layer"), which was determined in Ref. [98] as the point when $f_G(t)$ decreases to the value $f_{G0} - 2f_{G+}$. The solution is, in general, nonexponential, but gradually evolves, as $f_G(t)$ falls to values $\ll 1$, to the exponential decay with the time $\tau_{slow}^{(low\ V)}$ determined by (4.36).

In the case of $\tau_w < \tau_{rel}^{(G)} < \tau_{esc}^{(G)}$, the population of the excited level is low throughout the process, and the evolution of the ground-state occupancy is described by a simple exponential decay with the time $\tau_{slow}^{(high\ V)} \approx \tau_{esc}^{(G)}$ (4.37), not only for $f_{G0} \ll 1$ but also for an arbitrary f_{G0}.

Interestingly, there appears to be no clear sign in the reported experiments of separate electron and hole dynamics, and so the model needs to include only one sign of carriers to explain the experimental results. One way of interpreting this is to assume that the excitonic approximation $f_G^{(e)} = f_G^{(h)}$ holds reasonably well in QD SAs. Alternatively, one can deduce that the dynamics of one type of carriers is much slower than the other.

While there may be scope for further investigation and refinement of the model, for example, including escape from both the excited state(s) and the wetting layer, the general conclusions on the nature of the two major timescales involved to do with intradot relaxation and sweepout are convincing. They have also been recently confirmed by a more extensive numerical model involving both escape from the excited state into the WL (with subsequent escape into the barriers) and direct escape from the excited level into the barriers (see Chapter 5).

The nature of kinetic processes in the QD SA is presented schematically in Figure 4.3, which is the SA counterpart of Figure 2.1 for an amplifying section. Additional processes of escape through the WL, not described by (4.31) and (4.32), are shown with dashed lines.

The sweepout of carriers from the *excited* level(s) is considerably shorter than from the ground level since there is no interlevel carrier escape stage involved in the process, as will be discussed in Section 5.4.

The relaxation dynamics as described above is different from the case of bulk and QW materials, and has several significant implications for the use of QD SAs.

First, since the smallest observed relaxation time in QD SAs is on the order of 1 ps (as opposed to <100 fs in bulk and QW materials), the QD saturable absorbers would appear to act as purely *slow* absorbers when used with subpicosecond lasers (e.g., solid-state lasers), without the fast absorber component present in QWs and bulk materials. Their behavior is thus fully characterized by a saturation fluence, which will be discussed below, rather than instantaneous intensity. In semiconductor mode-locked lasers, whether in-plane or vertical external cavity, which generate pulses of a duration on the order of 1–10 ps, the pulse duration may be of the same order as the relaxation time, so the absorber is, technically speaking, neither fast nor slow. Full dynamic model has to be analyzed, as is the case with bulk and QW lasers.

Second, the "slow" relaxation processes in the QD SA, while classified as "slow" compared to subpicosecond pulse durations, are nevertheless faster than the

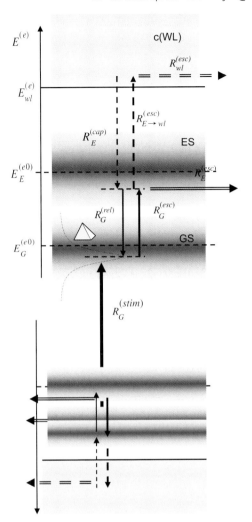

Figure 4.3 Schematic representation of level structure and the main electron kinetic processes in a dynamic model of a QD *saturable absorber* pumped with light resonant with GS transitions. Processes shown in solid lines are those included in the simplest model described in Equations 4.31 and 4.32. Shown in dashed lines are additional carrier escape mechanisms involving the wetting layer (and possibly higher excited states).

corresponding time constants in a QW absorber. Indeed, the measured sweepout times in QWs are typically not smaller than 5–6 ps, whereas 2–3 ps has been observed in QD SAs at high bias.

This fast response of QD SAs should prove useful in the development of devices for switching at frequencies above 1 THz. In particular, this provides significant promise for ultrafast electroabsorption modulators and for the optimization of saturable absorbers used for the passive mode locking of semiconductor lasers.

Incidentally, quantum dots suspended in glasses also exhibit fast carrier dynamics. The relaxation kinetics of their saturable absorption also exhibit biexponential characteristics, with fast and slow components where the fast component decreases with reduction in the radius of the dots [35]. Furthermore, a strong decrease in the relaxation time with increasing pump fluence (from several tens of picoseconds to several picoseconds) has been observed in CdSe, PbSe, and PbS QDs embedded in glass matrices [102–104]. The chemical composition, crystalline structure, and the surface states determine the characteristics of the nonlinear response of materials such as CdS, CdSe, CdTe, and CuCl [105–108], Cu_xS [109], and PbS [104, 110]).

4.3.3
Saturation Fluence

QD-based SESAMs exhibit lower saturation fluence than quantum well-based materials due to their delta-like density of states. This facilitates the self-starting of mode locking at modest pulse energies. This feature is particularly important in high-repetition rate lasers, where the optical energy available in each pulse is small. On the other hand, low saturation fluence is also crucial for average power due to limitations imposed by thermal issues and cavity stability as demonstrated with external vertical cavity surface emitting lasers, or VECSELs.

To investigate the physical processes contributing to the (all-optical) saturation in the QD SA, a theoretical approach very similar to that used for analyzing saturation in a QD amplifier can be used. Indeed, in a simple model, the QD SA absorption at a photon energy $\hbar\omega$ can be written similar to the gain of the amplifier (2.9), except for the opposite sign. For the ground-state absorption coefficient, we have

$$\alpha_G = A_G^{(a)} \varrho_G N_D \int (1 - f_G^{(e)}(E_G^{(e)}) - f_G^{(h)}(E_G^{(h)})) D_{\text{hom}}(\hbar\omega, E_G^{(e-h)}) D_{\text{inh}}(E_G^{(e-h)}) dE_G^{(e-h)}, \tag{4.41}$$

or, in shorthand notation,

$$\alpha_G = A_G^{(a)} \varrho_G N_D \sum_i (1 - f_{Gi}^{(e)} - f_{Gi}^{(h)}) D_{\text{hom},i}. \tag{4.42}$$

The coefficient $A_G^{(a)}$, proportional to the absorption cross section, is in the free particle model identical to that used in the amplifier: $A_G^{(a)} = A_G$. However, it has been experimentally noted [51] that a fully inverted QD amplifier shows a gain about three times smaller than the peak unsaturated absorption in the same sample. This implies $A_G^{(a)} > A_G$ and may be attributed to the Coulomb interaction between the electron and the hole, which is screened in the amplifier but is present in the absorber. This is consistent with the fact that Coulomb interactions are known to get more important in the optical properties of the material as the dimensionality of the material is reduced: for example, the free exciton peak in absorption is not observed in bulk materials at room temperatures but is observed in QWs. (It is possible also that what is observed in the experiments on the amplifier is not really full inversion, particularly

4.3 The Main Special Features of a Quantum Dot Saturable Absorber Operation

as in the real situation the excitonic assumption may not be a particularly good approximation, in which case the saturation of the current dependence of the peak gain is considerably slowed down as discussed in Section 2.2, Figure 2.7). The inhomogeneous broadening function in (4.41) is clearly the same as in (2.9) as the same electron and hole states are involved in transitions in the amplifier and the absorber. The homogeneous broadening may be somewhat different since the very low WL carrier densities in QD SAs remove one of the mechanisms (WL-dot carrier scattering) that causes dephasing of electron and hole states in QD SOAs and thus contributes to the homogeneous broadening). Thus, the homogeneous broadening of transitions may be somewhat smaller in absorbers than in amplifiers: $\Delta E_{hom}^{(a)} < \Delta E_{hom}$; as a rough estimate, a value of the same order may be taken.

Considering the case of an incident pulse short enough for the SA to be treated as a slow absorber (but still long enough for the electronic polarization to be adiabatically eliminated – from the previous section, this implies pulses several hundred femtoseconds long), we can describe the dynamics of dot population by rate equations similar to (2.16) but with relaxation and escape, as well as recombination, terms omitted (this is the stage preceding the dynamic analysis of the previous section and determining the initial condition f_{G0} for this analysis). Only the stimulated (absorption) term (2.64) is present in the analysis. In this case, given the absence of recombination and thermalization, the excitonic assumption $f_{Gi}^{(e)} = f_{Gi}^{(h)} = f_{Gi}$ is thoroughly justified: an act of absorption creates an electron–hole pair. Thus,

$$\frac{\partial f_{Gi}}{\partial t} = v_g A_G^{(a)} (1 - 2 f_{Gi}) D_{hom,i} S. \tag{4.43}$$

The dots that contribute to absorption are those with transition energies within $\sim \Delta E_{hom}^{(a)}$ around $\hbar\omega$. In general, from (4.43), the population of each of them would saturate with its own saturation fluence dependent on $D_{hom,i}$. To obtain an analytical estimate for F_{sat} of the entire absorber, one has, as in the analysis of QD SOA operation for the slow relaxation case (Section 3.3), to make the top-hat approximation (2.100) for the homogeneous broadening function:

$$D_{hom,i} = D_{hom}(\hbar\omega, E_{Gi}^{(e-h)}) = \frac{1}{\pi} \frac{\Delta E_{hom}^{\alpha}}{(\hbar\omega - E_{Gi}^{(e-h)})^2 + \Delta E_{hom}^{\alpha 2}}$$

$$\approx \begin{cases} \frac{1}{\Delta E_{hom}^{\alpha}} & \text{if } \left| \hbar\omega - E_{Gi}^{(e-h)} \right| < \frac{\Delta E_{hom}^{\alpha}}{2}, \\ 0 & \text{otherwise.} \end{cases} \tag{4.44}$$

Then, the absorption saturation is described by the formula (4.1) for a slow saturable absorber, with the unsaturated absorption being

$$\alpha_{uns} - \alpha_{ns} \approx A_G^{(a)} \varrho_G N_D D_{inh}(E_G^{(e-h)} = \hbar\omega) \tag{4.45}$$

and the saturation fluence given by

$$F_{sat} \approx \frac{\hbar\omega}{2X} \frac{\Delta E^\alpha_{hom}}{A_G^{(a)}} = \frac{1}{X} \frac{\Delta E^\alpha_{hom}}{2\sqrt{\pi}\Delta E_{inh}^{(G)}} \frac{\hbar\omega}{\sigma_G^{(a)}}, \qquad (4.46)$$

where in the latter equality we have introduced, similar to the amplifier case, the absorption cross section $\sigma_G^{(a)} = A_G^{(a)}/\sqrt{\pi}\Delta E_{inh}^{(G)}$, which may be different from the gain cross section $\sigma_G = A_G/\sqrt{\pi}\Delta E_{inh}^{(G)}$ due to Coulomb effects.

Expression (4.46) is approximate for a number of reasons. It uses the top-hat approximation for the homogeneous broadening function, neglects the contribution of $f_{Gi} > 0$ to the screening of the Coulomb electron–hole interaction (which would make $A_G^{(a)}$ to decrease with population rather than be constant, and thus contribute further to saturation), and, finally, neglects any optoelectronic contribution to the SA saturation – (4.46) describes only the all-optical part. However, it is a useful indication of the possibilities offered by the QD absorbers. Indeed, compared to the formula for QW absorbers, expression (4.46) for QD SAs contains a small factor $\Delta E^\alpha_{hom}/(2\sqrt{\pi}\Delta E_{inh}^{(G)})$ that comes from the relatively slow energy redistribution of carriers in QDs, meaning that the states that contribute most to the absorption are filled preferentially, enhancing the absorption saturation. This is similar to the observation made in Section 3.3 on the saturation of gain in QD SOAs being stronger when the pulse durations are shorter than the characteristic relaxation times. Estimating the cross section $\sigma_G^{(a)}$ compared to its QW or bulk counterpart is less straightforward. On the one hand, using the logic presented in Ref. [50], it may be expected to be somewhat smaller due to the reduced overlap between the electron and the hole wave functions in a QD compared to a QW; on the other hand, it may benefit from an increased Coulomb electron–hole interaction in a tightly confined nanostructure such as a QD.

Assuming the cross section, homogeneous, and inhomogeneous broadening energy in the SA the same as in the SOA (see Ref. [50] and Section 3.3), a numerical estimate from (4.46) gives $F_{sat} \sim 0.1$ J m^{-2} for $X = 1$.

The considerations above are centered on the *all-optical* mechanism of the absorption saturation in the SA. The *electro-optical* mechanism has received less attention for the time being; to obtain a theoretical estimate of its magnitude, the knowledge of the Stark effect in QD SAs is needed. Experimental data indicate that the Stark effect in QDs is somewhat weaker than in QWs; according to the measurements reported in Ref. [111], an applied electric field of about 200 kV m^{-1} leads to a variation in the *modal* absorption in the waveguide geometry of 10 cm^{-1} at the peak of the ground-state absorption. A crude estimate from this result (noting that the surface-normal absorption properties may be different from the waveguide geometry) gives $F_{sat}^{eo} >\sim 1$ J m^{-2} at $X = 1$, which is somewhat higher than the value for the all-optical mechanism. Theoretical models of QD SAs published to date tend to ignore the electro-optical saturation effect; however, experimental observations of the mode-locked QD laser dynamics being dependent on the load resistance of the SA circuit [91] suggest that at least in some experimental situations it might play some part.

Experimentally, saturation flux values on the order of $0.01\,\text{J}\,\text{m}^{-2}$ have been observed in QD SESAMs [112], where $X > 1$, and of 0.1–$0.5\,\text{J}\,\text{m}^{-2}$, in the traveling-wave waveguide geometry, where $X < 1$ [98]; these values compare very favorably with other all-optical absorbers. In Ref. [98], an increase in the saturation fluence with an increase in reverse bias was also reported, which may be due to the fact that the pulse durations used were only a few times shorter than the absorber recovery time (particularly at high bias voltages), meaning that the SA was not a true slow absorber.

5
Monolithic Quantum Dot Mode-Locked Lasers

5.1
Introduction to Semiconductor Mode-Locked Lasers

5.1.1
Place of Semiconductor Mode-Locked Lasers Among Other Ultrashort Pulse Sources

Solid-state lasers based on vibronic gain materials such as Ti:sapphire, Cr:forsterite, and Cr:YAG have so far been delivering the best performances in terms of femtosecond pulse durations, very high peak power, and low jitter [113]. For example, pulses as short as 5 fs can be directly produced from a Kerr-lens mode-locked Ti:sapphire laser [114] and up to 60 W of average power has been generated in femtosecond pulses from a thin-disk Yb:YAG laser [115]. However, these laser systems present intrinsic limitations that have been preventing their widespread use in industrial and medical applications [116]. Ultrafast solid-state lasers can be very expensive and cumbersome optical sources. Despite some efforts on the miniaturization of these sources, the footprint of these systems is still very significant and integration in ultracompact setups is virtually impossible. They are not electrically pumped, and a second laser system has to be used as an optical pump, which often needs a cooling system. Ultrafast solid-state lasers are multielement systems, comprising at least a crystal, lenses, output couplers, and mirrors, all critically aligned between them. Most of these lasers also incorporate intracavity and extracavity dispersion compensation in order to achieve femtosecond pulse durations. Solid-state lasers are thus complex to operate and optimize, requiring a highly skilled user. The crystals used in these lasers have usually low gain, and therefore the necessary minimum crystal length limits the obtainable pulse repetition frequency. Moreover, these lasers cannot be easily synchronized with an external electrical signal.

Semiconductor lasers cannot yet directly generate the sub-100 fs pulses routinely available from diode-pumped crystal-based lasers, but they represent one of the most compact and efficient sources of picosecond and subpicosecond pulses. Furthermore, the bias can be easily adjusted to determine the pulse duration and the optical power, thus offering, to some extent, electrical control of the characteristics of the

Ultrafast Lasers Based on Quantum Dot Structures: Physics and Devices.
Edik U. Rafailov, Maria Ana Cataluna, and Eugene A. Avrutin
Copyright © 2011 WILEY-VCH Verlag GmbH & Co. KGaA, Weinheim
ISBN: 978-3-527-40928-0

output pulses. These lasers also offer the best option for the generation of high repetition rate trains of pulses, owing to their small cavity size. Ultrafast diode lasers have thus been favored over other laser sources for high-frequency applications such as optical data/telecoms. Being much cheaper to fabricate and operate, ultrafast semiconductor lasers also offer the potential for dramatic cost savings in a number of applications that traditionally use solid-state lasers. The deployment of high-performance ultrafast diode lasers could, therefore, have a significant economic impact, by enabling ultrafast applications to become more profitable, and even facilitate the emergence of new applications.

The physical mechanisms underlying the generation of short pulses from diode lasers are very different from vibronic lasers. Semiconductors have a higher nonlinear refractive index than other gain media. The interaction of the pulse with the gain and the resulting large changes in the nonlinear refractive index lead to a phase change across the pulse, owing to the coupling between gain and index, which in most laser materials and constructions can be described via the linewidth enhancement factor (LEF) α_H. This results in significant self-phase modulation, imparting an up-chirp to the pulses, which combined with the positive dispersion of the gain material leads to substantial pulse broadening. This mechanism has been one of the major limitations in obtaining pulse durations on the order of 100 fs *directly* from the diode lasers, with picosecond pulses being the norm. Furthermore, a strong saturation of the gain also results in stabilization of pulse energy, which limits the average and peak power to much lower levels than in vibronic lasers. Output average power levels for mode-locked laser diodes are usually between 0.1–100 mW, while peak power levels remain between 10 mW–1 W. Only with additional amplification/compression setups can the peak power reach the kW level [117].

Until very recently, ultrafast laser diodes have been solely based on bulk and quantum well semiconductors. Since 2001 [118], quantum dot (QD) materials have been generating much interest because they offer specific advantages when integrated into ultrafast lasers. The investigations on the amplification of the femtosecond pulses and the ultrafast carrier dynamics of quantum dot structures imply that such structures could be used simultaneously as an efficient broadband gain medium and as a fast saturable absorber. In this section, we will show how mode-locked quantum dot lasers can successfully combine these two aspects and deliver picosecond and subpicosecond pulses.

5.1.2
Mode-Locking Techniques in Laser Diodes: The Main Principles

To generate ultrashort pulses from semiconductor lasers, three main techniques can be used: gain switching, Q-switching, and mode locking [11]. Mode locking (ML) is usually the preferred method because it can provide higher repetition rate pulses and shorter pulse durations than the other techniques. To achieve these goals, a variety of mode-locking techniques and device structures have been investigated and optimized [11]. There exist three main forms of mode locking: active, passive, and hybrid.

Active mode locking consists of the direct modulation of the gain section with a frequency equal either to the pulse repetition frequency in the cavity or to a subharmonic of this frequency. Equivalently, an electroabsorption segment of a multielement device can be modulated to produce the same effect. The main advantages of this technique are the resultant low jitter and the ability to synchronize the laser output with the modulating electrical signal, which is a fundamental attribute for optical transmission and signal processing applications. However, high repetition frequencies are not readily obtained through directly driven modulation of lasers because fast RF modulation of the drive current becomes progressively more difficult with increase in frequency.

The frequency limitation imposed by electronic driver circuits can be overcome by employing passive mode-locking techniques. This approach typically uses a saturable absorbing region in the laser diode, which plays a crucial role in shortening the circulating pulses, thus providing the shortest pulses of all three techniques, without the need for an applied RF modulation. In a saturable absorber, the loss decreases as the optical intensity increases. This feature acts as a discriminator between CW and pulsed operation and can facilitate a self-starting mechanism for mode locking. Most importantly, saturable absorption plays a crucial role in shortening the duration of the circulating pulses, as will be explained, in more detail below thus providing the shortest pulses achievable by all three techniques, and the absence of an RF source considerably simplifies the fabrication and operation.

Passive mode locking also allows high pulse repetition rates that are determined solely by the cavity length. The most traditional way of achieving this is to place the saturable absorber in the center of the laser cavity; this is known as *colliding pulse* mode (CPM) locking and produces pulses at *twice* the fundamental ML frequency (the inverse round-trip period) of the cavity. The technique has been extended to integer multiples of the fundamental repetition frequency, by positioning one absorber (asymmetric CPM) or several absorbers (multiple CPM) at fractions of the cavity length (see Ref. [12] for an overview). Some experimental studies also reported observation of the doubling of repetition frequency in ordinary ML laser constructions at high currents; this effect will be explained in Section 5.2.

The technique of hybrid mode locking is a combination of active and passive ML. In this case, the pulse generation may be seen as initiated by a modulation provided by an RF current imposed in the gain or absorber section, while further shaping and shortening are assisted by a saturable absorber section. This process results in high-quality pulses, synchronized with an external source.

5.1.3
Passive Mode Locking: The Qualitative Picture, Physics, and Devices

In a simple frequency domain picture for mode locking, the relative phases are of primary relevance. A physical model for passive mode locking can be alternatively and equivalently described in terms of the temporal broadening and narrowing mechanisms.

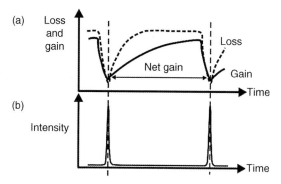

Figure 5.1 Mechanism of passive mode locking: (a) the loss and gain dynamics that lead to (b) pulse generation.

Upon start-up of laser emission, the laser modes initially oscillate with relative phases that are random such that the radiation pattern consists of noise bursts. If one of these bursts is energetic enough to provide a fluence that matches the saturation fluence of the absorber, it will bleach the absorption. This means that around the peak of the burst, where the intensity is higher, the loss will be smaller, while the low-intensity wings become more attenuated. The pulse generation process is thus initiated by this family of intensity spikes that experience lower losses within the absorber carrier lifetime.

The dynamics of absorption and gain play a crucial role in pulse shaping. In steady state, the unsaturated losses are higher than the gain. When the leading edge of the pulse reaches the absorber, the loss saturates more quickly than the gain, which results in a net gain window, as depicted in Figure 5.1. The absorber then recovers from this state of saturation to the initial state of high loss, thus attenuating the trailing edge of the pulse. It is thus easy to understand why the saturation fluence and the recovery time of the absorber are of primary importance in the formation of mode-locked pulses.

This temporal scenario can be connected to the frequency domain description of mode locking. The burst of noise is the result of an instantaneous phase locking occurring among a number of modes. The self-saturation at the saturable absorber then helps to sustain and strengthen this favorable combination, by discriminating against the lower power CW noise.

In practical terms, a saturable absorber can be monolithically integrated into a semiconductor laser, by electrically isolating one section of the device (Figure 5.2). By applying a reverse bias on this section, the carriers that are photogenerated by the pulses can be more efficiently swept out of the absorber, thus enabling the saturable absorber to recover more quickly to its initial state of high loss. An increase in the reverse bias serves to decrease the absorber recovery time, which will have the effect of further shortening the pulses. Alternatively, a saturable absorber can also be implemented through ion implantation on one of the facets of the laser. The defects generated in the semiconductor provide a channel for the nonradiative

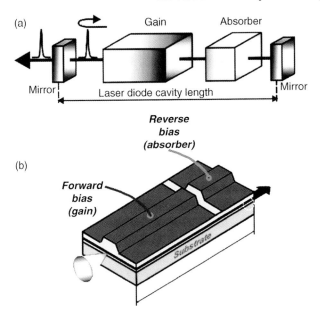

Figure 5.2 Schematic diagram of (a) the main components that form a two-section laser diode and (b) a monolithic two-section laser.

recombination of the photocarriers, thus enabling the recovery of the absorption at relatively short times [119].

5.2
Theoretical Models of Mode Locking in Semiconductor Lasers

Any model of mode-locked laser dynamics should account both for pulse shortening by modulation (active/hybrid ML) and/or saturable absorption (passive/hybrid ML) and for pulse broadening by saturable gain and cavity dispersion (including gain/loss dispersion and group velocity/phase dispersion). In addition, if spectral properties are to be accurately accounted for, self-phase modulation needs to be included in the model. Several approaches have been used so far, and most of them, with subsequent modifications, can be – and have been – applied to quantum dot lasers. In this section, we shall cover general principles of the approaches, while the specific QD features will be considered later.

5.2.1
Small-Signal Time Domain Models: Self-Consistent Pulse Profile

Conceptually the simplest, and historically the oldest, models of mode-locked lasers are *time domain lumped models*, based on the approximation that the pulse width is much smaller than the repetition period, and treating a hypothetic ring laser with unidirectional propagation. The amplification and gain/group velocity dispersion (GVD), which in reality are simultaneously experienced by the pulse, may then be approximately treated in two independent stages. This allows the substitution of the distributed amplifier in the model by a lumped *gain element* performing the functions of amplification and self-phase modulation. Mathematically, this element can be described by a nonlinear integral or integrodifferential operator acting on the complex pulse shape function Y(t), t being the *local* time of the pulse. The model was originally designed for solid-state and gas lasers, whose long lengths make for a round-trip time considerably longer than the pulse duration, so separate timescales are introduced explicitly for the pulse (short timescale) and relaxation period between pulses (long timescale).

The gain operator takes the form

$$\hat{G}Y(t) = \exp\left(\frac{1}{2}(1-i\alpha_{Hg})G(t)\right)Y(t), \quad (5.1)$$

with α_{Hg} the Henry linewidth enhancement factor (cf. Section 3.1) and $G(t) = \Gamma \int g(z,t)dz$ the total gain integrated over the length of the amplifying region. Neglecting dispersion and fast nonlinearities and using the fact that the gain recovery time $\tau_G \gg \tau_p$ (pulse duration), one may obtain an approximate explicit expression for G on the short timescale commensurate with the pulse duration (cf. Equation 3.12 in Section 3.1; note that in that section, we operated with the definition of gain as $G = \exp(G)$). In the case of $G \ll 1$, it takes the form

$$G(t) = G_-\exp(-U(t)/U_g). \quad (5.2)$$

Here, as in the analysis of amplifiers (cf. Section 3.1), $U(t) = \int_{-\infty}^{t} P(t')dt' = v_g\hbar\omega A_X \int_{-\infty}^{t}|Y(t')|^2 dt'$ is the pulse energy up to the time t, $G_- = \Gamma \int g(z, t \to -\infty)dz$ is the total amplification in the gain element at the time before the arrival of the pulse, $U_g = \hbar\omega A_X/\sigma_b = \hbar\omega A_X/dg/dN$ the saturation energy of the amplifier.

The saturable absorber, if any, is also considered a lumped element, described, in the simplest case, by an operator similar to (5.2):

$$\hat{Q}_S Y(t) = \exp\left(-\frac{1}{2}(1-i\alpha_{Ha})Q_S(t)\right)Y(t), \quad (5.3)$$

with the absorber linewidth enhancement factor α_{Ha}, and the dimensionless slow saturable absorption $Q_S = \Gamma \int a(z)dz$, in the ideal slow absorber approximation and with $Q \ll 1$, given by

$$Q_S(t) = Q_-\exp(-U(t)/U_a). \quad (5.4)$$

Here, again, the total initial absorption $Q_- = \Gamma \int a(z, t \to -\infty)dz$ and the absorber saturation energy is $U_a = \hbar\omega A_{X\alpha}/\sigma_\alpha = \hbar\omega A_{X\alpha}/d\alpha/dN$ (depending on the construction, the cross-section of the beam in the SA may be different from that of the amplifying section); as shown below, it is usually advantageous to have $A_{X\alpha} < A_{Xg} = A_X$.

Equations 5.2 and 5.3 take into account only true "slow" saturable gain/absorption and SPM, and treat the SA as an ideal slow absorber, neglecting, not only gain saturation but also absorber saturation during the pulse. (This implies that both the gain and the absorber recovery times should satisfy the condition $\tau_{g,a} \gg \tau_p$, τ_p being, as in Section 3.1, the pulse duration. In the case of a semiconductor laser, this is readily fulfilled in the case of the gain medium (τ_g, ~1 ns), but is easily strained in the case of the SA (τ_a, ~10 ps), which led to some modifications to the model, described below.) As mentioned above (Section 3.1), semiconductor saturable absorbers (though not necessarily the QD ones) tend to contain a subpicosecond component that acts as a "fast" saturable absorber even for short pulses of a duration of 1 ps, typically generated by semiconductor lasers. Some lumped time domain models [120] also include "fast" effects in the SA (SA nonlinearities) as an equivalent fast absorber characterized by an operator $\hat{Q}_F Y(t) = \exp(-\frac{1}{2} Q_F(t)) Y(t)$ and with an equivalent absorption

$$Q_F(t) = Q_{iF}(1 - \varepsilon_a |Y(t)|^2). \tag{5.5}$$

Then, the total absorption is

$$\hat{Q} Y(t) = \hat{Q}_S \hat{Q}_F Y(t). \tag{5.6}$$

Gain nonlinearities may, in principle, be included in the same way, although in the case of QD materials, a more accurate account of dynamics is preferable.

Finally, in the traditional form of a lumped model, the dispersion of material gain and refractive index, together with any artificial dispersive elements present in the cavity, such as a DBR grating, are combined in a lumped *dispersive element*. In the frequency domain, its effect on the pulse may be written as

$$\hat{D} Y^T(\omega) = e^{i\varphi_0} \left[\frac{1}{1 - i(\omega - \omega_p)/\omega_L} + D(\omega - \omega_0)^2 \right] Y^T, \tag{5.7}$$

where Y^T is the Fourier transform of the complex pulse shape $Y(t)$, ω_p and $\omega_L \ll \omega_p$ denote, respectively, the peak frequency and the bandwidth of the dispersive element (defined by the gain curve of the amplifier and the frequency selectivity of a grating element if it is present in the cavity), and ω_0 is the reference frequency as in the analysis of amplifiers. The value of ω_p may change during the pulse (due to gain curve changes with carrier density) that modifies the dispersive operator [121], although in the majority of papers on the subject, the effect is not included specifically. Variable φ_0 denotes the phase shift introduced by the element and D the equivalent dispersion (including the GVD of the passive waveguide and the effective dispersion of the external grating element, if any). To rewrite the operator (5.7) in the time domain, one may expand the first term around the reference frequency ω_0 noting that $|\omega_p - \omega_0| \ll \omega_0$. Then, after a standard transformation $(\omega - \omega_0) Y^T \div id/dt\, Y$, (5.7) becomes a differential operator; if the exponential is expanded keeping the first two terms, the operator is reduced to second order.

The dynamics of mode locking process are then described by cascading the operators and setting

$$Y_{i+1}(t) = (\sqrt{\kappa}\hat{G}\hat{Q}\hat{D})Y_i(t), \tag{5.8}$$

where i is the number of the pulse round-trip (determining the "slow" evolution of the mode-locking pulse), time t is on the fast timescale commensurate with the pulse duration, and the dimensionless parameter $\kappa < 1$ introduces the total (integrated) unsaturable intensity losses in the cavity, both dissipative and outcoupling. The model reflects the balance of the main processes affecting the pulse in a mode-locked laser in that the saturable absorption operator $\hat{Q}$ acts to narrow the pulse down, whereas the gain saturation $\hat{G}$ and the dispersion operator $\hat{D}$ act to broaden it; the result in the steady state is a constant pulse shape. The *stationary* mode-locking equation is thus obtained by writing out the condition that the shape of the pulse is conserved from one repetition period to the next. In the operator notation introduced above, this means

$$(\sqrt{\kappa}\hat{G}\hat{Q}\hat{D})Y(t) = e^{i\delta\psi}Y(t + \delta T), \tag{5.9}$$

where δT is the shift of the pulse or detuning between the repetition period and the round-trip of the "cold" cavity (or its fraction in case of locking at harmonics of the fundamental frequency), and $\delta\psi$ is the optical phase shift induced by the round trip. In between the pulses, on the slow timescale commensurate with the round-trip time, gain and SA are allowed to recover with their characteristic relaxation times, according to the rate equation for carrier density with $S = 0$, as was done in Section 3.3 for analyzing SOAs at high bit rates. This allows one to calculate the values of gain and saturable absorption at the onset of the pulse, given the pulse energy and repetition period (the only point at which this latter parameter enters a lumped time domain model).

In the approximation of no dispersion ($\hat{D} = 1$), the broadening of the pulse by gain saturation alone in the lumped model cannot compensate for the shortening by the absorption. The model in this approximation thus predicts the steady output in the form of a series of *infinitely short* (delta function-like) pulses; neither the pulse shape nor the duration can be analyzed in this approximation. However, it is possible to determine the total pulse *energy* and also analyze the stability of the solutions by requiring that net gain both immediately before the pulse and immediately after the pulse is smaller than 1:

$$\begin{aligned} G_- - Q_- - \ln \kappa &< 0, \\ G_+ - Q_+ - \ln \kappa &< 0. \end{aligned} \tag{5.10}$$

Here G_-, Q_- are (as in Section 3.3) total (integrated) gain and absorption immediately before the pulse and G_+, Q_+ are the values immediately after the pulse. This is known as New's theory of mode locking (G. New's original 1970s paper [122] on non-semiconductor lasers in which $\ln(1/\kappa) \ll 1$; however, (5.10) is also applied in the generalized version of the theory proposed by Vladimirov and coauthors [123, 124] and covered in more detail next).

Analytical approximations for the *pulse shape and duration* have been originally obtained if the pulse energy is smaller than $U_{G,A}$ and the gain and loss (saturable and

5.2 Theoretical Models of Mode Locking in Semiconductor Lasers

unsaturable) during one round-trip are small ($\ln(1/\kappa)$, G, $Q \ll 1$). Then, the exponentials in the formulas for the gain and loss operators may be expanded in a Taylor series keeping terms up to the second order in (5.2) and (5.4) (weak-to-moderate saturation of gain during the pulse):

$$\exp\left(-\frac{U(t)}{U_{g,\alpha}}\right) \approx 1 - \frac{U(t)}{U_{g,\alpha}} + \frac{1}{2}\left(\frac{U(t)}{U_{g,\alpha}}\right)^2, \tag{5.11}$$

and to the first order in (5.1) and (5.3) (small gain and loss):

$$\exp\left(\frac{1}{2}(1-i\alpha_{Hg})G(t)\right) \approx 1 + \frac{1}{2}(1-i\alpha_{Hg})G(t);$$

$$\exp\left(-\frac{1}{2}(1-i\alpha_{H\alpha})Q(t)\right) \approx 1 - \frac{1}{2}(1-i\alpha_{H\alpha})Q(t) \tag{5.12}$$

(the accuracy of the model can be improved by expanding these equations, too, to the second rather than first order).

Then, following the route pioneered by H. Haus in the first papers on mode locking in lasers of an arbitrary type [125] and later adapted specifically to diode lasers [121, 126], the mode-locking equation is rewritten as a complex second-order integrodifferential equation known as the master equation of mode locking, which permits an analytical solution of the form

$$Y(t) = Y_0 \exp(i\Delta\omega t)\left(\cosh\frac{t}{\tau_p}\right)^{-1+i\beta} \tag{5.13}$$

known as the *self-consistent profile* (SCP). The corresponding theoretical approach is known as the SCP or Haus's mode-locking theory. Assembling the terms proportional to the zeroth, first, and second power of $\tanh(t/\tau_p)$ in the mode-locking equation, one obtains three complex, or six real, transcendental algebraic equations [121, 126] for six real variables: pulse amplitude $|Y_0|$, duration measure τ_p, chirp parameter β, optical frequency shift $\Delta\omega = \omega - \omega_0$, repetition period detuning δT, and phase shift $\arg(Y_0)$ (which is not a measurable parameter, so in reality there are five meaningful equations). These equations, being nonlinear and transcendental, generally speaking, cannot be solved analytically, but still allow some insight into the interrelation of pulse parameters. For example, it can be deduced [121] that the pulse duration may be considerably shortened by the presence of a fast SA, and the achievable pulse durations are estimated about 10 times the inverse gain bandwidth, decreasing with increased pulse energy.

By requiring the net small-signal gain before and after the pulse to be negative (Equation 5.10), so that noise oscillations are not amplified, the self-consistent profile approach also allows to estimate the parameter range of the stable ML regime.

Some conclusions from the SCP approach that are borne out by more precise models (see below) are as follows:

First, arguably the most important parameter governing the mode-locked laser performance is the gain-to-absorber saturation energy ratio

$$s = \frac{U_g}{U_\alpha} = \frac{\sigma_a A_{Xg}}{\sigma_g A_{X\alpha}}. \tag{5.14}$$

A minimum value of $s > 1$ is needed to achieve mode locking at any range of parameters at all, and the range of stable mode-locking operation broadens with an increased s. Colliding pulse mode-locked configurations, linear or ring, not only increase the pulse stability but also lead to shorter pulses by increasing the parameter s.

Second, increasing the dispersion parameter D also increases the parameter range for ML, at the expense of broadening the pulses.

The SCP model has also been successfully used to explain, at least qualitatively, the shape of the pumping current dependence of the repetition frequency detuning from the "cold cavity" round-trip frequency [127] in passively mode-locked DBR LDs. Indeed, experimentally the detuning shows a minimum value in its dependence on the gain section current. Since the pulse energy is approximately proportional to the excess current over threshold, this experimental result is in at least qualitative agreement with the parabolic dependence on the pulse energy U_p predicted by the SCP model:

$$\frac{\Delta f_{\text{rep}}}{f_{\text{rep}}} \approx \frac{1}{4} \left(-(Q_- - sG_-) \frac{U_p}{U_\alpha} + \frac{1}{4} Q_- \left(\frac{U_p}{U_\alpha} \right)^2 \right). \tag{5.15}$$

When applied more quantitatively, however, the self-consistent pulse profile model is not too accurate and cannot adequately describe details of pulse shape and spectral features – indeed, the pulse shape given by expression (5.13) and the corresponding spectrum (both of which have the shape of a hyperbolic secant) are always symmetric, which, in general, need not be, and often is not, the case in practice.

This is to do with a large number of approximations involved in the self-consistent profile approach, which have been progressively removed by various researchers at the expense of making the model more complex and requiring numerical rather than semianalytical analysis of the pulse profile – while still keeping the model lumped.

First, achieving the self-consistent profile requires that the relaxation of gain and absorber during the pulse is negligible, so that the gain and absorber operators can be written in the form (5.2) and (5.3). As mentioned above, this is a safe assumption in semiconductor lasers in the case of gain media, whose characteristic response times are on the order of 1 ns, but for the semiconductor absorbers, which have response times down to a few picoseconds, the approximation is strained. The obvious upgrading of the model is then to characterise the relaxation of the saturable absorption Q by a characteristic recovery time τ_A:

$$\frac{dQ(t)}{dt} = -X(Q)Q \frac{P(t)}{U_\alpha} - \frac{Q_0 - Q}{\tau_\alpha},$$

$$X(Q) = \frac{1 - \exp(-Q)}{Q}. \tag{5.16}$$

Here, Q_0 is the unsaturated total absorption (at repetition periods $T_{rep} \gg \tau_\alpha$, $Q_- = Q_0$), and, in the case of small absorption (Q_0, $Q \ll 1$) treated above, the geometric factor X is a constant $X = 1$ (note that the more general expression in (5.16) is equivalent to Equation 4.28 for the traveling wave geometry absorber).

Equation 5.16 is then used with (5.3) instead of (5.4), which it obviously reproduces in the limiting case of $X = 1$, and $\tau_\alpha \gg \tau_p$, or $\tau_\alpha \to \infty$ on the short timescale $t \sim \tau_p$. Unfortunately, even this apparently minor modification to the model means that a closed form solution in the form of (5.13) is no more possible even with Q_0, $Q \ll 1$ ($X = 1$), and the iteration-type procedure (5.8) has to be repeated *numerically* until a steady-state profile that satisfies (5.9) is found.

Studies with such a modified SCP model found that even with $X = 1$ in (5.16) (small gain/absorption case) and even with absorber recovery times a few times greater than the pulse duration, the finite τ_p makes some difference to the results, noticeably shortening the pulse, making its shape less symmetric, and affecting boundaries of stable ML regime [128])

5.2.2
Large-Signal Time Domain Approach: Delay Differential Equations Model

Having sacrificed the analytical solution in order to improve the accuracy of the model, one notes that both the assumption of weak-to-moderate pulse saturation during the pulse and, even more so, that of small gain and absorption per pass may become more tenuous in semiconductor lasers than the assumption of ideally slow absorption – in fact, in edge-emitting lasers, the small-gain one is almost always completely inapplicable since at least one of the laser facets is usually uncoated (or even AR coated to reduce the reflectance to 0.05–0.1) to increase the output power, so the outcoupling losses are significant. Then, it makes sense to abandon the expansions (5.11) and (5.12) in the fully numerical procedure and use the full exponential form of (5.1)–(5.5), as well as the more accurate full expression for X in (5.16), thus moving from a small-signal SCP model to a *large-signal iterative model* (see, for example, Ref. [129]). This also means that the fast nonlinearities of gain and absorption, and possibly part of the dispersion, may be included directly into the gain and absorber operators.

Even in its large-signal form and with the finite absorber (and gain, if necessary) relaxation time taken into account, the iterative procedure (5.8) is still somewhat artificial in that it requires a trial pulse shape to start with, and explicitly separates the timescale into the short timescale of the pulse and the long timescales of the repetition period. Moreover, if the time window of the pulse is taken as much smaller than the repetition period (which is the standard thing to do if the repetition period is much longer than the pulse), any instability to do with secondary pulses arising far away from the main pulse may be missed by the model. In semiconductor lasers, neither of these assumptions is well justified, as the pulse may be only about an order of magnitude shorter than the repetition period, so that the separation of scales is not as justified as in lasers of other types, and the chaotic instabilities with several competing pulse trains are a very real threat.

An elegant solution to these modeling limitations is offered in the form of the most sophisticated and the most realistic of the lumped models of mode-locked lasers. In this form of the lumped approach, the two different scales for pulse analysis are, in general, abandoned, and the iteration procedure (5.8) substituted by a *delay* one. In its most general form, this procedure may be written as

$$Y(t) = (\sqrt{\kappa}\hat{G}\hat{Q}\hat{D})Y(t-T_{RT}), \qquad (5.17)$$

where T_{RT} is the round-trip of the cold cavity and t is still the local time of the pulse.

A particularly useful form of this model is obtained if the dispersion operator $\hat{D}$ is expanded as a differential one. A very simple form of such an expansion has been derived by Vladimirov et al. [123, 124] who showed that for a bandwidth-limiting element with a Lorentzian spectrum

$$\hat{D}Y^T(\omega) = \left[\frac{1}{1-i(\omega-\omega_p)/\gamma}\right] Y^T \qquad (5.18)$$

(i.e., neglecting group velocity dispersion), assuming without much loss of generality that the peak gain frequency ω_p coincides with one of the laser resonator modes, and taking it as the reference optical frequency, we can rewrite (5.17) as

$$Y(t) = -\gamma^{-1}\frac{\partial Y(t)}{\partial t} + (\sqrt{\kappa}\hat{G}\hat{Q})Y(t-T_{RT}). \qquad (5.19)$$

Equation 5.19 is a delay differential one, and the model thus becomes the *delay differential equation*, or DDE, *model* of mode locking in semiconductor lasers.

The operators $\hat{G}$ and $\hat{Q}$ can be calculated using (5.1)–(5.4) (in this version of the DDE model, no fast absorption is present); the integrated absorption Q is found from Equation 5.16, and for the integrated gain G, a similar equation is written. Assuming that the pulse in the unidirectional cavity treated by the model passes the absorber before the amplifier, the equation takes the form

$$\frac{dG(t)}{dt} = -[\exp(G(t))-1]\exp(-Q(t))\frac{P(t)}{U_g} + \frac{G_0-G(t)}{\tau_g}, \qquad (5.20)$$

where, as in the theory of amplifiers, G_0 is the unsaturated gain determined by the pumping conditions.

Equations 5.19, (5.16), and (5.20) are a closed system suitable for a detailed numerical simulation of both stationary and dynamic behavior of passive mode locking. They can also be fairly easily adapted to allow numerical analysis of *hybrid* mode-locking behavior. As shown in Ref. [124], the DDE model also allows significant *analytical* progress, similar to one achieved with classical New and Haus's models as described above, but for a more general case of large single-pass gain and absorption. As this approach is more relevant for most semiconductor laser constructions than the classical SCP one (and has been adapted to QD lasers as explained below), we shall present it here in more detail, following [124]. In the analytical procedure, the slow (relaxation of gain and absorption between pulses) and fast (evolution during the pulse) stages of laser dynamics are still, as in the traditional SCP model, treated separately. Considering the slow stage results in equations connecting the gain and absorption before and after the pulse, similar to the case of analyzing the bit rate dependence of amplifier performance (Section 3.1), we get

$$G_- = G_0 - (G_0 - G_+)\exp(-T_{RT}/\tau_g) \quad (5.21)$$

and

$$Q_- = Q_0 - (Q_0 - Q_+)\exp(-T_{RT}/\tau_a). \quad (5.22)$$

At the fast stage, again as in the high bit rate amplifier analysis, the relaxation terms are omitted, and so (5.16) and (5.20) take the form

$$\frac{dG(u)}{du} = -[\exp(G(u))-1]\exp(-Q(u)); \quad \frac{dQ(u)}{du} = -s(1-\exp(-Q(u))), \quad (5.23)$$

where u is the dimensionless energy within the pulse, $u(t) = U(t)/U_g$, $U(t) = \int_{-\infty}^{t} P(t')dt' = v_g\hbar\omega A_{Xg}\int_{-\infty}^{t}|Y|^2(t')dt'$, again as in the amplifier analysis. Introducing $u_p = U_p/U_g$, $U_p = U(t \to \infty)$ as the total dimensionless pulse energy (using the time of minus infinity on the short timescale, meaning the time before the pulse, and plus infinity, covering the entire time of substantial pulse energy, that is, the entire pulse duration), we can integrate (5.23) to get another set of equations connecting the prepulse and postpulse gain and absorption:

$$Q_+ = Q(u_p) = \ln[1 + \exp(-su_p)(\exp(Q_-)-1)], \quad (5.24)$$

$$G_+ = G(u_p) = -\ln\left[1 - \frac{1-\exp(-G_-)}{[\exp(-Q_-)(\exp(su_p)-1)+1]^{1/s}}\right]. \quad (5.25)$$

The pulse energy itself may be calculated from (5.19), by taking the modulus square of both sides of the equation and integrating over the pulse. The result can be expressed as

$$\gamma^{-2}v_g\sigma_g\int_{-\infty}^{\infty}\left|\frac{\partial Y}{\partial t}\right|^2 dt + u_p = \kappa\ln\frac{\exp(G_-)-1}{\exp(G_+)-1}. \quad (5.26)$$

In general, the integral on the left-hand side cannot be calculated analytically. Two particular cases when this is possible have been analyzed in Ref. [124].

The first is the case of a model without spectral filtering, when the integral can be set to zero. As noted by the authors of Ref. [124], this is a fairly crude approximation, as in fact the value of the integral does not disappear even in the limit of infinitely wide gain dispersion curve ($\gamma \to \infty$). Indeed, the integral is over the time of the pulse, and as such roughly proportional to the pulse duration. In the lumped theories of mode locking, this duration scales as γ^{-1} meaning that $|\partial Y/\partial t|^2 \propto \gamma^2 A^2$, so that the integral remains finite as $\gamma \to \infty$. In fact, as mentioned above, the theory with $\gamma \to \infty$ cannot predict the pulse shape or duration, leading to pulses collapsing to a delta function shape. The total pulse *energy*, however, can be estimated *approximately* by neglecting the integral on the left-hand side of (2.38) and thus obtaining an equation for u_p in the form

$$u_p = \kappa \ln \frac{\exp(G_-) - 1}{\exp(G_+) - 1}. \qquad (5.27)$$

Equations 5.21, 5.22, 5.24, 5.25, and 5.40 form a closed system of five (nonlinear and transcendental) equations for the five unknowns: $G_\pm$, $Q_\pm$, and u_p. The authors of Ref. [124] identified this system as the *generalized New's model*, as it does not include spectral filtering (as the original New's model) but, unlike this model, does include arbitrarily large gain and absorption per pass. The (numerical) solution gives the dependence of pulse energy (though neither duration nor peak power) on pulse parameters, represented by the unsaturated gain (which is related to pumping current) and absorption (which is related to the reverse bias applied to the absorber). The other fundamental absorber parameter also dependent on the reverse bias, the absorber lifetime, enters the calculations only through the relaxation equation (5.22) and does not influence the results from this model at all if $\tau_\alpha \ll T_{RT}$ (in which case, obviously, $Q_- \approx Q_0$).

The solution to this nonlinear algebraic equation system can then be substituted into the inequalities (5.10) to analyze the stability boundaries of the ML operating range with respect to the leading edge and trailing edge instability. The curves, of course, can be calculated only numerically; however, the authors of Ref. [124] noted that the leading edge and trailing edge instability boundaries met at the codimension-2 point lying on the linear threshold line ($G_0 - Q_0 - \ln \kappa = 0$). This point can be calculated explicitly and is

$$Q_0 = \ln \frac{\kappa(s-1)}{s\kappa - 1}; \quad G_0 = \ln \frac{s-1}{s\kappa - 1}. \qquad (5.28)$$

This means that the condition $s > 1$, derived in the traditional SCP approach for the case of small gain and loss per period, needs to be generalized in the case of arbitrary losses in the cavity as

$$s\kappa > 1 \qquad (5.29)$$

(in a more realistic construction, an extra geometric factor would also be required).

In the case of G_0, Q_0, $\ln \kappa \ll 1$, Equations 5.24 and 5.25 simplify to (5.2) and (5.4), and as a result the equation for the pulse energy simplifies to

$$Q_- \frac{\exp(su_p) - 1}{s} - G_-(\exp(u_p) - 1) - u_p \ln \kappa = 0, \qquad (5.30)$$

which is the equation for u_p that features in the original New's theory of mode locking.

In the dispersionless approximation, the linewidth enhancement factor of the gain and absorber sections does not influence the stability criteria, which will not be the case in the general analysis.

The second case where full (semi-) analytical solution of the DDE model possible is when the dispersion is taken into account, but the saturation of gain and absorption during the pulse (though not necessarily gain and absorption themselves) is assumed to be small. The authors of Ref. [124] called this the generalized Haus model. In this case, a steady-state solution is sought in the form similar to (5.9); in our notations, $Y(t+T_{RT}) = e^{-i\delta\psi} Y(t-\delta T)$. Then, from (5.19),

$$\gamma^{-1}\frac{\partial Y(t-\delta T)}{\partial t} + Y(t-\delta T) = F(u(t))Y(t), \quad (5.31)$$

$$F(u) = \sqrt{\kappa}\exp(G(u)(1-i\alpha_{Hg}) - Q(u)(1-i\alpha_{Hq}) - i\delta\psi).$$

The "complex net gain" $F(u)$ can be written out explicitly, substituting expressions (5.24) and (5.25) (with u instead of u_p) for $G(u)$ and $Q(u)$.

Next, assuming that the single-pass pulse shift is significantly smaller than the pulse duration and that the saturation of both the gain and the absorption during the pulse is weak enough ($u(t) < u_p \ll 1/s$) – the latter being the underlying assumption of Haus's theory – both sides of (5.31) can be expanded in Taylor series up to the second-order terms in their respective arguments:

$$Y(t-\delta T) \approx Y(t) - \frac{\partial Y}{\partial t}\delta T + \frac{1}{2}\frac{\partial^2 Y}{\partial^2 t}\delta T^2, \quad (5.32)$$

and generalizing the expansions (5.11) and (5.12) of the original Haus's theory,

$$F(u) \approx F_0 + F'_0 u + \frac{1}{2}F''_0 u^2; \quad F_0 = F|_{u=0}; \quad F'_0 = \left.\frac{\partial F}{\partial u}\right|_{u=0}; \quad F''_0 = \left.\frac{\partial^2 F}{\partial u^2}\right|_{u=0}. \quad (5.33)$$

Then, following [124], we obtain an equation in the form

$$\delta T\left(\frac{\delta T}{2} - \gamma^{-1}\right)\frac{\partial^2 Y}{\partial^2 t} - (\gamma^{-1} - \delta T)\frac{\partial Y}{\partial t} + \left(F_0 - 1 + F'_0 u(t) + \frac{1}{2}F''_0 u^2(t)\right)Y = 0. \quad (5.34)$$

Recalling the definition of $u \propto \int |Y|^2 dt$, one identifies this second-order nonlinear integro-differential equation as the generalization of the *master equation of mode locking* in Haus's theory, which admits solutions of the same form (5.13) as the original master equation. Six equations are then obtained for six real parameters: peak pulse power, duration, time shift δT, optical frequency shift $\Delta\omega$, phase shift per round-trip $\delta\psi$, and the chirp parameter β.

By substituting the solutions into the conditions (5.10), stability limits in the generalized Haus form can be obtained. In general, these depend on the linewidth enhancement factors; however, for direct comparison with other models, the case of $\alpha_{Hg} = \alpha_{H\alpha} = 0$ is useful. Results of such analysis, reproduced from Ref. [124], are plotted in Figure 5.3. In the plot, the subscript N refers to results from New's model, generalized (solid lines) or standard (dashed lines); and the subscript H, to those from Haus's model (calculated with zero linewidth enhancement factors). The filled/empty dots are the leading/trailing instability boundaries calculated by numerical

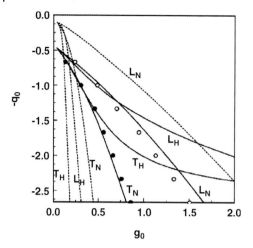

Figure 5.3 Stability boundaries of ML with respect to leading (L) and trailing (T) edge instabilities, calculated semianalytically in traditional and generalized New's (N) and Haus's (H) models. In the calculations, $s = 25$, $T_{RT}/\tau_\alpha = 1.875$, $\tau_\alpha/\tau_g = 0.0133$, $\kappa = 0.1$. Reproduced with permission from Ref. [124].

integration of the model. In this numerical integration, the gain and absorber operators are treated on a continuous timescale, without the need to introduce separate timescales for pulse and the free relaxation period as in the iterative procedure. As seen in the figure, standard Haus and New's models are extremely inaccurate in predicting the instability boundaries (with the range predicted by New's model being too wide and that from Haus's model, too narrow, as noted also in Ref. [128]). The generalized Haus's model gives good agreement within its validity limits at low currents/unsaturated gain values, while the generalized New's model gives very good agreement with numerical simulations at all parameter values (there are some modest deviations, which will be discussed in more detail below). Thus, the large-signal nature of the DDE model is proven to be a very important advantage over the classical mode-locking theories.

Apart from allowing some analytical progress in the limiting cases, the DDE model also allows the use of numerical techniques that have been developed for the analysis of delay differential equations, in particular of numerical packages that allow a full bifurcation analysis of delay differential equations. Such a study was indeed performed in Ref. [124], comprising the full (in)stability analysis of the stationary solution of the DDE. The stationary solution (the steady-state light–current characteristic of the laser) itself is found by seeking the steady-state light output in the form of $Y(t) = Y_{0s} \exp(i \Delta \omega_s t)$. Substituting this into the original Equations 5.16, 5.19, and 5.20 gives the steady-state amplitude and frequency in the parametric form

$$\kappa \exp(G_s(Y_0) - Q_s(Y_0)) - \Delta \omega_s^2 = 0,$$
$$\Delta \omega_s \gamma^{-1} + \tan[\Delta \omega_s T_{RT} + (\alpha_{Hg} G_s(Y_0) - \alpha_{H\alpha} Q_s(Y_0))/2] = 0. \tag{5.35}$$

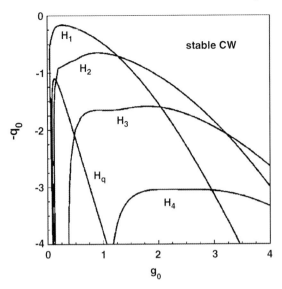

Figure 5.4 Bifurcation analysis of the steady-state solutions of the DDE model. Parameters used: $\gamma\tau_a = 33.3$; $\alpha_{Hg,\alpha} = 0$; the rest as in Figure 5.3. Reproduced with permission from Ref. [124].

Equation 5.39 is a transcendental trigonometrical equation and thus has an infinite set of formal solutions, corresponding to the cavity modes. The steady-state solution, as usual in laser theory, is the one with the lowest value of the threshold gain $G_s(Y=0)$, in other words, the closest to the peak of the gain spectrum. Figure 5.4, adapted from Ref. [124], shows the results of a bifurcation analysis of this solution using the numerical package DDEBIFTOOL. The line H_1 indicates the *Andronov–Hopf bifurcation* (transition from a steady state to a periodically oscillating solution with an amplitude smoothly increasing from zero as the controlling parameter, for example, the unsaturated gain in this case, increases beyond a critical value), corresponding to oscillations at the fundamental mode-locking frequency. ML is predicted at a certain range of conditions (unsaturated gain and absorption) above threshold, whereas at high enough unsaturated gain (or current) and low enough absorption, CW lasing is expected to be stable. The line H_q indicates the Andronov–Hopf bifurcation corresponding to passive Q-switching, or self-sustained pulsation, instability, which essentially means the well-known relaxation oscillations in the laser being turned from decaying to self-sustained pulsations by the positive feedback provided by the saturable absorber. The frequency of these oscillations is determined mainly by the unsaturated gain, gain cross section, gain relaxation time, and losses in the cavity, and is typically on the order of 1 GHz, or about an order of magnitude below the ML frequency. Thus, at low frequencies and with high enough amount of saturable absorption in the cavity, the ML pulse train is expected to be modulated by the self-pulsing envelope. The lines H_m, $m > 1$, show the bifurcations corresponding to a solution oscillating at the *m*th *harmonic* of the fundamental ML frequency. At high enough values of unsaturated absorption, there are ranges of

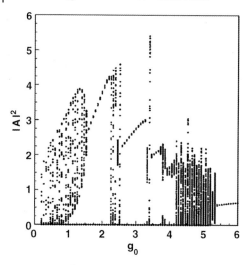

Figure 5.5 Bifurcation diagram obtained by direct numerical implementation of a DDE model. $Q_0 = 4$, the other parameters as in Figure 5.3. Reproduced with permission from Ref. [124].

G_0 (or current) in which ML at higher harmonics is predicted to be stable, but ML at fundamental harmonic is not.

These predictions are confirmed by a full numerical integration of the DDE model (Figure 5.5), showing the extrema of the laser intensity time dependence calculated for different values of the pumping parameter $g_0 = (\tau_a/\tau_g)G_0$. For each unsaturated gain, the initial transient is omitted before the start of registering signals. At low values of g_0 (and thus current), the laser exhibits a regime when the ML pulse power is modulated by passive Q-switching envelope, originally with almost 100% modulation depth (Figure 5.6a). As the pumping parameter increases, the Q-switching modulation gradually decreases in amplitude, and eventually the modulation regime undergoes the backward bifurcation, moving to a stable ML regime (this corresponds to the border of the trailing edge instability in Figure 5.3). Within the area of stable ML, the fundamental round-trip frequency, a train of short pulses is observed as in Figure 5.7a, with amplitudes increasing with G_0. At still higher pumping, the laser dynamics sees areas of harmonic mode locking at the second and third harmonic of the fundamental ML frequency Figure 5.7b,c (obviously with a lower pulse power than in the case of fundamental frequency ML, to preserve the total power), separated by narrow areas of unstable operation. Finally, the ML breaks up completely with the onset of chaotic modulation of the pulse power, with multiple pulse trains competing in the cavity, as in Figure 5.6b (the regimes separating fundamental frequency ML and harmonic ML areas are similar). Eventually, the system undergoes a transition to CW single-frequency operation in agreement with the bifurcation diagrams of Figures 5.4 and 5.5.

An interesting result obtained in Ref. [124] is that while the conditions (5.10) of negative net gain before and after the pulse are useful indications of the stability ranges of mode-locked operation, the onset of instabilities in numerical simulations

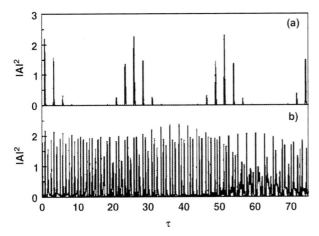

Figure 5.6 Illustration of the aperiodic regimes in Figure 5.5: combined mode-locking/Q-switching regime at $G_0 = 50$ (a) and chaotic pulse competition regime at $G_0 = 350$ (b). Reproduced with permission from Ref. [124].

does not coincide with those limits *exactly*. This may be caused in part by the omission of gain dispersion in the analytical study and in part by the neglect of absorber relaxation during the pulse (though the pulses simulated in Ref. [124] were about an order of magnitude shorter than T_{rt}, and several times shorter than τ_a, which was

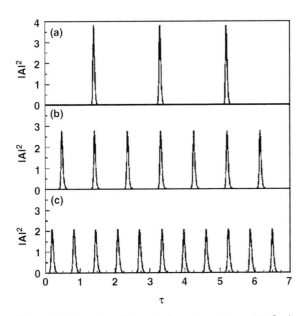

Figure 5.7 Illustration of the periodic regimes in Figure 5.5: fundamental frequency mode locking at $G_0 = 150$ (a) and first- and second-harmonic ML $G_0 = 225$ (b) and 270 (c). Reproduced with permission from Ref. [124].

about $1/2\,T_{rt}$, so this was probably not a very significant factor). However, there is also a genuine physical reason for the discrepancy, in that not all small fluctuations in a ML laser were found in [124] to grow into full-scale instabilities even if a window of positive gain preceded the ML pulse. Instead, stable ML operation was shown to be possible for a range of parameter (unsaturated gain and absorption) values such that before the pulse the fast absorption had recovered to its unsaturated value, but the slower gain continued recovery, leading to a window of positive net gain preceding the pulse (some previous studies using modifications of Haus model for semiconductor lasers indicated the possibility of positive net gain at the *trailing* edge of a stable ML pulse, see, for example, Ref. [124]).

As we shall discuss (see Section 5.4), the possibility of stable ML operation, despite a positive net gain window, is confirmed by more accurate traveling wave simulations, including those performed specifically for QD lasers.

One of the consequences of this effect is that the onset of instabilities may be expected to be sensitive to perturbations such as spontaneous noise. The effect of spontaneous emission was indeed studied analytically and numerically in Ref. [124], with the noise introduced as a delta-correlated random term on the right-hand side of (5.19). It was concluded that while the onset of Q-switching oscillations (trailing pulse edge instability) is a dynamic process independent of noise, the onset of the chaotic envelope instability (leading edge instability) is strongly affected by the noise, with an increase in the noise narrowing the window of stable mode locking. This is fully confirmed by the more complex traveling wave models described below.

The DDE model when used as a numerical tool is not only a fully large signal but also self-starting: it does not require a trial pulse to start with and can reproduce the emergence of mode-locking pulse train from randomly pulsing light output that is seen as the laser crosses the threshold condition. It can also include spontaneous emission/noise sources, as discussed above. Thus, the DDE model removes most of the shortcomings traditionally associated with lumped models of mode-locked lasers and presents a relatively simple yet very powerful tool for analyzing the qualitative tendencies of their behavior. As illustrated above, it combines analytical and numerical methods and possibilities very naturally within the same framework and, as will be discussed in more detail later, correctly predicts virtually all the dynamic regimes and tendencies observed in a real laser.

There remain some limitations to linking the DDE approach directly to the performance of a specific practical laser construction. First, the model as studied in Ref. [124] does not account for fast absorber saturation and fast gain nonlinearity (though it may be possible to include them, at least in some approximation, and in the QD case, the explicit introduction of fast nonlinearities may not be necessary, as will be discussed later; instead, separate rate equations for dot and reservoir populations are used). Second, Equations 5.16 and 5.20 for gain G and absorption Q integrated over the length of the amplifier/absorber element are accurate only if both the gain and the absorption have a simple linear dependence on the carrier densities in the corresponding elements, which is in itself an approximation. For an arbitrary $g(N)$ dependence, an accurate expression of the type of (5.20) for the total gain cannot, in general, be written, unless the variation of N along the propagation direction is weak.

Third, the geometry of the system analyzed in a DDE model as described above remains, as in most lumped models, somewhat artificial in that Equations 5.1 and 5.3 are, strictly speaking, valid only in a hypothetical unidirectional ring cavity (which is what can be expected, given the similarity of these equations to those written for traveling wave (unidirectional) amplifiers (Section 3.1), and the dispersion, which in a real laser is experienced by the pulse at least partly at the same time as gain/absorption, is assigned to a separate dispersive element.

There is one class of mode-locked semiconductor lasers for which these limitations may be relatively easily removed, and the delay differential model, with some modifications, may thus be expected to give a fully *quantitative*, as well as qualitative, description of the laser behavior. These are *vertical external cavity surface emitting lasers* (VECSELs), which consist of an amplifying (gain) chip and a SESAM chip (see Section 4.2), separated by a free space propagation path (with collimating optics). From the electromagnetic/wave propagation point of view, the two chips have a qualitatively similar structure in that both are *very short* (~1 μm) asymmetric Fabry–Perot resonators (with one of the facet reflectances of the resonator, the outer one, approaching one, and the other, the inner one, being between about 0.3 for an uncoated facet and zero for an ideally AR coated facet), containing semiconductor amplifying or absorbing layers. The main difference between them is that the gain chip is forward biased and thus provides gain rather than absorption (it also tends to be longer than the SESAM in the propagation direction). As the resonators are short, the characteristic travel time of light wave through each of them is smaller than the ML pulse duration, making the lumped element formalism a very natural one for their description. The gain and absorption operators have to be modified to take into account the resonator nature of the amplifier and absorber sections. These operators can also include the dispersion of the gain and resonators, so that the separate dispersion operator is not needed. Also, since in the surface-normal geometry both the gain and the absorption paths are short, no integration over length is needed for calculating G and Q, and so generic nonlinear $g(N)$ and $\alpha(N)$ dependences may be (and have been) used in a lumped element model.

Appropriate modifications are also required to the rate equations for the gain and absorption (or, more accurately, for the carrier densities in the gain and SA sections from which G and Q are then calculated), which is achieved, essentially, using the resonant enhancement factor X in the SESAM cavity instead of the traveling wave expression $X(Q) = [1-\exp(-Q)]/Q$ used in (5.16) and likewise modifying the geometric factor X_g for the gain section from the traveling wave value of $X_g(G) = [\exp(G)-1]/G$ implicitly used in (5.20).

A model based on an approach of this type, though presented in a somewhat different form, was successfully used to analyze the dynamics of external cavity VECSELs, with the predicted pulse duration and stability ranges matching the experimentally observed ones, not only qualitatively but also with a reasonable numerical agreement [130].

The model used in Ref. [130] employed expressions specific to QW media to describe carrier density dependences of gain/absorption and refractive index, but the electrodynamic part of the model should be relatively straightforwardly applicable to QD lasers as well, with the appropriate description of gain and absorption in QDs substituted.

MIXCEL constructions, with the QW gain and QD absorber layers located in one chip, could be described by a similar, possibly even somewhat simpler, model, with the single-chip reflectance containing the effects of both the gain and the absorption.

In the case of edge-emitting lasers, the situation is more difficult. While the delay differential model does give a very good qualitative description of the laser performance and has been successfully used to predict the dynamic regimes of the laser, using the unidirectional expressions for gain and absorption is not quantitatively accurate in this case because the pulse passes through both the amplifier and the absorber twice, not once, with the reflected pulse traveling through areas in which gain or absorption has been already partially saturated by the incident pulse, and possibly partially recovered. An attempt at taking into account a realistic edge-emitting tandem laser geometry with an end reflector by introducing averaging of gain/absorption over the corresponding sections with reflection taken into account was made by Khalfin et al. [129] (essentially by introducing resonant enhancement factors similar to the factor X for vertical cavity elements, but with only one side reflector), and used in a large-signal iterative (*not* delay differential) model of the type represented by (5.8). Similar forms of operators could also be used in a numerical DDE model. However, such an approach has its own inaccuracies since, unlike the case in the VECSEL chips, the travel time of the pulse through the edge-emitting amplifier section is greater than the pulse duration, so instantaneous averaging of gain over the section length (particularly in the case of the longer amplifier section) is not justified. It would appear that this limitation cannot be adequately addressed in a lumped element model, at least without introducing multiple delays and making the model almost as complicated as the fully traveling wave model described in the next section.

5.2.3
Traveling Wave Models

The most accurate and realistic, if also the most computationally intensive, approach to simulating edge-emitting mode-locked lasers is offered by *distributed time domain*, or *traveling wave*, models, which treat the propagation of an optical pulse through a waveguide medium with space as well as time resolution. As described in Section 2.4, the model then starts with decomposing the optical field in the laser cavity into components propagating right and left in the longitudinal direction (say, z)

$$Y(r,t) = \Phi(x,y)(Y_f \exp(i\beta_{ref}z) + Y_b \exp(-i\beta_{ref}z))\exp(-i\omega_{ref}t) \quad (5.36)$$

with Φ being the transverse/lateral waveguide mode profile and ω_{ref} and

$$\beta_{ref} = n(\omega_{ref})k_{ref} = n(\omega_{ref})\omega_{ref}/c \quad (5.37)$$

being the reference optical frequency and the corresponding wave vector, respectively. This results in a reduced equation for slowly varying amplitudes $E_{R,L}$, similar to Equation 2.51, which has the form

$$\pm\frac{\partial Y_{f,b}}{\partial z}+\frac{1}{v_g}\frac{\partial Y_{f,b}}{\partial t}=\left(\frac{1}{2}(\hat{g}_{mod}-\alpha_{int})+ik_{ref}\widehat{\Delta\eta}_{mod}\right)Y_{f,b}+iK_{bf,fb}Y_{b,f}+F_{spont}(z,t), \quad (5.38)$$

and is directly solved numerically without the partially analytical integration involved in deriving (5.1) and (5.3). At first sight, the equation looks like a first-order differential one, to which it does indeed reduce if the dispersion of gain and refractive index is neglected and the equation is rewritten in the simplified form of

$$\pm\frac{\partial Y_{f,b}}{\partial z}+\frac{1}{v_g}\frac{\partial Y_{f,b}}{\partial t}\approx\left(\frac{1}{2}(\hat{g}_{mod}-\alpha_{int})+ik_{ref}\widehat{\Delta\eta}_{mod}\right)Y_{f,b} \quad (5.39)$$

(cf. Equation 3.11 for the traveling wave amplifier), where g is the gain (at z-values within SA sections, the gain value g is naturally changed to $-\alpha$, α being the saturable absorption coefficient) and α_{int} is the internal unsaturable loss due to losses in cladding layers and waveguide imperfections. The simplified Equation 5.39 contains all the *first-order* effects needed to simulate pulse propagation and shaping: gain, saturable absorption, and self-phase modulation.

The gain and saturable absorption coefficients for QW (and bulk) materials are most often parameterized as functions of the carrier density and, through the gain and absorption compression coefficients ε_g and ε_α, on the photon densities (cf. Equations 3.1 and 3.20 for semiconductor amplifiers and (4.5) and (4.14) for SAs):

$$g^{bulk,QW}=\frac{g_{lin}(N)}{1+\varepsilon_g S}; \quad \alpha^{bulk,QW}=\frac{\alpha_{lin}(N)}{1+\varepsilon_\alpha S}, \quad (5.40)$$

where the subscript lin means the (quasi-)linear gain or absorption that does not depend on light intensity *directly* but only indirectly through N, and $S=|Y_f|^2+|Y_b|^2$. The dependences $g_{lin}(N)$ and $\alpha_{lin}(N)$ are, in the simplest version of the model, taken as linear (for absorption in most type of materials and for gain in bulk lasers, as in (3.1)) or logarithmic (for gain in QW lasers) or, in more accurate implementations, calculated microscopically with varying degrees of rigor. The compression factors $\varepsilon_{G,\alpha}$ also may be either introduced phenomenologically or calculated microscopically for the two main types of optical nonlinearities in bulk and QW lasers: spectral hole burning and dynamic carrier heating. As pulse duration decreases, the finite relaxation times of the nonlinearities become important. To take those into account, Equation 5.40 can be substituted by phenomenological relaxation equations [131]

$$\frac{d\alpha}{dt}=\frac{1}{\tau_{nl}^{(a)}}\left(\frac{\alpha_{lin}}{1+\varepsilon_\alpha S}-\alpha\right); \quad \frac{dg}{dt}=\frac{1}{\tau_{nl}^{(g)}}\left(\frac{g_{lin}}{1+\varepsilon_g S}-g\right), \quad (5.41)$$

where $\tau_{nl}^{(g,a)}$ are the (subpicosecond) characteristic relaxation times of the gain and absorption nonlinearities, respectively. This is particularly important for ultrahigh-speed mode-locking constructions operating at frequencies of hundreds of GHz. Some authors choose not to introduce $\varepsilon_{G,\alpha}$ at all due to heating, instead include microscopic analysis of carrier temperature dynamics [132] and gain–carrier temperature dependence into the model. Indeed, the duration of mode-locked laser pulses is typically shorter than the critical one determined by the condition (3.21), meaning that the accurate account of fast nonlinearity may be no less important in determining the pulse parameters than accurate account of (quasi-)linear gain g(N). This will be even more significant in the case of QD lasers, with their strong spectral hole burning effects.

The dynamic correction $\Delta\beta = \Delta\eta_{\text{mod}} k_{\text{ref}}$ to the propagation constant, in bulk and QW lasers, is usually related to gain by means of a single parameter, Henry's linewidth enhancement factor α_H (with different values used for the gain and absorber sections). The exact forms of implementing it vary in choosing the reference, one possible way being $\Delta\beta = \Delta\beta_{\text{SPM}} = -\alpha_H(g-g_{\text{th}})$, the latter parameter being the threshold value of peak gain.

The terms present in (5.38) but not in (5.39) describe second-order effects in the sense that they are not absolutely essential for obtaining mode-locking behavior as such in the theory or simulations. They are, however, needed to predict the correct pulse parameters and the correct range of operating points and laser parameters in which the mode-locking behavior is observed. In particular, the gain dispersion represented by the operator nature of gain $\hat{g}_{\text{mod}}$ as in (2.51) (and to a certain extent group velocity dispersion, represented by the operator nature of the modal refractive index $\widehat{\Delta\eta_{\text{mod}}}$) is very important in determining the *stability* range of ML. As mentioned before, in the lumped model, no stable mode-locking with a finite pulse width can be simulated in the absence of dispersion. In the distributed model, most models not including gain dispersion cannot predict stable mode-locking either. Some authors [132] reported stable mode locking with finite pulse widths simulated without the dispersion term, but the model of [132] included finite relaxation times of nonlinearities, which may have had a side effect of introducing *effective* dispersion.

An important point in any numerical implementation of (5.38) is thus the implementation of the dispersion operators, particularly of gain dispersion in the operator $\hat{g}$. In mode-locked laser constructions realized so far (including QD ones), the spectrum of mode-locked lasers, although quite broad, is still usually significantly narrower than that of gain/absorption, meaning that only the top of the gain curve needs to be accurately represented. Therefore, in the studies reported so far, the dispersion has usually been approximated by either a parabola or a Lorentzian curve, which can be implemented numerically in time domain using several different approaches. One involves a convolution integral formula $\hat{g}E = g\omega_L \int_0^\infty E(t-\tau) \exp(-\omega_L\tau)d\tau$, with ω_L being the gain linewidth parameter (which can be made complex, the imaginary part representing the static or dynamic shift of the gain maximum with respect to the reference frequency; the preintegral factor then contains only the real part of ω_L). The simplest numerical implementation of the formula is through an *infinite response digital filter*, which, however, acts both on the phase and on the amplitude of the signal, effectively introducing some group velocity dispersion that cannot be controlled separately from gain dispersion. Thus, higher order digital filters have been used by some researchers [133] to simulate parabolic gain profile without affecting the phase of light. Some other approaches to implementing dispersion involve rewriting the integral relation between the input and the output of the spectral filter representing gain in a small segment of a laser in a differential form, as in Equation 5.19, or introducing a separate differential equation for gain polarization [134].

The dispersive correction $\widehat{\Delta\beta} = k_{\text{ref}} \widehat{\Delta\eta}_{\text{mod}}$ to the (real part of) the propagation constant is usually less important than gain dispersion. Still, particularly in cases when very short (subpicosecond) pulses may be expected in the simulation, the operator $\widehat{\Delta\beta}$ may include an additional term describing GVD of the structure [135]

$$\widehat{\Delta\beta} Y = k_{\text{ref}} \widehat{\Delta\eta}_{\text{mod}} Y = \Delta\beta_{\text{SPM}} Y - \frac{\beta_2}{2} \frac{\partial^2 Y}{\partial^2 t^2}, \quad (5.42)$$

where the first term describes the self-phase modulation effects and the second the GVD, with $\beta_2 = (1/c)(dn_g/d\omega)$ the first-order group velocity dispersion coefficient and n_g being the group velocity refractive index of the laser waveguide. From numerical simulations [135], GVD affects the parameters of picosecond pulses significantly for the dispersion values of $dn_g/d\omega \sim 10^{-14}$ s and higher. Thus, this term is usually negligible in most QW lasers (where the GVD magnitude is estimated as $dn_g/d\omega = 10^{-16} - 10^{-15}$ s) and indeed is omitted in most models of mode-locked laser diodes published to date. It may, however, become important for QD lasers since the large difference in the experimentally observed mode-locking repetition frequencies for ground- and excited-state ML suggests the presence of stronger GVD in these lasers than in other types of lasers (the relative difference in the repetition frequency $F_{\text{rep}} = v_g/2L$ for the ground- and excited-state ML is $\Delta F_{\text{rep}}/F_{\text{rep}} \approx \Delta n_g/n_g \approx 0.05$, the wavelength difference between ground- and excited-state transitions being about 60 nm, suggesting $dn_g/d\omega > 10^{-14}$ s).

The only difference between (5.38) and (2.51) are the terms containing the forward–back and back–forward propagating coupling constants $K_{\text{fb,bf}}$. These terms need to be included only if a *Bragg grating* is present at the position z (and time t); the constants K_{fb} and $K_{\text{bf}} = -K_{\text{fb}}^*$ are, in general, complex due to both refractive index and gain/absorption grating being possible. In the context of a mode-locked laser, accounting for a grating may be needed if the laser construction contains a *distributed Bragg reflector* (DBR) section or accounting for *standing wave-induced gratings or short-scale spatial hole burning*, mainly in the saturable absorber sections. At the time of writing, DBR constructions have not yet been studied for mode-locked quantum dot lasers (though might become of interest, for example, for controlling or stabilizing the operation at either the ground- or the excited-state wavelength). Standing wave-induced effects, on the other hand, have been predicted to be significant in QD lasers (though not in the context of mode locking), due to suppressed diffusion in QD materials [136] and may merit inclusion in a comprehensive ML laser model. In the context of non-QD lasers, standing wave-induced grating exists both due to the carrier density being increased in the antinodes of the standing wave within the SA, thus decreasing the local absorption [131] and forming an absorption (and possibly refractive index due to self-phase modulation) grating, and due to the fast nonlinearities responding to the standing wave. Assuming the Henry factor of the absorber is small, as was done in Ref. [131], the grating is a purely absorption grating, smoothed down by diffusion. The magnitude of the periodic carrier density modulation is then given by an equation

$$\frac{d}{dt} N_{\text{grat}}(z,t) = -\frac{N_{\text{grat}}(z,t)}{\tau_{\text{grat}}} + \frac{v_g}{2} \alpha (Y_f^* Y_b + Y_b^* Y_f), \qquad (5.43)$$

where the grating relaxation time in QW materials is mainly determined by ambipolar diffusion with the coefficient D_a, $1/\tau_{\text{grat}} = (1/\tau_\alpha) + (16\pi^2 D_a n_g^2/\lambda^2)$, where λ is the lasing wavelength in vacuum. An estimate gives a value about 2–3 ps for the ambipolar diffusion coefficient of $2\,\text{cm}^2\,\text{s}^{-1}$ typical for III–V materials. Note that in quantum dots, the diffusion term will be absent and so the grating will relax with the same characteristic times as the average absorption in the SA (see Section 4.3), and the induced carrier density grating may be expected to be stronger than in QW materials. From the magnitude of carrier density modulation, the coupling in QWs can be estimated as $K_{\text{fb}} \approx \sigma_\alpha N_{\text{grat}}(z,t)(1-i\alpha_{\text{H}\alpha})$; in QDs, it would be necessary to consider separate QD and WL population with corresponding partial refractive indices.

The final term in (5.38) is the random noise source that leads the model to self-start and is essential for modeling noise and pulse jitter.

At the laser facets, standard reflection/transmission boundary conditions are imposed on $Y_{f,b}$; thus, unlike the delay differential model, the traveling wave one accounts accurately for the laser geometry.

The traveling wave equations are coupled with coordinate-dependent rate equations for the relevant populations. In the context of QW and bulk lasers,

$$\frac{d}{dt} N(z,t) = \frac{J(z,t)}{ed} - N\left(BN + \frac{1}{\tau_{\text{nr}}} + CN^2\right) - v_g \text{Re}(Y_f^* \hat{g} Y_f + Y_b^* \hat{g} Y_b), \qquad (5.44)$$

where J/ed is the pumping term, J is the current density, e is the elementary charge, d is the active layer thickness, B is the bimolecular recombination constant, τ_{nr} is the nonradiative recombination rate, and C is the Auger recombination rate. Carrier capture dynamics can be taken into account by adding an extra equation for carrier densities in the contact layers, but its significance for non-QD mode-locked lasers can be expected to be very modest in all cases except where direct current modulation is involved. For quantum dot lasers, similar equations would describe the *total* N, including both the dot and the wetting layer population; for full description of laser dynamics, energy-resolved rate equations for different size groups of dots (transition energies), similar to Equations 2.16–2.18 and 2.69–2.71, may be required.

Lasing spectra are, as usual in time domain modeling, calculated by fast Fourier transforming the calculated temporal profiles, discarding the initial turn-on transient to describe steady-state mode locking.

Traveling wave models are very powerful and general and their use is not restricted to ML edge-emitting lasers, though the essentially distributed nature of these devices, with an emphasis on timescales shorter than the round-trip, makes ML lasers an ideal showcase for the strength of such models. They have indeed been extensively and successfully used by a number of research groups to analyze mode-locked lasers and design particular laser constructions [137]. Several commercial simulators of laser diodes include traveling wave approach of some form in their solvers, which have also been applied to mode locking [137], though not yet to analysis of QD lasers, to the best of our knowledge.

The main limitation of distributed time domain models is the considerable requirements they impose on the computing time and memory, mainly due to the fact that time and space steps are usually related in these models and need to be sufficiently short (typical spatial step being 1–5 μm) to faithfully reproduce the pulse characteristics. This is exacerbated in a rigorous QD laser model due to the need to model the dynamics of separate groups of dots within the inhomogeneously broadened gain spectrum by separate, energy-resolved equations (Equations 2.69–2.71 or equivalent). Simplified approaches may, however, be possible for express analysis as described below.

5.2.4
Frequency and Time–Frequency Treatment of Mode Locking: Dynamic Modal Analysis

An alternative (or complementary) approach to the time domain analysis of mode locking is offered by the technique of *modal analysis*, static or dynamic [138]. In this approach, instead of (5.36), one uses an expansion of the optical field in the form

$$Y(r,t) = \Phi(x,y) \sum_k Y_k(t) u_k(z) \exp\left[-i\left(\int \omega_k dt + \varphi_k(t)\right)\right], \quad (5.45)$$

where E_k, u_k, ω_k, and φ_k are, respectively, the amplitude, mode profile, frequency, and phase of the longitudinal mode k. The fast laser dynamics (on the order of, and shorter than, round-trip, that is, on the pulse scale) is then contained in the instantaneous values of amplitudes and phases, whereas the slow dynamics (on the scale longer than round-trip) is determined by the dynamics of complex mode amplitudes $Y_k = E_k(t)\exp(i\varphi_k(t))$ that satisfy complex rate equation:

$$\frac{\partial}{\partial t} Y_k = \left[\frac{1}{2} g_k^{net} - i(\Omega_k - \omega_k)\right] Y_k + \sum_m \Delta g_m^{coupl} Y_{k+m} + F_k^{(L)}(t), \quad (5.46)$$

where g_k^{net} and Ω_k are the net gain and cold cavity eigenfrequency of the kth mode, Δg_m^{coupl} the mode coupling parameters that include contributions both due to four-wave mixing in gain and SA (passive/hybrid ML) sections (this contribution is nonlinear and proportional to the intensity of light oscillating at m intermodal intervals $(\sum_k Y_k Y_{k+m}^*)$) and due to modulation (active/hybrid ML) [138]. This approach is less powerful and universal than the traveling wave one (and, to a degree, less than the large-signal DDE model, though on the other hand, the modal analysis has the advantage over the DDE model in that it can account for laser cavity geometry, including colliding pulse and compound-cavity harmonic mode locking). Its main limitations are (i) the inherent assumptions of weak-to-modest nonlinearity, modulation, and dispersion, and (ii) the explicit introduction of slow and fast timescales obstructing the analysis of regimes such as clock recovery. The advantage of the modal expansion, on the other hand, is that the time steps can be much longer (and the number of variables can be smaller) than in the traveling wave model making it particularly efficient in analyzing, say, long-scale dynamics of external locking of DBR hybridly mode-locked lasers. It also has the logical advantage of describing steady-state ML as a steady-state solution and is particularly quite suitable for lasers with spectrally selective elements such as compound cavities.

Of the approaches outlined above, the delay differential equation model, the traveling wave approach, and the dynamic modal analysis have all been applied, either directly or with appropriate modifications discussed below, to quantum dot mode-locked lasers. In the next section, we shall summarize the main predictions of the generic models with emphasis on features that may be important for QD lasers.

5.3
Main Predictions of Generic Mode-Locked Laser Models and their Implication for Quantum Dot Lasers

5.3.1
Laser Performance Depending on the Operating Point

The first result of all the modern mode-locking theories, confirmed by the experiments, is that the dynamics of notionally mode-locked semiconductor lasers are quite rich and can show, apart from stable mode locking, a number of other dynamic regimes. Here, we shall briefly discuss the general trends in their dependence on the laser parameters.

One of the most important features in the dynamic map of operating regimes of a mode-locked laser is the self-sustained pulsations or passive Q-switching instability at low currents. As shown in Figure 5.5, produced by the DDE model, the range of currents, or unsaturated gain, values in which this regime is observed, increases with the amount of saturable absorption in the laser (which, in a given laser construction, either QW or QD, may be varied to some extent with reverse bias due to electro-absorption). The other model parameter affected by the reverse bias is the *absorber lifetime* τ_a, which is known to decrease approximately exponentially with the reverse bias; this concerns both QW structures and, to a degree, the QD ones where at least one of the characteristic time constants depends exponentially on voltage. The dependence of the Q-switching range on this parameter is not straightforward; the Q-switching range tends to be broadest at a certain absorber recovery time, on the order of the round-trip time though somewhat longer. At longer τ_a, the SP range slowly decreases; however, it also decreases as τ_a is decreased, and at τ_a of fraction of the round-trip, Q-switching instability can disappear, leaving an area of stable mode locking. This is illustrated by Figure 5.8, produced using a traveling wave model and showing approximate borders of different dynamic regimes for a representative Fabry–Perot laser operating at 40 GHz.

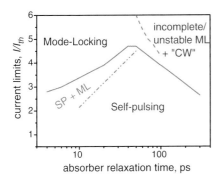

Figure 5.8 Schematic diagram of regimes in a generic QW mode-locked laser operating at the repetition rate of 40 GHz.

This means, first, that care needs to be taken when interpreting the bias voltage effects on the performance of either QW or QD mode-locked laser, as both the unsaturated absorption, the *s*-parameter, and the absorber recovery time are likely to be affected. The effect on the latter is probably the most significant though, as the dependence of τ_a on voltage is quite strong (exponential), while the effect on the unsaturated absorption appears, from measured threshold currents, to be more modest. Second, it means that for the same absorber parameters, longer lasers with longer repetition periods are less likely to suffer from the Q-switching instability, which needs to be borne in mind when analyzing the dynamics of QD lasers (due to the relatively low gain, these often have to be quite long if stable operation at the ground-level wavelength band is desired).

The lower current (or unsaturated gain) limit of the self-pulsing instability may be positioned either below or above the low boundary of mode locking itself, depending on the gain and absorber saturation energies (*s*-parameter) and the absorber recovery time. If the boundary for mode locking is below that for self-pulsing (which tends to happen in long lasers, when τ_a is significantly smaller than T_{RT} but not small enough to completely eliminate self-pulsing), then the stable mode-locking range is split in two by the self-pulsing area, with an area of stable ML seen below the Q-switching limit at currents just above threshold. The area is narrow, however, and the pulse powers generated in this regime are typically rather low. If, on the other hand, the boundary for mode locking is above that for self-pulsing (which tends to be the case for shorter lasers or longer absorber relaxation time, when $T_{RT} > \tau_a$, then an area of pure self-pulsing, with noisy/chaotic filling of pulses, is seen at small-to-modest excess currents above threshold, as in Figure 5.5; as the current is increased, the pulses acquire a regular structure and the combined ML/SP regime develops.

Figure 5.8 also shows the other main instability of ML, the leading edge, or chaotic, instability. As was shown by the DDE model, this is pushed up by the increased amount of absorption in the laser. Shortening absorber relaxation time also decreases the risk of this instability.

At long τ_a and high currents, the laser operates in a chaotic regime, which gradually evolves into some type of CW or quasi-CW operation. Whether or not true single-frequency CW operation predicted by the DDE is achieved depends on the length of the laser and the gain bandwidth; longer lasers (with a repetition frequency ~10 GHz) with broader gain tend to not reach true CW, instead operating in a chaotic quasi-CW regime with a narrow spectrum including only a few modes.

Within the stability range, the pulse amplitude grows with pumping current, as illustrated by Figure 5.9, calculated by a traveling wave simulation. Notice that qualitatively the figure is quite close to the bifurcation behavior seen from Figure 5.5, except that in the numerical model, and with the different set of parameters, only the second rather than third harmonic operation is predicted. The pulse duration tends to grow with current as well, although in some simulations, a decrease in pulse duration

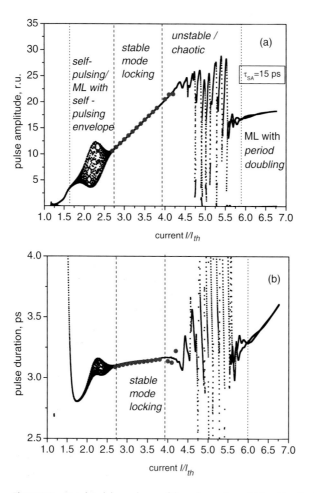

Figure 5.9 Simulated dependence of the pulse amplitude (a) and duration (b) on the pumping current – simulations using a traveling wave model with typical values for a QW laser. Solid line: obtained by slowly ramping the current; filled circles: time-averaged steady-state results for a fixed current.

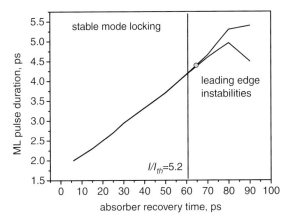

Figure 5.10 Typical simulated dependence of pulse duration on the absorber recovery time – simulations using a traveling wave model.

with current is also predicted, particularly in long lasers, in the area of ML below the Q-switching range if one is observed.

The dependence of the pulse duration on the *absorber recovery time*, within the stability range, is shown in Figure 5.10. As seen in the figure, to achieve stable ML, the absorber relaxation time needs to be below a certain critical value; longer τ_a produce instabilities. Within the stable ML range, a decrease in τ_a tends to shorten the pulses both due to the effects of partial absorber relaxation during the pulse and, probably more significantly, due to the fact that the slow relaxation of the absorber leads to the absorber being always partially saturated, thus reducing the initial absorption Q_-.

5.3.2
Main Parameters that Affect Mode-Locked Laser Behavior

Before moving on to discuss the specific features of mode-locking behavior in QD lasers, it is instructive to briefly overview the effects of the main laser parameters on mode locking in semiconductor lasers in general, as predicted by the models discussed above and confirmed by experiments. The effects of the operating point (unsaturated gain and absorption and the absorber relaxation time) have already been partly discussed. Among other important parameters that affect the ML properties are the following:

i) **The s-factor or the absorber-to-gain saturation energy ratio.** While it is well known that the increase in s facilitates mode locking, it is not immediately intuitively clear whether an increased s helps ML stability, as it is known [139] that the passive Q-switching regime, which is one of the instabilities threatening mode locking, is also facilitated by an increase in s. However, the results from both DDE and traveling wave simulations show that in fact it is the stable ML range that is increased with *s* at the expense of the Q-switching (or self-pulsing) range.

ii) **Gain and group velocity dispersion parameters**. Most models of mode-locked laser operation predict that without gain dispersion, stable mode locking with a finite pulse duration is impossible, and so the gain dispersion, or width of gain curve, represented by the parameter γ in the DDE or ω_L in the traveling wave model, should play an important role in determining the pulse width and stability. Within the range of gain dispersion typical in mode-locked semiconductor lasers, which usually corresponds to ω_L of the order of tens of meV (or nm) and does not change too much with operating conditions or construction, gain dispersion is not the most drastic factor limiting the pulse width; however, achieving a broad gain spectrum is still desirable. This may be one of the advantages of QD active media, as discussed below.

Group velocity dispersion, like gain dispersion, acts to broaden the pulses in the case of a normal dispersion (which is usual in semiconductor lasers). As already discussed, the effect of this parameter is modest on non-QD semiconductor lasers; however, with stronger GVD possible in QD lasers, some account of this effect may be necessary.

iii) **The gain suppression and absorber compression coefficients**. Pulses generated by mode-locked lasers tend to be of picosecond duration, below the critical pulse width for fast gain saturation to dominate in pulse amplification and shaping (see Section 3.1 and Equation 3.21). Thus, the gain compression (and, similarly, absorber compression) effects and the coefficients describing them (if introduced) may be expected to play a significant part in ML properties.

In practice, the effect of these parameters is twofold. First, gain compression tends to broaden ML pulses, with absorption compression having the opposite effect. Second, and in some regards more important, an increase in gain compression stabilizes ML operation, suppressing Q-switching instability (the latter can be easily shown by rate equation analysis of Q-switched lasers [140]). Again, fast absorber saturation has the opposite effect.

In the context of mode-locked QD lasers, the effect of nonlinearities on both pulse duration and stability is expected to be significant. In particular, the stronger gain compression is likely to be a significant factor contributing to the Q-switching instabilities being much less pronounced in QD lasers than they are in other semiconductor laser sources (the situation with absorption compression is more difficult since, as described above (Section 4.3.2), the distinction between fast and slow saturable absorption is blurred in QD materials and so the notion of absorption compression may not be relevant for short pulses).

iv) **The gain and absorber linewidth enhancement factors**. The linewidth enhancement factors have a very modest effect on pulse energy for a given current and absorption, but a more noticeable one on amplitude and duration. They do not significantly affect the onset of the Q-switching instability (the lower current or unsaturated gain limit of ML stability), but very strongly influence the upper limit of ML stability associated with the irregular envelope and pulse competition. ML behavior is the most stable when the gain and absorber linewidth enhancement factors are not too different from each other. According to the DDE model predictions, the most stable operating point (which also corresponds to the highest pulse amplitude and lowest duration) is for $\alpha_{Hg} = \alpha_{H\alpha}$; however, traveling wave and modal analysis predict that the best-quality mode locking is achieved with $\alpha_{Hg} > \alpha_{H\alpha}$; the discrepancy is likely to be caused by the different geometry of the long amplifier and the shorter absorber. The main parameter determined by the linewidth enhancement factors is the *chirp* (dynamic shift of the instantaneous frequency) of the pulse. Passively mode-locked pulses tend to be up-chirped (with the instantaneous optical frequency increasing toward the end of the pulse) when the absorber saturation factor $\alpha_{H\alpha}$ is small and the chirp is mainly caused by α_{Hg}. With a certain combination of α_{Hg} and $\alpha_{H\alpha}$ (typically $\alpha_{Hg} > \alpha_{H\alpha}$), an almost complete compensation of chirp is possible; with $\alpha_{H\alpha} > \alpha_{Hg}$, the pulse is typically down-chirped [141]. As up-chirp is observed more frequently than down-chirp in experiments, one may conclude that typical values of $\alpha_{H\alpha}$ are significantly smaller than α_{Hg}. The situation in QD lasers in this respect may be expected to be rather complicated since, as discussed in Chapter 2, describing their behavior with a single Henry factor is not always possible or accurate; this will be discussed in more detail below.

5.4
Specific Features of Quantum Dot Mode-Locked Lasers in Theory and Modeling

Quantum dot multilayers are essentially semiconductor materials, if rather specific ones, and thus QD lasers share some features and parameters with semiconductor lasers of other types. These include strong inhomogeneous broadening leading to relatively broad gain spectra, the short ($\sim$100 fs) polarization relaxation times T_2 making it possible to eliminate polarization adiabatically (meaning that QD lasers, like other semiconductor lasers, are usually classified as *class B lasers* from the point of view of generic laser theory), the scales of the recombination times ($\sim$1 ns), the saturable absorber recovery times (units to tens of picoseconds), round-trip time and photon lifetimes, the significant single-pass gain and absorption making small-signal models not very accurate, and so on. Thus, with some reservations, some properties of mode-locked QD lasers can be deduced from the generic mode-locked semiconductor laser models described above. However, there are also significant features of QD lasers, particularly in the context of short-pulse generation, which are different from other semiconductor lasers. These come essentially from the discrete energy-level structure of the dots and the independent dynamics of different dots, and include the importance of capture/escape and interlevel relaxation dynamics in the mode-locking properties, the stronger nonequilibrium distribution effects (such as hole burning) than in semiconductor lasers of other type, the extra caution needed when using the concept of the linewidth enhancement factors (particularly in the SA) for analyzing pulse chirp, and, finally, the possible dual-peak (ground and excited state) nature of the gain and saturable absorption spectra, and thus the possibility of switching or coexistence of two wavelength bands of lasing/mode locking. Thus for more accurate analysis, the generic models need to be modified to include the specific features of QD materials, as will be discussed next.

5.4.1
Delay Differential Equation Model for Quantum Dot Mode-Locked Lasers

Of all the theoretical approaches to analyzing mode-locked lasers presented above, the DDE model offers by far the best returns for its computational complexity: in its classical form, it has only three dynamic variables, three equations – one delay differential and two ordinary differential – and an absolute minimum of free parameters, yet is a large-signal model, can describe all the main dynamic regimes, and predict all the main tendencies correctly. Thus, it also promises a powerful, if not necessarily the most accurate, way of treating the specific features of QD mode-locked lasers.

A DDE model of QD ML lasers was presented by Viktorov et al. [49]. The dynamics of light is described by exactly the same equation (5.19) as that used for lasers of other types; substituting the explicit form of gain and absorption operators, the equation is

$$Y(t) = -\gamma^{-1}\frac{\partial Y(t)}{\partial t} + \sqrt{\kappa}\exp\left[\frac{1}{2}(G(t-T_{RT})(1-i\alpha_{Hg})\right.$$
$$\left.-Q(t-T_{RT})(1-i\alpha_{H\alpha}))\right]Y(t-T_{RT}), \qquad (5.47)$$

where all the parameters have the same meaning as in the original (5.19). The laser is assumed to be operating at the ground-level transitions. In the original version of the model, neither the inhomogeneous broadening nor the excited-level dynamics is explicitly taken into account, and the excitonic (dot neutrality) assumption is made. Then, gain and absorption are calculated using simple equations obtained from (2.2) and (2.3):

$$Q \approx Q_0(1-2f_\alpha), \qquad (5.48)$$

$$G \approx G_p(2f_g-1), \qquad (5.49)$$

where $G_p = 2\sigma_G^g N_D L_g$ and $Q_p = 2\sigma_G^\alpha N_D L_\alpha$ are the peak gain at full population inversion and the unsaturated absorption, respectively (σ_G^g and σ_G^α being the gain/absorber cross sections as in (2.2) and (2.3), the factor $\varrho_G = 2$ has been substituted for the ground-level spin degeneracy factor, and L_g and L_α are gain and absorber section lengths, respectively). f_α and f_g are space-averaged ground-state occupancies in the gain and absorber sections (following [49], we drop the G subscript as only one level is considered).

For the occupancies, rate equations of the type (2.37), and with the stimulated term (2.57) (negative for gain, positive for absorption) averaged over the cavity length in the same way as in (5.16) and (5.20), can be written out. In our notations, they are

5.4 Specific Features of Quantum Dot Mode-Locked Lasers in Theory and Modeling

$$\frac{df_g}{dt} = R_g^{(cap)} - R_g^{(esc)} - \frac{f_g}{\tau_g} - [\exp(G(t)) - 1]\exp(-Q(t))\frac{P(t)}{U_g}, \quad (5.50)$$

$$\frac{df_\alpha}{dt} = R_\alpha^{(cap)} - R_\alpha^{(esc)} - \frac{f_\alpha}{\tau_\alpha} + s[1 - \exp(-Q(t))]\frac{P(t)}{U_g}, \quad (5.51)$$

where the power is connected to the amplitude as $P(t) = v_g \hbar \omega A_{Xg}|Y|^2(t)$, the saturation optical energies of the gain and absorber sections are $U_{g,\alpha} = v_g \hbar \omega A_{Xg,\alpha}/\sigma_{g,\alpha}$, and the relative parameter $s = U_g/U_\alpha$ is introduced. Note that since the dependences (5.48) and (5.49) of gain and absorption G and Q on the level occupancies f_g and f_α are linear, the use of integrated quantities in rate equations is mathematically self-consistent.

The capture and escape rates, in the simplest version of the model, are calculated as in (2.39):

$$R_{g,\alpha}^{(cap)} = \frac{N_{g,\alpha}^{(wl)}}{2N_D} \frac{1 - f_{g,\alpha}}{\tau_{cap}^{(g,\alpha)}}; \quad R_{g,\alpha}^{(esc)} = \frac{f_{g,\alpha}}{\tau_{esc}^{(g,\alpha)}} f_{wlg,\alpha}' \quad (5.52)$$

(assuming the wetting layer density is normalized to the same volume as the dots). In the published version of the model, the complementary distribution function of the WL layer was taken as $f_{wl}' = 1$ in both gain and absorption sections since buildup of excessive WL population in lasers is less likely than in amplifiers (an increase in current, which in an amplifier directly and significantly increases N_{wl} and thus affects f_{wl}', does not have this effect in lasers above threshold; instead, optical power is increased).

The average wetting layer populations in the gain and absorber sections are given by

$$\frac{dN_{g,\alpha}^{(wl)}}{dt} \approx \frac{N_{g,\alpha}^{(wl0)} - N_{g,\alpha}^{(wl)}}{\tau_{g,\alpha}^{(wl)}} - 2N_D(R_{g,\alpha}^{(cap)} - R_{g,\alpha}^{(esc)}), \quad (5.53)$$

where $N_{g,\alpha}^{(wl0)}$ are unsaturated values of the corresponding carrier density; in the gain part, $N_g^{(wl0)} = (I/eV_{wl})\tau_g^{(wl)}$ is the carrier density pumped in the absence of lasing, and in the SA section, $N_\alpha^{(wl0)} \ll N_g^{(wl0)}$ (zero in the first approximation, though the authors of Ref. [49] used a small but finite value). The time $\tau_g^{(wl)} \sim 1$ ns represents spontaneous (or, more accurately, all nonstimulated) recombination in the wetting layer of the amplifier section.

The interpretation of the time $\tau_\alpha^{(wl)}$ and its relation to the relaxation time τ_α of the dot absorber population is less straightforward. As the energy levels of carriers in the absorber are much lower than those in the WL, and the escape time increases strongly (approximately exponentially) with the level depth for both thermionic and tunneling escape, the most accurate representation of the realistic situation is to make "direct" escape time long ($\tau_\alpha > \sim 1$ ns) but assign a relatively short ($\tau_\alpha^{(wl)} \sim 1$–10 ps) relaxation (escape) time to the WL absorber carriers, which was indeed done in Ref. [49]. This results (Figure 5.11) in the two-stage dynamics of the type discussed in more detail in Section 4.3.2 – indeed, the analysis there referred to an upper energy level in a dot,

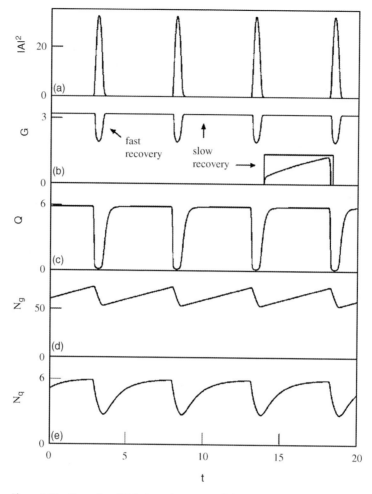

Figure 5.11 Dynamics of light intensity, gain, and absorption simulated by the delay differential model of a quantum dot laser. Reproduced with permission from Ref. [49].

but the same dynamics will result if the upper state is in the wetting layer – this would be even better consistent with the exponential dependence of $\tau_a^{(wl)}$ on the applied voltage, well known for carrier sweepout from QWs as discussed in Section 4.2.

As can be expected, recovery of *gain* after each ML pulse also shows two distinct stages: the fast stage due to the capture processes and the slow stage due to escape and recombination. In other terms, the fast stage approximately corresponds to reestablishment of quasi-equilibrium between dots and the wetting layer in the amplifier section, broken by the pulse, whereas the slow stage is the joint dynamics of quasi-equilibrium populations.

The relative magnitude of these two stages in terms of gain variation can be very different if a significant reservoir of wetting layer carriers exists in the gain section, as was the case in the simulations of Ref. [49]. In that case, the initial depletion of f_g and thus of G by the pulse is virtually completely replenished by the fast supply of carriers through capture from the WL reservoir. Then, between the pulses, gain is almost constant. This is particularly true if, as in Ref. [49], the carrier escape rate may be taken to be much smaller than the capture rate. Then, the quasi-equilibrium dot occupancy $f_g^{QE}(N_g^{(wl)}) \approx (1 + (\tau_{cap}^{(g)}/\tau_{esc}^{(g)})(N_{QD}/N_g^{(wl)}))^{-1}$ is very close to unity for both prepulse and postpulse WL carrier density values, and so after the recovery of f_g to its quasi-equilibrium value, subsequent variations in the quasi-equilibrium gain $G^{QE} \approx G_p(2f_g^{QE}(N_g^{(wl)})-1)$ are very weak. In the simulations of Ref. [49], the slow recombination-related stage of recovery consisted only of gain variations a few *orders of magnitude* smaller than those associated with the fast recovery stage (Figure 5.11); with different parameter values, the difference could be not quite as striking but the tendency of the fast recovery rate being more significant than the slow one is generic. This is different from the dynamics of semiconductor lasers of any other type and has significant consequences for the dynamic regimes of the laser.

One of these consequences is strong suppression of all types of instabilities in QD lasers. Indeed, the two-stage dynamics of gain and absorption makes it easier for the system to satisfy the conditions (5.10) of the net round-trip gain being negative both before and after each ML pulse. This, first, explains the experimentally observed fact that self-pulsation or passive Q-switching instability at low currents is either not observed at all in QD mode-locked lasers or observed only in a very narrow range of parameters (although the short recovery time compared to round-trip should also play a part in this) and, second, predicts that the chaotic pulse competition instability at high currents is either not observed at all or moved to unphysically high currents. To a degree, this could be expected from the start. Indeed, the picosecond capture times describing the fast response of QD active materials (as opposed to subpicosecond carrier–carrier and carrier–phonon times that govern fast nonlinearities in lasers of other type) lead (see Sections 2.4 and 3.3) to a larger gain compression coefficient than that seen in lasers of other types, and the role of gain compression in suppressing Q-switching in lasers with a saturable absorber is well established (see Section 5.3.2).

With a variation in parameters, the DDE model does predict instabilities in the laser behavior [142]. At high currents, the appearance of secondary pulses (pulse doubling) is predicted in qualitative agreement with experiments, whereas at low currents, an area of Q-switching instability, also observed in experiments, is predicted (Figure 5.12); however, compared to the case of non-QD lasers (cf. Figures 5.5 and 5.9), the current range in which this instability is observed is very narrow – in full agreement with experimental observations.

The same similarity between the increased capture time and an increased nonlinearity also explains the fact that an increase in the capture time broadens the ML pulses, as does the increase in the gain suppression coefficient in a generic ML laser model discussed in Section 5.3.2. In a model with inhomogeneous broadening taken into account (which is difficult to do accurately with the DDE model), the carrier capture time would also have the opposite effect of increasing the spectral hole burning, thus broadening the lasing spectra and potentially shortening the pulse [52]; the relative magnitude of the two effects would depend on the

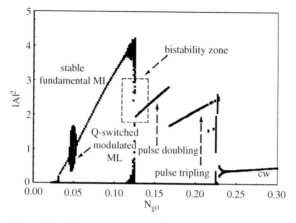

Figure 5.12 Bifurcation diagram of a quantum dot mode-locked laser simulated by the delay differential model. $T_{RT} = 50$ ps, $\kappa = 0.34$, $\gamma^{-1} = 0.5$ ps; $\tau = 1$ ns, $G_p = 3.4$, $Q_0 = 2.1$; $\alpha_{Hg,\alpha} = 2.2$; $s = 15$. Reproduced with permission from Ref. [142].

homogeneous and inhomogeneous broadening parameters. A relatively narrow width of the spectral hole assumed in the frequency domain modeling of Ref. [52] led to a net narrowing of the pulse in the simulated *active* mode locking with increased capture time, but the broadening due to gain compression is likely to be more significant than the spectral hole burning effect in a more realistic construction and with realistic parameter values.

Another feature explained by the model of Ref. [49] is the possibility of a *hysteresis/bistability* between the lasing and the nonlasing states at the switch-on of mode-locked QD lasers, indeed observed in a number of experiments. A qualitative explanation may be partly due to the fact that the strongly saturating curve $G^{QE}(f_g^{QE})$ makes for a very low effective gain cross section dG/dN in QD lasers (N being the total carrier density, including both dots and wetting layer). Bistability at threshold in semiconductor lasers is known to be governed by the ratio $(dQ/dN)/(dG/dN)$ and by the absorber recovery time τ_α or, in the case of a QD absorber, the slower recovery time

$$\tau_\alpha^{(slow)} \approx \tau_\alpha^{(wl)}(V)\left(1 + \frac{\tau_{esc}^{(\alpha)}}{\tau_{cap}^{(\alpha)}}\right) \tag{5.54}$$

(adapting Equation 4.36 for the situation treated here); the width of the hysteresis loop increases with both parameters. Normally, τ_α on the order of 1 ns, too long for mode locking, is needed to achieve bistability; however, as the decreasing dG/dN can make the ratio $(dQ/dN)/(dG/dN)$, in theory, arbitrarily large, bistability becomes possible even with the relatively small τ_α values and thus becomes observable simultaneously with mode locking.

DDE model of mode locking in QDs is capable of reproducing some of the important trends of laser pulse behavior with operating conditions. In particular, it is predicted by both DDE and traveling wave models (see Figure 5.10) that the ML pulse can be shortened significantly with the decreased absorber recovery time. In the case

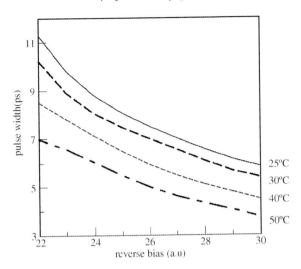

Figure 5.13 Temperature dependence of the mode-locking pulse duration simulated by the delay differential model of a quantum dot laser. $\kappa = 0.2$, $G_p = 2.4$, $Q_0 = 13.5$; $\alpha_{Hg,\alpha} = 2$; $s = 30$. Reproduced with permission from Ref. [143].

of a QD absorber, the part of the absorber recovery time is played by the slow time $\tau_\alpha^{(slow)}$. Since the escape time from a QD is related to the capture time by the relation (2.42), $\tau_{esc}/\tau_{cap} \propto \exp(E_{wl} - E_{Gi}/k_B T)$, it is clear that the absorber recovery will be sped up considerably by an increase in the laser temperature, leading in turn to a decrease in the ML pulse duration. This has indeed been predicted by a DDE model similar to that presented above (Figure 5.13), with even a reasonable quantitative agreement with experiments [143] (in Ref. [143], instead of (2.42), it was assumed that both the capture and the escape times scale with temperature as the inverse Bose–Einstein distribution, $(\tau_{esc}; \tau_{cap} \propto \exp(E_{wl} - E_{Gi}/k_B T) - 1)$, which contains the same exponential factor as (2.42)).

The version of the DDE model presented above is the simplest and the most computationally efficient. A number of modifications may be made to it in order to improve both its accuracy and its versatility (within the limitations of a DDE model as discussed in Section 5.2.2). The easiest of them is inclusion of the effect of the carrier density in the wetting layer (QW) on self-phase modulation of ML pulses. As mentioned in Ref. [49], this can be achieved by introducing the corresponding factor in Equation 5.47, which then becomes

$$Y(t) = -\gamma^{-1}\frac{\partial Y(t)}{\partial t}$$
$$+ \sqrt{\kappa} \exp\left[\frac{1}{2}(G(t-T_{RT})(1-i\alpha_{Hg}) - Q(t-T_{RT})(1-i\alpha_{H\alpha}) \quad (5.55)\right.$$
$$\left. - i\psi_{WL}(t-T_{RT}))\right] Y(t-T_{RT}),$$

where self-phase modulation due to the WL is

$$\psi_{WL}(t) = \frac{2\pi}{\lambda} \left.\frac{\partial n}{\partial N^{(wl)}}\right|_g L_g N_g^{(wl)}(t) - \frac{2\pi}{\lambda} \left.\frac{\partial n}{\partial N^{(wl)}}\right|_\alpha L_\alpha N_\alpha^{(wl)}(t), \qquad (5.56)$$

where $\partial n/\partial N^{(wl)}$ is the carrier density derivatives of the modal refractive index in the gain and SA sections, and $L_{g,\alpha}$ is the section length.

In this form, the DDE model has been applied to analysis, not only of purely passive but also of hybrid mode locking of QD lasers achieved experimentally by means of modulating the voltage at the SA section [144]. Simulations were used to predict the locking cone (the minimum modulation voltage needed for stable hybrid ML versus the frequency detuning between the modulation signal and the free running passive ML) of the laser. Since the Stark effect in QDs is believed to be weaker than in QWs (see Section 4.3.3), the only effect of the AC voltage on the SA properties included in the model was the variation in the recovery time described by the exponential fit (4.34); no direct modulation of unsaturated absorption was included. Nevertheless, the locking cone simulated in Ref. [144] was (at least at low values of the modulation amplitude that corresponds to the experimentally measured locking cone of <30 MHz) remarkably similar to that described by previously published traveling wave simulations of generic (QW) lasers [137], in which, on the contrary, the absorption magnitude was modulated by the voltage and the recovery time was kept constant. In both cases, the locking cone was rather linear in shape and also asymmetric, making slowing the laser down easier than speeding it up, in agreement with experimental observations. Mathematically, the transition between the locked and the unlocked operations in the case of a low modulation amplitude/small detuning was identified as a *saddle-node bifurcation* (as was previously found in time–frequency simulations (dynamic modal analysis) of a QW laser [138]). At larger modulation amplitudes, the nature of the locking stability limit changes to an Andronov–Hopf bifurcation, which is accompanied by a strong increase in the locking range asymmetry (this does not seem to have been observed in any experiments, possibly due to the practical limitations on the modulation amplitude achievable under the experimental conditions). In agreement with the experiments (and also with the published traveling wave simulations of generic (QW) lasers [137]), the locking range is wider for broader pulses/larger dispersion in the system, which leads to a lower quality factor of the ML oscillations.

So far, only application of the DDE model to GS mode locking has been discussed. At the next level of complexity, the model can also be relatively easily modified to take into account two-level dynamics including both excited- and ground-state transitions; in this case, one can introduce an extra equation for the light resonant with the excited-level transitions in the form

$$Y_E(t) = -\gamma_E^{-1} \frac{\partial Y_E(t)}{\partial t}$$

$$+ \sqrt{\kappa_E} \exp\left[\frac{1}{2}\left(G_E(t-T_{RT}^{(E)})(1-i\alpha_{Hg}^E) - i\alpha_{Hg}^{GE} G(t-T_{RT}^{(E)})\right.\right.$$
$$\left.\left. - Q_E(t-T_{RT}^{(E)})(1-i\alpha_{H\alpha}) + i\alpha_{H\alpha}^{GE} Q(t-T_{RT}^{(E)}) - i\psi_{WL}^{(E)}(t-T_{RT}^{(E)})\right)\right] Y_E(t-T_{RT}^{(E)}),$$
$$(5.57)$$

5.4 Specific Features of Quantum Dot Mode-Locked Lasers in Theory and Modeling

where almost all parameters have the same meaning as in (5.47), the super- or subscript E meaning excited level, and cross-phase modulation has been introduced with the cross-level phase modulation coefficient $\alpha_{H\alpha}^{GE}$ describing the effect of *ground* states on the light resonant with the *excited*-state transitions. Equation 5.47 for the light resonant with ground-state transitions would have to be modified accordingly:

$$Y(t) = -\gamma^{-1}\frac{\partial Y(t)}{\partial t} + \sqrt{\kappa}\exp\left[\frac{1}{2}\left(G(t-T_{RT})(1-i\alpha_{Hg}) - i\alpha_{Hg}^{EG}G_E(t-T_{RT})\right.\right.$$
$$\left.\left. -Q(t-T_{RT})(1-i\alpha_{H\alpha}) + i\alpha_{H\alpha}^{EG}Q_E(t-T_{RT}) - i\psi_{WL}(t-T_{RT})\right)\right]Y(t-T_{RT}),$$
(5.58)

where the cross-level phase modulation $\alpha_{H\alpha}^{EG}$ describes the effect of *excited* states on the light resonant with the ground-state transitions.

Note that the round-trip times may be easily made different for the two levels in accordance with experimental observations; this is considerably more cumbersome to implement in a traveling wave model. The spatially averaged populations of the excited state in the gain and absorber sections would be given by the equations similar to (5.50) and (5.51) with slightly revised kinetic terms:

$$\frac{df_{gE}}{dt} = \sum_{v=wl,G}(R_g^{(v\to E)} - R_g^{(v\to E)}) - \frac{f_{gE}}{\tau_{gE}} - [\exp(G_E(t))-1]\exp(-Q_E(t))\frac{P_E(t)}{U_{gE}},$$
(5.59)

$$\frac{df_{\alpha E}}{dt} = \sum_{v=wl,G}(R_\alpha^{(v\to E)} - R_\alpha^{(v\to E)}) - \frac{f_{\alpha E}}{\tau_{\alpha E}} + s_E[1-\exp(-Q_E(t))]\frac{P_E(t)}{U_{gE}}.$$
(5.60)

The power and saturation energy for the excited state are introduced in the same way as for the ground state ($P_E(t) = v_g\hbar\omega_E A_{xg}|Y_E|^2(t)$; $U_{g,\alpha} = v_g\hbar\omega A_{xg,\alpha}/\sigma_E^{(g,\alpha)}$), the gain and absorption cross sections $\sigma_E^{(g,\alpha)}$ are likely to be somewhat smaller than $\sigma_G^{(g,\alpha)}$ due to the smaller overlap of the electron and hole wave functions in the excited state; on the other hand, the twice greater degeneracy of the excited state will make for a higher peak gain/unsaturated absorption: $G_{pE} = 4\sigma_E^g N_D L_g$ and $Q_{pE} = 4\sigma_E^\alpha N_D L_\alpha$ (substituting the degeneracy level $\varrho_E = 4$) with the saturation described the same way as for the ground level:

$$Q_E \approx Q_{0E}(1-2f_{\alpha E}); \quad G_E \approx G_{pE}(2f_{gE}-1).$$
(5.61)

The capture and escape rates in the two-level model, in general, would be governed by equations similar to (2.25). For the ground state (omitting the G subscript), we have the rates of direct capture from the wetting layer and the escape in the opposite direction

$$R_{g,\alpha}^{(wl\to G)} = \frac{N_{g,\alpha}^{(wl)}}{2N_D\tau_{wl\to G}^{(g,\alpha)}}(1-f_{g,\alpha}); \quad R_{g,\alpha}^{(G\to wl)} = \frac{f'_{wl(g,\alpha)}}{\tau_{G\to wl}^{(g,\alpha)}}f_{g,\alpha},$$
(5.62)

and of the two-stage capture and escape via the excited state:

$$R_{g,\alpha}^{(E\to G)} = \frac{2f_{(g,\alpha)E}}{\tau_{E\to G}^{(g,\alpha)}}(1-f_{g,\alpha}); \quad R_{g,\alpha}^{(G\to E)} = \frac{1-f_{(g,\alpha)E}}{\tau_{G\to E}^{(g,\alpha)}}f_{g,\alpha}, \tag{5.63}$$

where the subscripts "initial → final" show the initial and final state of the transition. For the excited state, the capture from the wetting layer and the escape in the opposite direction are

$$R_{g,\alpha}^{(wl\to E)} = \frac{N_{g,\alpha}^{(wl)}}{4N_D}\frac{1-f_{(g,\alpha)E}}{\tau_{cap,E}^{(g,\alpha)}}; \quad R_{g,\alpha}^{(E\to wl)} = \frac{f'_{wl(g,\alpha)}}{\tau_{E\to wl}^{(g,\alpha)}}f_{(g,\alpha)E}. \tag{5.64}$$

Then, for the ground-state occupation, Equations 5.50 and 5.51 still hold, with

$$R_{g,\alpha}^{(cap)} = R_{g,\alpha}^{(wl\to G)} + R_{g,\alpha}^{(E\to G)}; \quad R_{g,\alpha}^{(esc)} = R_{g,\alpha}^{(G\to wl)} + R_{g,\alpha}^{(G\to E)}. \tag{5.65}$$

The wetting layer densities interact with both QD states, so (5.66) becomes

$$\frac{dN_{g,\alpha}^{(wl)}}{dt} \approx \frac{N_{g,\alpha}^{(wl0)} - N_{g,\alpha}^{(wl)}}{\tau_{g,\alpha}^{(wl)}} - 2N_D(R_{g,\alpha}^{(wl\to G)} - R_{g,\alpha}^{(G\to wl)}) - 4N_D(R_{g,\alpha}^{(wl\to E)} - R_{g,\alpha}^{(E\to wl)}). \tag{5.66}$$

Note that in the SA sections, $\tau_{g,\alpha}^{(wl)} < \tau_{\alpha E} \ll \tau_{\alpha G}$, as the wetting layer carriers are able to escape into the barriers using both thermionic and tunneling mechanisms whereas carriers in the dots can only tunnel out (or escape to the WL), and it is known that the escape of carriers from a QW through both thermionic and tunneling mechanisms depends very strongly (exponentially) on the level depth, and the same can be expected in a QD.

A model in this form may be used to gain some further insight into the dynamics of ML QD lasers including operation at the excited-level wavelength. However, it is likely that such dynamics will be influenced by spectral hole burning, inhomogeneous broadening, and the geometric arrangement of the amplifier and absorber. Accurate inclusion of any of these into a DDE model is not straightforward; however, some progress toward this is possible with some approximations. In a recent paper [145], a modified DDE model was applied to analyze dual-wavelength (GS and ES) short-cavity (40 GHz) mode-locked QD laser. Separate kinetic equations (of the type (5.59) and (5.60)) have been written for populations of different size groups of dots. Since closed form averaging over the length in such a model is not possible, the spatially averaged factors $[\exp(G_{G,E}(t))-1]\exp(-Q_{G,E}(t))$ in the equation for gain section populations f_l of the lth group of dots were substituted by just G_l, the gain coefficient for the corresponding energy, calculated using an equation similar to (2.9) (with the excitonic approximation). The model was thus not an entirely accurate large-signal one; despite this, good qualitative agreement between experimental and simulated behavior was achieved. At low currents, ground-state ML, producing pulses 3 ps long, was predicted with large enough SA voltage, with low ES populations and no ES lasing. As the GS gain saturated at higher currents, ES lasing emerged, which for large enough SA bias was accompanied by mode locking. Stable dual-wavelength mode locking was registered, in agreement with experiments, within a relatively

5.4 Specific Features of Quantum Dot Mode-Locked Lasers in Theory and Modeling | 141

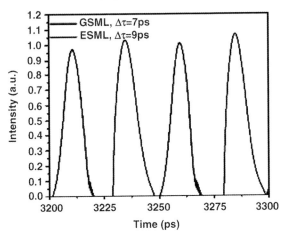

Figure 5.14 Typical temporal traces of ground- and excited-state lasing predicted by the multipopulation version of the delay differential model. Reproduced with permission from Ref. [145].

narrow current range ($I_G/I_{th} = 3.4$–3.7) and for SA bias voltages higher than 7 V. The ground- and excited-state pulse trains generated during the stable dual-wavelength mode-locked operation (Figure 5.14) were, in general, out of phase with each other, with a significantly different repetition rate in agreement with experiments. As seen in the figure, the individual pulses were broadened for both GS ($\tau \approx 6.5$–7 ps) and ES pulse ($\tau \approx 9$ ps) compared to the GS-ML, due to the high gain current value needed for the appearance of lasing from the ES. At gain currents beyond $3.8\,I_{th}$, ES-only ML was observed while at even higher current values (i.e., $I > 4.2\,I_{th}$), it gave way to CW ES-only lasing. Some minor differences existed between the model and the experiment in describing the unstable lasing between the distinct operating regimes.

For a more accurate account of both spatial and spectral inhomogeneities, *traveling wave model* studies of ML QD lasers become important, particularly when a laser, and operating conditions, capable of operating at either of the two levels, or both levels simultaneously, is to be investigated.

5.4.2
Traveling Wave Modeling of Quantum Dot Mode-Locked Lasers: Effects of Multiple Levels and Inhomogeneous Broadening

In a traveling wave model of a QD ML laser, propagation of light is described, as in a non-QD laser diode, in terms of equations for the slow amplitudes of light propagating forward and backward in the cavity (5.38):

$$\pm \frac{\partial Y_{f,b}}{\partial z} + \frac{1}{v_g}\frac{\partial Y_{f,b}}{\partial t} = \left(\frac{1}{2}(\hat{g}_{mod} - \alpha_{int}) + ik_{ref}\widehat{\Delta\eta}_{mod}\right) Y_{f,b} + iK_{bf,fb} Y_{b,f} + F_{spont}(z,t).$$

(5.67)

As discussed in Section 2.4, several degrees of approximation are possible when calculating the dynamics of gain and refractive index (complex dielectric permittivity) of a QD active layer.

A) In the most rigorous, *ultrabroadband*, version of the model, the slow amplitudes Y could, in principle, describe the light resonant with both ground- and excited-state transitions, with a common reference frequency introduced as in (5.36). Gain and refractive index operators are then introduced in a way similar to (2.69), but with a summation of contributions from both levels, and consist of a resonant contribution that includes all the transitions in the dots and a nonresonant contribution that describes the effect of the wetting layer on the light resonant with transitions between confined levels in the dots (ground and excited), as well as any background dispersion in $\widehat{\Delta\eta}_{\text{mod}}$:

$$(\hat{g}_{\text{mod}} + i2k_{\text{ref}}\widehat{\Delta\eta}_{\text{mod}})Y_{f,b} = 2\frac{n_l N_D^{(2d)}}{w_{\text{mod}}^{(G,E)}} \sum_l (\hat{\sigma}_{\text{mod},l}^{(G)} + \hat{\sigma}_{\text{mod},l}^{(E)})Y_{f,b} + ik_{\text{ref}}^{(G,E)}\widehat{\Delta\eta}_{\text{mod}}^{\text{nr}} Y_{f,b}.$$

(5.68)

The first term, the resonant contribution, is then determined as a sum over all dot groups (sizes): $\sum_l \hat{\sigma}_{\text{mod},l}^{(G,E)} Y_{f,b} \leftrightarrow \int dE_{(G,E)l}^{(e-h)} D_{\text{inh}}^{G,E}(E_{(G,E)l}^{(e-h)}) \hat{\sigma}_{\text{mod},l}^{(G,E)} Y_{f,b}$. The effective complex cross-section operators $\hat{\sigma}_{\text{mod},l}^{(G)}$ include the Lorentzian broadening of the corresponding transitions, and the Lorentzian can then be implemented in time domain, for example, using infinite impulse response filters:

$$\hat{\sigma}_{\text{mod},l}^{(G,E)(g,\alpha)} Y_{b,f}(z,t) = \hat{\sigma}_{\text{mod},l}^{(G,E)(g,\alpha)}(E_{(G,E)l}^{(e-h)})Y_{b,f}(z,t)$$

$$= \frac{\sigma_{G,E}^{g,\alpha}}{T_2^{(G,E)}} \int_0^\infty d\tau \exp\left[\left(-\frac{1}{T_{2g,\alpha}^{(G,E)}} - i\frac{\Delta E_{(G,E)l}^{(e-h)} - \hbar\omega_{\text{ref}}}{\hbar}\right)\tau\right]$$

$$\times (f_{G,E}^{(e)}(E_{(G,E)l}^{(e)}, z, t-\tau) + f_G^{(h)}(E_{(G,E)l}^{(h)}, z, t-\tau) - 1)Y_{b,f}(z, t-\tau),$$

(5.69)

or alternatively using additional dynamic equations, as discussed in Section 5.2.3. Within the gain sections, the gain cross section $\hat{\sigma}_{\text{mod},l}^{(G,E)g}$ is used in the formula; within the saturable absorber section(s), the absorption cross section is substituted, $\hat{\sigma}_{\text{mod},l}^{(G,E)\alpha}$, and with the negative statistical factor the gain turns to saturable absorption: $\hat{g}_{\text{mod}}^{(G,E)} \to -\hat{\alpha}_{\text{mod}}^{(G,E)}$.

Such a broadband implementation has the advantage of introducing all the amplitude and phase effects on the same footing and faithfully describing both the gain and the refractive index dispersions, as the complex Lorentzian linewidth broadening means that the resonant contributions to gain and refractive index automatically satisfy Kramers–Kronig relations and are self-consistent. However, such a rigorous approach also has some disadvantages, partly for the fundamental reason that the derivation of (5.38) and the introduction of the slow amplitudes Y rely on the slow-wave condition $(\partial Y_{f,b}/\partial t) \ll \omega_{ref} Y_{f,b}$ (2.49). The optical frequency difference $\Delta\omega_{GE} = 2\pi c(\lambda_E^{-1} - \lambda_G^{-1})$, $\lambda_{G,E}$ being the wavelengths in vacuum, between the excited-level and the ground-level transitions is only about an order of magnitude smaller than the optical frequency itself; thus, the approximation would be rather strained for at least one of the levels. On the more practical side, stringent constraints are also placed on the time steps in numerical simulations: steps on the order of $\Delta t \ll (\Delta\omega_{GE})^{-1}$ will be needed to describe such a potentially broad spectrum, which is extremely taxing on computational resources.

B) A more practical and efficient, if somewhat less rigorous, approach is to treat the reference frequency (5.36) and the slow amplitudes in (5.38) as referring only to ground-level transitions, define a separate reference frequency $\omega_{ref}^{(E)}$ for the light resonant with excited-state transitions, and introduce slow waves for the light resonant with the excited state using this separate reference frequency, thus obtaining separate traveling wave equations for the light resonant with the two levels. The slow waves are thus introduced via

$$Y^{(G,E)}(\mathbf{r}, t) = \Phi^{(G,E)}(x, y)(Y_f^{(G,E)} \exp(i\beta_{ref}^{(G,E)} z) + Y_f^{(G,E)} \exp(-i\beta_{ref}^{(G,E)} z))\exp(-i\omega_{ref}^{(G,E)} t), \quad (5.70)$$

where both the transverse/lateral waveguide mode profiles $\Phi^{(G,E)}$ and the reference wave vectors $\beta_{ref}^{(G,E)} = n(\omega_{ref}^{(G,E)})k_{ref}^{(G,E)} = n(\omega_{ref}^{(G,E)})\omega_{ref}^{(G,E)}/c$ are different for the two levels, alongside the frequencies $\omega_{ref}^{(G,E)}$. The traveling wave equations are then

$$\pm \frac{\partial Y_{f,b}^{(G,E)}}{\partial z} + \frac{1}{v_g^{(G,E)}} \frac{\partial Y_{f,b}^{(G,E)}}{\partial t} = \left(\frac{1}{2}(\hat{g}_{mod}^{(G,E)} - \alpha_{int}) + ik_{ref}^{(G,E)} \hat{\Delta n}_{mod}^{(G,E)}\right) Y_{f,b}^{(G,E)}$$
$$+ iK_{bf,fb}^{(G,E)} Y_{b,f}^{(G,E)} + F_{spont}^{(G,E)}(z, t) \quad (5.71)$$

(where the superscript G has been introduced in (5.36) to describe ground-state transitions and a similar equation with a superscript E written for excited-state transitions). Note that, strictly speaking, not only the gain but also the group velocity of light should be different for ground- and excited state-transitions, as evidenced by the experiments.

To calculate the complex dielectric permittivity (or complex gain/absorption), an approach similar to Equations 2.69 and 5.68 is then used:

$$(\hat{g}_{mod}^{(G,E)} + i2k_{ref}^{(G,E)} \hat{\Delta n}_{mod}^{(G,E)})Y_{f,b}^{(G,E)} = 2\frac{n_l N_D^{(2d)}}{w_{mod}^{(G,E)}} \sum_l \hat{\sigma}_{mod,l}^{(G,E)} Y_{f,b}^{(G,E)} + ik_{ref}^{(G,E)} \hat{\Delta n}_{mod}^{(G,E)nr} Y_{f,b}^{(G,E)}. \quad (5.72)$$

The first term in (5.72) again contains the resonant (in this case, same-level) contribution to the complex dielectric permittivity, while the second term describes the nonresonant part, which in the multipopulation models with separate slow amplitudes for the two levels is both due to the effect of the ground-level states on the light resonant with the excited state and vice versa (cross-level dielectric permittivity contribution) and due to the effect of the wetting layer carrier densities on the refractive index at both ground- and excited-state wavelengths.

Note that the effective modal width $w_{mod}^{(G,E)}$ is, strictly speaking, different for the two levels (as the wavelengths are noticeably different, this would lead to different waveguide confinement), as is the background waveguide refractive index.

In the calculation of the resonant term, the summation is over the groups of dots of different sizes/different energies but for the same (ground or excited) state as in (2.69) ($\sum_l \leftrightarrow \int dE_{(G,E)l}^{(e-h)} D_{inh}^{G,E}(E_{(G,E)l}^{(e-h)})$), thus including the inhomogeneous broadening, while the Lorentzian homogeneous broadening is implemented in time domain, for example, as in (5.69),

$$\hat{\sigma}_{mod,l}^{(G,E)(g,\alpha)} Y_{b,f}^{(G,E)}(z,t) = \hat{\sigma}_{mod,l}^{(G,E)(g,\alpha)}(E_{(G,E)l}^{(e-h)}) Y_{b,f}^{(G,E)}(z,t)$$

$$= \frac{\sigma_{G,E}^{g,\alpha}}{T_2^{(G,E)}} \int_0^\infty d\tau \exp\left[\left(-\frac{1}{T_{2g,\alpha}^{(G,E)}} - i\frac{\Delta E_{(G,E)l}^{(e-h)} - \hbar\omega_{ref}^{(G,E)}}{\hbar}\right)\tau\right]$$

$$\times (f_{G,E}^{(e)}(E_{(G,E)l}^{(e)}, z, t-\tau) + f_G^{(h)}(E_{(G,E)l}^{(h)}, z, t-\tau) - 1) Y_{b,f}^{(G,E)}(z, t-\tau)$$

(5.73)

(or using any equivalent method, as discussed in Section 5.2.3). The superscripts g,α, as in the previous section, represent the gain and SA sections; within the saturable absorber section(s), again, $\hat{g}_{mod}^{(G,E)}$ is substituted by $-\hat{\alpha}_{mod}^{(G,E)}$ calculated using the same formulas, with the statistical factor negative, and the absorber cross section and dephasing times used. Note that the dephasing time $T_2 = \hbar/\Delta E_{hom}$ and the inhomogeneous broadening ΔE_{inh} are both, in principle, different for ground and excited states, though in the absence of exact information and observing that the dephasing processes are fundamentally the same for both levels, it may be acceptable to use the same T_2. In addition, as mentioned in Section 2.4, it is probable that T_2 is different in gain and SA sections ($T_{2\alpha}^{(G,E)} > T_{2g}^{(G,E)}$) since one of the dephasing processes is the collision of carriers in dots with WL carriers, which are significantly less pronounced in the SA sections).

If the *cross-level* contributions from the excited-state transitions to the complex dielectric permittivity of the slow wave $Y_{b,f}^{(G)}$ and of the ground-state ones to that of the slow wave $Y_{b,f}^{(E)}$ were treated on the same footing as the same-level contributions, then the introduction of separate reference frequencies would not really reduce the computational complexity of the model (though may have slightly improved its accuracy, depending on the step). The simplification comes if the cross-level terms are included in the *nonresonant* contribution to the complex dielectric permittivity and treated more approximately than the gain and self-phase modulation from the same (resonant) level. In fact, in the simplest approximation, the cross-level contributions to *gain/absorption* can be neglected completely; a slightly more accurate approach will be mentioned below. The *refractive index/self-phase modulation* contributions, which have a longer frequency range and are thus more significant, can be included in the nonresonant contribution to the refractive index operator, which will take the form

$$\widehat{\Delta\eta}_{mod}^{G(E)nr} Y_{f,b}^{G(E)} = \frac{\partial n_{mod}^{G(E)}}{\partial N^{(wl)}}\bigg|_g N_g^{(wl)}(t) Y_{f,b}^{G(E)} + \widehat{\Delta\eta}_{mod}^{EG(GE)} Y_{f,b}^{G(E)} - \frac{\beta_2}{2k_{ref}^{(G,E)}} \frac{\partial^2 Y_{f,b}^{G(E)}}{\partial^2 t^2}.$$

(5.74)

Here, the first term describes the effect of the wetting layer carriers, the second represents the cross-level modulation between ground and excited levels, and the third includes both the wetting layer and the nonresonant energy-level (ground for $Y_{f,b}^E$, excited for $Y_{f,b}^G$) contributions to the group velocity dispersion; the contribution of the resonant level (ground for $Y_{f,b}^G$, excited for $Y_{f,b}^E$) is taken into account self-consistently by the operator (5.73). The cross-modulation term, in the simplest form, may neglect any hole burning and gain dispersion in the nonresonant state and assume cross-phase modulation factors $\alpha_{Hg,\alpha}^{EG(GE)}$ as in (5.57) and (5.58), precalculated, for example, at the reference frequencies.

$$\widehat{\Delta\eta}_{mod}^{EG(GE)} \approx \Delta\eta_{mod}^{EG(GE)} = -\alpha_{Hg}^{EG(GE)} g_{E(G)}.$$

(5.75)

A slightly more accurate approximation, along the lines of the simplified multilevel approach proposed in Section 2.4, is to use the real energy distribution in the nonresonant level, but still not consider dispersion/operator nature of the nonresonant contribution (or consider it included in the dispersion term in (5.72)), for example, using the approach of Equation 2.58 and setting

$$\Delta\eta_{mod}^{EG(GE)} = \frac{A_{E(G)} \varrho_{E(G)} N_D}{k_{ref}} \sum_l (f_{E(G)l}^{(e)}(E_{E(G)l}^{(e)}) + f_{E(G)}^{(h)}(E_{E(G)l}^{(h)}) - 1) D_{hom}^{''E(G)}(\hbar\omega_{ref}^{G(E)}, \Delta E_{E(G)l}^{(e-h)}),$$

(5.76)

where the complementary homogeneous broadening function (the imaginary part of the complex Lorentzian) is in the form of (2.59):

$$D_{hom}^{''G(E)}(\hbar\omega, \Delta E_{G(E)}^{(e-h)}) = \frac{1}{\pi} \frac{\hbar\omega - \Delta E_{G(E)}^{(e-h)}}{(\hbar\omega - \Delta E_{G(E)}^{(e-h)})^2 + (\Delta E_{hom}^{G(E)})^2}.$$

(5.77)

The cross-level *gain* (or absorption) contribution in such an approximation need not necessarily be neglected and can be calculated along the same lines, but with the real part $D_{hom}^{G(E)}(\hbar\omega, \Delta E_{G(E)}^{(e-h)}) = (1/\pi)(\Delta E_{hom}^{G(E)}/(\hbar\omega - \Delta E_{G(E)}^{(e-h)})^2 + (\Delta E_{hom}^{G(E)})^2)$ of the complex Lorentzian:

$$\Delta g_{mod}^{EG(GE)} = A_{E(G)} \varrho_{E(G)} N_D \sum_l (f_{E(G)l}^{(e)}(E_{E(G)l}^{(e)}) + f_{E(G)}^{(h)}(E_{E(G)l}^{(h)}) - 1) D_{hom}^{E(G)}(\hbar\omega_{ref}^{G(E)}, \Delta E_{E(G)l}^{(e-h)}).$$

(5.78)

As mentioned above, since the real part of the Lorentzian decays faster (as $1/\Delta\omega^2$ rather than $1/\Delta\omega$) with its argument – the frequency detuning – and as the frequency difference between the levels is larger than the homogeneous broadening of either level so that $\left|\hbar\omega_{\text{ref}}^{G(E)} - \Delta E_{E(G)}^{(e-h)}\right| \gg \Delta E_{\text{hom}}^{E(G)}$, so the cross-level gain (absorption) contribution is likely to be less significant than that of the cross-level phase modulation.

The dynamics of wetting layer, ground-, and excited-state populations can be described by the rate equations similar to those used in the two-level DDE model, except the occupation factors are local rather than averaged, resolved in dot group (size) index l (which determines the ground- and excited-state transition energies $\Delta E_{G,E}^{(e-h)}$), and with the stimulated recombination (or absorption) terms modified to reflect homogeneous and inhomogeneous broadening. Thus, the ground- and excited-state occupations satisfy the modified equations (5.50, 5.51, 5.59, and 5.60)

$$\frac{df_{(g,\alpha)G,l}}{dt} = \sum_{v=wl,E}(R_{(g,\alpha)l}^{(v \to G)} - R_{(g,\alpha)l}^{(G \to v)}) - \frac{f_{(g,\alpha)G,l}}{\tau_{gE}} - R_{G(g,\alpha)l}^{(\text{stim})}, \tag{5.79}$$

$$\frac{df_{(g,\alpha)E,l}}{dt} = \sum_{v=wl,G}(R_{(g,\alpha)l}^{(v \to E)} - R_{(g,\alpha)l}^{(E \to v)}) - \frac{f_{(g,\alpha)E,l}}{\tau_{gE}} - R_{E(g,\alpha)l}^{(\text{stim})}, \tag{5.80}$$

with the stimulated terms for the ground and excited states written in a form similar to (2.71):

$$\begin{aligned}R_{G,E(g,\alpha)l}^{(\text{stim})}(z,t) &= R_{E(g,\alpha)}^{(\text{stim})}(E_{(G,E)l}^{(e-h)}, z, t) \\ &= v_g \text{Re}[Y_f^{(G,E)*} \hat{\sigma}_{\text{mod}}^{G,E(g,\alpha)}(E_{(G,E)l}^{(e-h)}) Y_f^{(G,E)} + Y_b^{(G,E)*} \hat{\sigma}_{\text{mod}}^{G,E(g,\alpha)}(E_{(G,E)l}^{(e-h)}) Y_b^{(G,E)}]\end{aligned} \tag{5.81}$$

The relaxation rates in (5.79) and (5.80) are still given by (5.62)–(5.64), treated as spatially resolved equations, and the dynamics of the wetting layer described by the spatially resolved version of (5.66). In the DDE model, the excitonic (dot neutrality) approximation was made; in the more complex traveling wave model, it may be sensible to write the equations for electrons and holes separately – though good agreement with some experiments has been obtained using the excitonic approximation too.

Again, for excited-level and wetting layer states in the SA section, fast relaxation terms are introduced, describing excited-level carrier tunneling out of the dots and the combined tunneling and thermionic escape from the WL.

If standing wave-induced gratings are to be included in the calculations, we need to consider the spatial modulation of all the populations with the spatial periodicity of the standing wave (half wavelength in the waveguide). The amplitudes of this modulation, $\Delta f_{(g,\alpha)(E,G),l}^{(\text{gr})}$ and $\Delta N_{g,\alpha}^{(\text{wl,gr})}$, the superscript gr standing for grating, are determined by the same kinetic equations as the values $f_{(g,\alpha)(E,G),l}$ and $N_{g,\alpha}^{(\text{wl})}$, with two differences. First, the standing wave-induced population modulation (grating) in the wetting layer (but not in the dots) is smoothed down by the diffusion (and the pumping term is absent) so that the equation becomes

$$\frac{d\Delta N_{g,\alpha}^{(\text{wl,gr})}}{dt} \approx -\frac{N_{g,\alpha}^{(\text{wl})}}{\tau_{g,\alpha}^{(\text{wl,gr})}} - 2N_D(R_{g,\alpha}^{(\text{wl}\rightarrow G;\text{gr})} - R_{g,\alpha}^{(G\rightarrow\text{wl;gr})}) - 4N_D(R_{g,\alpha}^{(\text{wl}\rightarrow E;\text{gr})} - R_{g,\alpha}^{(E\rightarrow\text{wl;gr})}),$$

$$\frac{1}{\tau_{g,\alpha}^{(\text{wl,gr})}} = \frac{1}{\tau_{g,\alpha}^{(\text{wl})}} + \frac{16\pi^2 D_a n_g^2}{\lambda^2}$$

(5.82)

where D_a is the ambipolar diffusion coefficient in the QW wetting layer, and the capture and escape rates are calculated using (5.62)–(5.64) but with $\Delta f_{(g,\alpha)(E,G),l}^{(\text{gr})}$ and $\Delta N_{g,\alpha}^{(\text{wl,gr})}$ substituted instead of $f_{(g,\alpha)(E,G),l}$ and $N_{g,\alpha}^{(\text{wl})}$, respectively. Second, the stimulated recombination terms in the equations for $f_{(g,\alpha)(E,G),l}$ take the form

$$R_{G,E(g,\alpha)l}^{(\text{stim,gr})}(z,t) = \frac{v_g}{2}\text{Re}\left[Y_b^{(G,E)*}\hat{\sigma}_{\text{mod}}^{G,E(g,\alpha)}(E_{(G,E)l}^{(e-h)})Y_f^{(G,E)} + Y_f^{(G,E)*}\hat{\sigma}_{\text{mod}}^{G,E(g,\alpha)}(E_{(G,E)l}^{(e-h)})Y_b^{(G,E)}\right].$$

(5.83)

The gain/absorption and refractive index grating can then be calculated from $\Delta f_{(g,\alpha)(E,G),l}^{(\text{gr})}$ and $\Delta N_{g,\alpha}^{(\text{wl,gr})}$ in the same way as the total gain, absorption, and refractive index are calculated from $2f_{(g,\alpha)(E,G),l}-1$ (assuming the excitonic approximation) and $N_{g,\alpha}^{(\text{wl})}$. Some simplifications (e.g., neglect of hole burning in the carrier distributions in energy) may be justified, as the grating terms are less important, and calculating them accurately is less critical, than in the case of the total gain and absorption.

C) Even with separate reference frequencies for ground- and excited-state transitions, the multipopulation model remains computationally quite demanding, particularly if a considerable number of size groups l are to be considered. One way of introducing the next level of approximation is to use the approach of Equation 2.73:

$$f_{G,E}^{(e)}(E_{G,E}^{(e,h)}) = f_{QE}^{(e)}(E_{G,E}^{(e,h)}) + \Delta f^{(e)}(E_{G,E}^{(e,h)}),$$

(5.84)

with the first term describing the quasi-equilibrium population determined by the total electron/hole density including both the dots and the WL assuming a Fermi distribution and the second term $\Delta f^{(e,h)}(E_{G,E}^{(e,h)})(|\Delta f^{(e,h)}(E_{G,E}^{(e,h)})| \ll 1)$ being the fast-relaxing nonequilibrium correction describing the spectral hole burning and the associated nonlinear correction to the gain (2.75):

$$g_G(\hbar\omega) = g_G^{(QE)}(\hbar\omega) + \Delta g_G(\hbar\omega).$$

(5.85)

Nonequilibrium corrections to gain and refractive index can then be calculated along the lines of Equations 2.77, 2.86, 2.89, and 2.94, which can be relatively easily generalized for the case of ground and excited levels and gain and absorber sections. This procedure does not need to resolve separate energy states, with the inhomogeneous broadening included through the dependence of the slow quasi-equilibrium gain and absorption peak, spectral linewidth, and spectral position on the carrier density (all of which can be precalculated before dynamic simulations).

The magnitude of the fast response to the light intensity (i.e., the spectral hole burning-related gain compression coefficient $\varepsilon = \varepsilon_{SHB}$ and the gain relaxation time $\tau_{rel}^{(g,e)}$) can also be precalculated as functions of carrier density using equations of the type (2.89) and (2.96)–(2.97). Strictly speaking, these equations describe the *electronic* nonequilibrium contribution to gain. If an excitonic approximation is made, the relaxation times may be assumed the same for electrons and holes; in the opposite approximation, the hole distribution can be assumed equilibrium since their energy levels are closely spaced and the energy relaxation may be expected to be fast, and only the electron distribution to hole burning needs to be considered.

The wetting layer contributions can be added as in (5.76).

The spectrally dependent nature of the spectral hole burning may be included by introducing the spectral width of the spectral hole $\Delta\omega_{SHB} \approx \frac{1}{\hbar}\Delta E_{hom}$. Then, instead of the instantaneous relations (2.96) or (2.99), the nonlinear part of the gain due to spectral hole burning may be described (assuming the electron contribution dominates the nonlinearity) by an operator $\widehat{\Delta g}_{shb}$ given by

$$\widehat{\Delta g}_{shb} Y_{f,b} = \Delta\omega_{shb} \int_0^\infty \exp(-\Delta\omega_{shb}\tau)\Delta g_{shb}(t-\tau)Y_{f,b}(t-\tau)d\tau, \quad (5.86)$$

with the integrand evaluated from

$$\frac{d}{dt}(\Delta g_{shb}) \approx -\frac{1}{\tau_{rel}^{(g,e)}}\left(\Delta g_{shb} + \Delta\Omega'_g \int_0^\infty \frac{\exp(-\Delta\Omega_g\tau)\left(\varepsilon g_G^{(QE)}(|Y_f|Y_f^* + |Y_b|Y_b^*)\right)}{1+\varepsilon(|Y_f|^2+|Y_b|^2)}\bigg|_{t-\tau} d\tau\right). \quad (5.87)$$

In the limit of $\tau_{rel}^{(g,e)} \to 0$, $\Delta\omega_{shb} \to \infty$, $\Delta\Omega_g \to \infty$, (5.86) and (5.87) reproduce (2.99). The finite relaxation time $\tau_{rel}^{(g,e)}$ in (5.87) introduces the dynamics of hole burning into the model, and the finite $\Delta\omega_{shb}$, and thus the integral in (5.86), represents in time domain the finite width of the spectral hole (the integral in (5.87) takes into account linear gain dispersion). Computationally, the integrals can be implemented using, say, an infinite impulse response filter in the same way as the linear gain bandwidth is taken into account.

In the case of saturable absorbers, with the intralevel relaxation time comparable to the sweepout time by the external field, no quasi-equilibrium distribution in energy is formed; therefore, the entire saturable absorption should be described by the formulas of the type (5.86) and (5.87):

In the long lasers with high-power pulses generated, the assumption $|\Delta f^{(e)}(E_{G,E}^{(e)})| \ll 1$ (and thus $|\Delta g| \ll g^{(QE)}$) may become somewhat tenuous, however, particularly in the absorber section, so ideally the calculations will need to be validated against the multipopulation model for a particular structure.

D) Finally, a population model with no account of inhomogeneous broadening at all, in other words, a traveling wave version of the DDE approach, is a fast and versatile, if not the most accurate, approach to dynamics of QD lasers.

Mathematically, the kinetic part of such a model essentially consists of solving Equations 5.70–5.83 but without a summation in the dot energy (size) group l.

The dependence of gain on the ground-state population may be calculated microscopically or fitted to experimental data. It is worth noting that, according to simulations, even with a significant inhomogeneous broadening, a simple linear dependence (2.4) of the peak ground-state gain on the ground-state carrier population (if the excitonic approximation is used), or on the sum of the electron and hole ground-state populations (if the excitonic approximation is not made), remains a fairly good description of the gain–carrier population relation in the gain section. This considerably simplifies a model. The gain *dispersion curve* is described either by a single digital filter or with a separate equation for polarization, with the characteristic energy of the notionally "homogeneous" spectral broadening in the model artificially increased to include approximately the effect of *inhomogeneous* broadening on the ground and excited transitions gain spectrum, too, as is usual in the case of modeling non-QD lasers. The most natural reference frequencies for the ground- and excited-state absorption are in this case determined by the transition energies for the corresponding levels, so one can set $\Delta E_{(G,E)}^{(e-h)} = \hbar \omega_{\text{ref}}^{(G,E)}$ in Equation 5.73.

While the details of the laser behavior, such as the peculiarities of the spectral hole burning and their effect on the laser properties, may not be captured accurately by such a model, it can be trusted – on the basis of the success of the DDE model as described above, which also does not include inhomogeneous broadening – to give as much useful information as the DDE model itself, and more, because the geometry of the laser is treated accurately in the traveling wave model.

The computational complexity of models C and D is approximately the same, the former having the advantage of describing, albeit somewhat approximately, the inhomogeneous broadening and spectral hole burning, and the latter, of presenting a more accurate large-signal analysis, which is useful in long lasers and/or at high operating powers, when population variation after even a single pulse ML may strain the approximation $|\Delta f^{(e)}(E_{G,E}^{(e)})| \ll 1$. Another advantage of the single-population model with a full account of capture/escape dynamics is more straightforward account of the excited-state dynamics, with equations of the same type (albeit with very different parameter values) used for modeling ground- and excited-state dynamics in both gain and SA sections.

Most of the models discussed above have indeed been used to simulate ML QD laser dynamics.

Recently, multipopulation traveling wave models (of the types A and B) have been extensively investigated by Rossetti, Bardella, and Montrosset [146, 147]. It was found [147] that only a set of several (typically seven) size groups was sufficient to accurately reproduce the gain spectrum in the time domain model.

The simulations were, in fact, more sophisticated than those described above as they took into account ground and *two* excited levels (first and second), the higher of which, while not contributing significantly to the optical properties of the active layer, served as one of the channels of escape of carriers from the saturable absorber and a part of the capture chain in the gain section.

With lasers operating entirely at the ground level (which tends to happen at longer lengths and/or larger aggregate dot density per unit area), stable mode locking at $F_{rep} < 20$ GHz was observed in a broad range of parameters: gain section current of about 200–350 mA (150 mA being the operating threshold) and absorber section voltage in excess of about 6 V. The amplitude and duration of the ML pulse both increased with current (Figure 5.15) as in most other passive mode-locking simulations and in agreement with most experimentally observed data, with the simulated

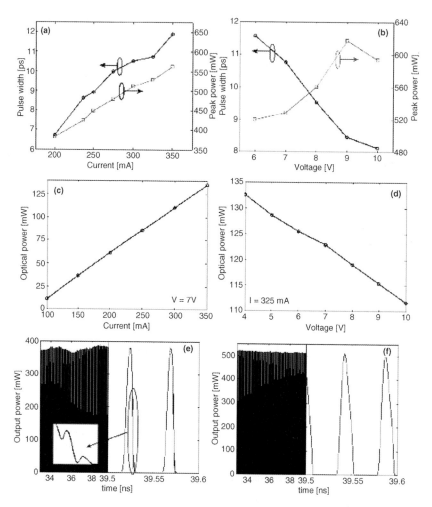

Figure 5.15 Pulse duration and peak power dependence on (a) the bias current in the gain section at a fixed absorber voltage of $V_{SA} = -7$ V; and (b) the reverse bias on the absorber at a fixed current of $I_G = 325$ mA in a mode-locked QD laser diode; the corresponding current (c) and voltage (d) dependences of the average power; (e and f) time traces of ML output for $V_{SA} = -7$ V and $I_G = 200$ and 325 mA, respectively. Note the trailing edge instability at the lower current. All the results simulated by a traveling wave multipopulation model. Reproduced with permission from Ref. [147].

values close to those observed experimentally. At low currents, low frequency envelope was observed in the simulations. Simulations (Figure 5.15e) identify it as trailing edge instability, in common with the self-pulsations typically observed at low currents in most ML laser diodes. At modest currents (and high enough reverse bias), stable mode locking (Figure 5.15f) is seen, with pulses rather broad (several picoseconds in duration) and of a strongly asymmetric shape not seen in lasers of other type, which may be attributed to the effects of inhomogeneous broadening and slightly dephased dynamics of different dot groups.

By integrating the gain over propagation distance in the traveling time frame of the pulse, the authors analyzed the pulse round-trip gain in the traveling wave model and found that the pulses remained stable even though, as the bias increased, a window of positive net gain opened around the pulse.

Finally, at high bias, an irregular instability manifested itself again, with satellite pulses, initially stable and then unstable, appearing in the pulse train.

The effect of the SA bias voltage on the laser dynamics, presented in Figure 5.15b,d, was consistent with the effective absorber recovery time being shortened by the voltage. With the voltage magnitude below a certain critical value (about 6 V), no stable mode locking was observed. As the voltage increased from 6 to 10 V, decreasing the absorption recovery time, the pulse width of stable mode-locking pulses decreased by about 50% (from 12 to 8 ps), in agreement with experiments.

The traveling wave model also made it possible to analyze the interplay of the level dynamics and the influence of the excited state on the mode-locked emission pulses, which in the case of long lasers are always emitted in the ground-state band. Figure 5.16 shows the evolution of gain and absorption spectra at two points inside the laser, one in the gain section and one in the SA, at an equal distance from the facets. As can be seen in the figure, the depletion of the gain by the pulse is significantly weaker than the saturation of the absorption. This is the result of the gain saturation energy being higher than the SA saturation energy, which in turn stems not only from the position of the pulse photon energy in the gain and absorption spectra but also from the strongly inhomogeneous nature of the absorption saturation (see Section 4.3). The shape of the gain and absorption spectra implies that the effect of crossgain/absorption from the excited-level transitions on the pulse propagation is very small in both gain and absorber sections.

The multipopulation traveling wave model was also used to successfully interpret the dynamics of gain and absorption sections in lasers whose design makes operation at the excited level possible. These are essentially lasers with a relatively short cavity length, increasing the threshold for ground-state lasing.

Unlike the multipopulation DDE model [145], which to date has only predicted switching from GS to ES ML, the traveling wave model has been used to analyze switching both from excited- to ground- and from ground- to excited-state mode locking, with an area of ground- and excited-state ML coexisting, depending on the density of dots in the laser and the operating point. The subtle balance of gain and absorption, sensitive to the reverse bias applied to the SA section, and even to the redistribution of current in the absorber circuit (see Section 4.2.1), makes both the GS to ES and the ES to GS switching scenarios possible. In the latter case, the mechanism

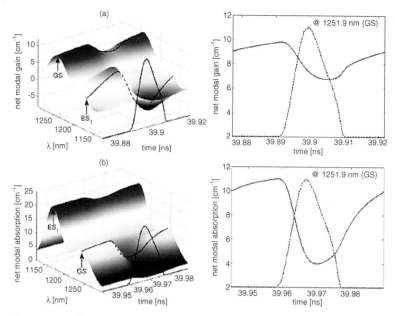

Figure 5.16 *Left*: Simulated dynamics of gain (a) and absorption (b) spectra, at a point 20 μm away from the corresponding facet. *Right*: details of gain (a) and absorption (b) dynamics at the center pulse wavelength of 1251.9 μm (in the GS band). Reproduced with permission from Ref. [147].

was shown to be as follows. At the onset of lasing, the saturation of the ground-state gain with current and the high outcoupling losses make it difficult for the ground-state gain to reach the threshold condition, and the laser starts operating at the excited-state wavelength, since gain saturation at the excited-state wavelength happens at higher currents. During the laser operation, however, carriers are created in the saturable absorber section. Originally, they have energies in the excited-state band, but as the excited- to ground-state relaxation time in the SA is comparable to the time of sweepout by the reverse bias field, a fraction of photocarriers relax to the ground state. Thus, the ground-state band of the saturable absorber is effectively pumped by excited-state band light, leading to GS absorption saturation. Eventually, as power is increased, the threshold condition is satisfied for the ground-state transition, which can lead either to the switching from the ES to GS operation or to coexistence of both levels.

It should be noted that good agreement with experiments was achieved in the above simulations, though the excitonic approximation $f^{(e)}_{(g,\alpha)(G,E)l} = f^{(h)}_{(g,\alpha)(G,E)l}$ was made, the group velocity was assumed the same at both levels, and induced gratings were not taken into account, meaning that the influence of the effects "missed out" can be expected to be relatively modest.

Simulations using *single-population models*, without taking into account inhomogeneous broadening (model of the type D according to the above classification), have been performed by several groups, a recent example being the study reported in Ref. [148]. Only ground-level resonant light was studied in Ref. [148] (along with other single-population models reported so far), with only ground-level states contributing to the gain as in Equations 2.2 and 2.3. The excitonic approximation was also made. The excited state entered the model through kinetic equations for the absorber section only, as described in Section 4.3. In the gain section, capture was considered to be effectively straight from the wetting layer to the ground state (the excited state may be said to have been eliminated adiabatically).

A number of predictions of the multipopulation models are also confirmed by the single-population one. Namely, self-pulsation instability has been observed in a narrow current range at low pumping currents [148]. At modest currents, good quality mode locking is seen, and at high currents, pulse breakup is observed, with trailing pulses following the main pulse, as in DDE modeling, leading eventually to an instability. The breakup is preceded with the (stable) trailing pulses showing as shoulders of the main ML pulse, reminiscent of the irregular and very asymmetric pulse shapes produced by the multipopulation model.

One result of the multipopulation model that the single-population one in the form of [148] had difficulties reproducing was the increase in pulse duration with current seen both in multipopulation model and in experiments. Instead, pulse narrowing with increased current was registered in the simulated dynamics. In Ref. [148], this was attributed partly to the neglect of spatial and spectral hole burning effects and partly to the effect of the excited-state dynamics. Indeed, more recently the same team showed that more careful analysis of excited-state effects, with the two-stage relaxation included in both the gain and the SA sections and with the excited- to ground-state relaxation time taken as comparable to the WL excited-state time, allowed this trend to be reversed and pulse broadening with current to be obtained in simulations. Thus, correct prediction of this feature is possible in a properly formulated single-population model.

It has to be pointed out that with carefully adjusted phenomenological parameters, some features (e.g., the pulse shapes under certain operating conditions) of QD mode-locked lasers (operating within the ground-level band) can be reproduced with some success also by models that contain no specific features of QD dynamics at all and are just generic laser models of the type discussed in Section 5.2, of a traveling wave type or a DDE type (e.g., in Ref. [149]). It is not clear, however, how generic such predictions are.

5.4.3
Modal Analysis for QD Mode-Locked Lasers

Dynamic modal analysis may be a way of introducing the effect of inhomogeneous broadening and spectral hole burning on mode locking in QD lasers without needing to introduce spatial resolution. When applied to quantum dots, it consists of analyzing pulsations of dot and wetting layer populations at the intermodal interval (round-trip frequency), and the harmonics of this frequency, due to the beating between modes in the laser spectrum, and using the amplitude of those pulsations to calculate the complex coupling terms that lead to synchronization of modes. Dynamic modal analysis has not yet been applied to passive mode locking of QD lasers, but in Ref. [52], the approach was applied to a hypothetic *actively* mode-locked construction, predicting picosecond pulses with durations decreasing with the modulation strength. The limitation of the model is its essentially small-signal nature, relying on the modulation of all the dot populations being weak. This assumption is, by and large, supported by the results of time domain simulations of the *gain* section dynamics, but is much more strained in the SA section, with the modulation of absorption being almost 100% in some cases; therefore, it is not clear how accurate modal analysis would be in the case of realistic *passively* mode-locked QD lasers.

5.5
Advantages of Quantum Dot Materials in Mode-Locked Laser Diodes

5.5.1
Advantages of QD Saturable Absorbers

Bearing in mind that the presence of a saturable absorber is required for passive and hybrid mode locking, all the advantages of broadband absorption, ultrafast absorption recovery, and low saturation fluence for quantum dots as used in saturable absorbers can be exploited quite generally in QD mode-locked lasers. It is worth mentioning that the required specifications of a saturable absorber may be more stringent if the absorber is integrated into a semiconductor mode-locked laser. First of all, the optical peak power is usually much lower than in mode-locked vibronic lasers. Moreover, in a monolithic configuration, the usual ridge waveguide geometry of the two-section laser imposes a similar mode area on both the gain and the absorber sections, which inhibits a flexible optimization of the saturation fluence in the absorber. To overcome this limitation, the use of a flared waveguide geometry has been investigated in a QD monolithic laser, which recently allowed the generation of 360 fs pulses [150].

Importantly, it has also been observed that the saturation power is at least 2–5 times smaller for a QD saturable absorber than for a QW-based counterpart when integrated into a monolithic mode-locked laser [151], which appears to tally up well with the estimates and reasoning in Section 4.3.2. In this paper, the authors pointed out that saturation would further depend on the density of dots, reverse bias, and inhomogeneous broadening.

5.5.2
Broad Gain Bandwidth

Quantum dot semiconductor structures are particularly exciting materials for the generation and amplification of femtosecond pulses because one of the key parameters that influence the emission spectrum and the optical gain in QD devices is the spectral broadening associated with the distribution of dot sizes. The extremely broad bandwidth available in QD mode-locked lasers offers potential for generating sub-

100 fs pulses provided all the bandwidth could be engaged coherently and dispersion effects suitably minimized.

Indeed, it has been shown that there is usually some gain narrowing/filtering effects in mode-locked QW lasers [152]. With the inhomogeneously broadened gain bandwidth exhibited by QDs, there is support for more bandwidth and this can oppose the effect of pulse broadening that may arise from spectral narrowing. In addition, due to the particular nature of QD lasers, many possibilities open up in respect of the exploitation of ground-state (GS) and excited-state (ES) bands, thereby enabling multiple wavelength-band switchable mode locking [153–155] and dual-wavelength mode locking involving widely separated spectral bands [145, 156, 157]. On the other hand, the interplay between GS and ES can be deployed in novel mode-locking regimes [158]. Furthermore, by using an external cavity, it is possible to set up mode-locked sources that can be tunable across the GS or ES transition bands [159].

The inhomogeneous nature of the gain can also bring some advantages regarding the generation of low jitter pulses. In fact, the inhomogeneously broadened gain may lead to a lower mode partition noise, enabling the generation of individual comb components with reduced relative intensity noise [160].

However, high levels of inhomogeneous broadening also entail a number of disadvantages due to the resulting low levels of gain. In order to achieve higher optical power levels, the cavity length needs to be longer, which implies a higher level of dispersion and thus further broadening of the pulse duration.

5.5.3
Low Threshold Current

As devices, QD laser diodes have the advantage of requiring a very low threshold current to initiate lasing [20]. This attribute also applies to operation in the mode-locking regime because most QD lasers exhibit mode-locked operation right from the threshold of laser emission. (Bistability between the nonlasing state and the onset of lasing/mode locking might be present, as has been shown experimentally [118, 161] and numerically [49].) A low threshold current is clearly advantageous because this can represent a device that is compatible as an efficient and compact source of ultrashort pulses where the demand for electrical power can be very low.

Furthermore, having a low threshold avoids the need of higher carrier densities for pumping the laser and this implies less amplified spontaneous emission (ASE) and reduced optical noise in the generated pulse sequences.

5.5.4
Low Temperature Sensitivity

Due to the discrete nature of their density of states, QD lasers exhibit low temperature sensitivity, making them excellent candidates for many applications, for example, in optical communications. If the need for thermoelectric coolers can be avoided, then there is scope for low-cost, compact, lightweight, and lower power systems. More importantly, QD lasers can also show resilience to temperature effects in mode-locked operation. Indeed, stable mode-locked operation from a quantum dot laser has been demonstrated at temperatures up to 80 °C [162], and more recently it has been shown that the duration of the pulses generated actually decreases with temperature [163], which can be attributed to a faster absorber recovery time with increasing temperature – most notably due to faster thermionic escape [164].

5.5.5
Suppressed Carrier Diffusion

Owing to the clustered nature of QDs, carrier diffusion in the gain material is strongly suppressed, compared to QW materials [165]. This feature facilitates the fabrication of laser devices, as there are no or few defects created in the sidewalls of mesas, with the consequent reduction in the number of centers of nonradiative recombination at the surface or interface of the active material. This feature has allowed routine processing that involves etching through the active layer, without imposing a heavy penalty on the threshold current of laser devices [166, 167]. The ease of integrating deeply etched structures opens up wide and appealing opportunities for integrating a greater range of structures into QD mode-locked lasers, such as photonic bandgap structures, chirped mirrors for dispersion compensation, and distributed feedback (DFB) structures. The fabrication process of etching through the active layer brings related and significant advantages for efficient fiber coupling in narrow stripe lasers because the stronger index guiding of the optical mode that is possible results in a quasi-symmetric far field [168]. Moreover, better light confinement potentially could also help to saturate the saturable absorber more easily.

Etch-through fabrication has another advantage in respect of the direct modulation of the laser. This is important in hybrid mode locking where an RF electrical signal is injected either into the absorber or into the gain section. It has been demonstrated that an etch-through mesa exhibits a much lower parasitic capacitance than a shallow-etched mesa and so this enhances the electrical performance of the laser diode at high frequencies [168]. This feature is of key importance in active and hybrid mode-locked configurations. The etch-through approach has been pursued recently in mode-locked lasers as well, particularly when these are to be used under hybrid mode-locking operation [169, 170].

Finally, it has been stated that another positive aspect of the lower diffusion of carriers, when associated with low linewidth enhancement factor, is that it reduces the likelihood of beam filamentation in contrast to quantum well lasers [171, 172]. This affords some additional potential for upgrading the optical power of QD mode-locked lasers without degrading beam quality – something that has been proven difficult for QW lasers. Indeed, QD mode-locked lasers with ridges as wide as 24 or 30 μm are possible [173, 174], and a tapered waveguide mode-locked laser with a 100 μm wide front facet has been demonstrated [161]. The absence of carrier diffusion to the end facets may also concur with the higher threshold of optical catastrophic damage. This implies that QD lasers can withstand higher values of current injection than their QW counterparts, without degradation of the facet mirrors and thus higher power levels can be achieved [175, 176].

5.5.6
Lower Level of Amplified Spontaneous Emission

Pulse generation in semiconductor ultrafast lasers can be affected by timing jitter, in particular in passive mode-locked devices, due to the lack of synchronization with an electrical signal [177]. ASE is the primary cause of jitter because it generates random fluctuations in photon density, gain, and index of refraction, which contribute to variations in the round-trip time [11]. Longer devices are particularly prone to such problems, as there is a buildup of spontaneous emission along the length of the active material. QD-based materials could make a difference in this respect, as these nanostructures exhibit overall lower levels of ASE [61], compared to QW, due to the discrete nature of their energy levels. On the other hand, because QD lasers usually have a lower threshold current, the amount of carriers needed to start laser emission is also reduced, which consequently also leads to a smaller number of carriers being involved in nonstimulated emission [178].

5.5.7
Linewidth Enhancement Factor

One of the main motivations for the enthusiastic investigation of QD materials in the past few years has been the theoretically predicted potential for very low values of LEF, owing to the symmetry of the gain associated with QD structures. The possibility of a low LEF is very attractive for a number of performance aspects, such as weak self-phase modulation effects, lower frequency chirp in directly modulated lasers, lower sensitivity to optical feedback effects, and suppressed beam filamentation.

The first measurement of LEF in an InAs QD laser diode in 1999 resulted in a value of 0.1 below threshold [179], which was particularly exciting because it implied the possibility of lower chirp, and Fourier transform-limited pulses. In the past years, there have been some additional experimental demonstrations of low LEF, but limited to a narrow range of bias conditions (below or close to threshold). In practical devices and under bias conditions, it has been observed that the LEF increases significantly as the operating current is increased [180, 181], which in turn may contribute to an increase in the frequency chirp and duration of the pulses generated by mode-locked QD lasers [79]. In fact, LEF can be significantly higher for QD materials than for QW, and a giant LEF of 60 has been measured in a QD laser, the highest ever to be measured in a semiconductor laser [182]. The usual approach to generate shorter and transform-limited pulses is then to operate two-section mode-locked lasers very close to threshold, and at a high reverse bias [173]; however, this limits the output power and constrains the exploitation of their typically high differential efficiency in the generation of high-power ultrashort pulses.

The LEF increase in QD lasers has been attributed to the increasing asymmetry of the gain, as the carrier population is slowly building in the higher levels with increasing current.

Above laser threshold in the GS, the carrier population in the ES is not clamped. As there are a limited number of available GS states which the ES carriers can relax to, they start to accumulate in the ES levels, as the current increases. A "carrier pileup" starts to occur – a unique phenomenon in the domain of semiconductor lasers. The carrier pileup induces a change in the refractive index leading to an increase in the LEF [182].

The spectral dependence of LEF above threshold in QD lasers has recently been characterized by Kim and Delfyett [183], together with the investigation of the spectral dependence of pulse duration and linear chirp. This work has shown that the LEF is reduced when the laser operates ~10 nm offset of its gain peak (~1264 nm). The effect is even more dramatic on the anti-Stokes side, where the LEF reduces to ~1, as compared to 6 close to the gain peak. On the Stokes side, the LEF is also reduced to 4. This has a direct impact on the pulse duration and linear chirp, as investigated in a tunable mode-locked QD laser. While keeping the bias conditions intact, the laser was tuned in the spectral region around the gain peak (from 1260 to 1285 nm). The spectral bandwidth on the anti-Stokes side (1260–1275 nm) was found to be broader (~8 nm) than the bandwidth on the Stokes side (1280–1285 nm). Similar to the LEF, the output pulse duration was reduced on both the Stokes and the anti-Stokes side. On the other hand, the linear chirp was found to double from 3 to 6 ps nm^{-1}, when the wavelength was tuned from 1260 to 1285 nm. Such results suggest that the carrier-induced refractive index change is smaller on the anti-Stokes side. As the authors point out, there are also other additional factors that could contribute to this behavior, such as effects from the excited states and inhomogeneous broadening [184, 185].

5.6
Ultrashort Pulse Generation: Achievements and Strategies

5.6.1
Monolithic Mode-Locked Quantum Dot Lasers

The first demonstration of a QD mode-locked laser was reported in 2001 by Professor Luke Lester's group at the University of New Mexico, with ~17 ps pulse durations at 1.3 μm and repetition rate of 7.4 GHz, using passive mode locking [118]. Hybrid mode locking at the same wavelength was demonstrated in 2003 by the Photonic Systems research group at the University of Cambridge [174], reporting an upper limit estimation of 14.2 ps for the shortest pulses measured, at a repetition rate of 10 GHz. Later in 2004, the same group demonstrated Fourier transform-limited 10 ps pulses at a 18 GHz repetition rate, using passive mode locking [173]. In 2004, for the first time the generation of subpicosecond pulses directly from a QD laser was demonstrated, where the shortest pulse durations of 390 fs were observed (Figure 5.17), without any form of pulse compression [186, 187]. These pulses were generated by a two-section passively mode-locked QD laser. Owing to the excellent electrical characteristics of the device, it was possible to apply very high values of current and reverse bias (up to 10 V), which provided some latitude for exploring a wider range of these parameters. (It is important to recall that until then mode-locked

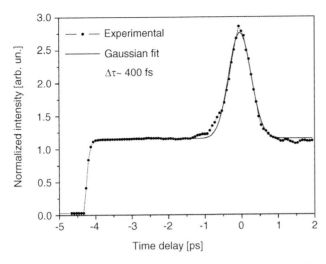

Figure 5.17 Intensity autocorrelation trace of a 400 fs pulse generated by a QD mode-locked laser, indicating a contrast ratio close to 3 : 1.

lasers were being operated close to threshold, and with relatively low values (<3 V) of reverse bias). Mode-locked operation was observed over a broad range of injection currents from above laser threshold (∼30 mA) up to 360 mA and over a relatively wide range of reverse bias levels on the absorber section (from 4.5 to 10 V). The deduced time–bandwidth product of about 1 indicated some residual frequency chirp in these pulses. The broad spectrum (14 nm) and the ultrashort pulse durations measured suggest that the generation of pulses in the sub-100 fs domain may yet be possible from relatively simple QD laser configurations.

Given that this laser was operating significantly above threshold, the power level was relatively high, with typical output powers of up to 60 mW in the CW and up to 45 mW in mode-locked regimes. This substantially exceeded all the previously reported values of output power for this type of lasers. It thus became evident that a wider range of operating conditions could be usefully explored to improve the performance of such QD lasers. This became possible as the material growth and device processing were further refined for increasing numbers of QD layers.

The generation of subpicosecond pulses was reported later by several groups [81, 161]. In this paper, Thompson and coauthors also confirmed that the pulse duration decreases exponentially with increasing reverse bias (up to 8 V) [161]. The main factor contributing to this decrease is the corresponding exponential decrease in the absorber recovery time as the reverse bias is increased, as was later demonstrated in a pump–probe investigation in Ref. [99]. Nevertheless, there are several limitations that restrict the amount of "useful" reverse bias that can be applied to the saturable absorber. For higher values of reverse bias (usually >8–10 V), it is commonly observed that the pulse duration does not decrease further, may even start to increase with increasing reverse bias [169], or the mode locking may cease to be stable and in fact cease to exist. There are several possible reasons for this. In the first instance, the presence of quantum-confined Stark effect may lead to a redshift of the absorption spectrum with increasing reverse bias [150, 188]. As a result, if the absorption maximum therefore no longer overlaps with the emission spectrum, the efficiency of the absorber as a pulse shaper becomes smaller. In addition, for a very high reverse bias, the optical power can become so low that the pulse has not enough energy to bleach the absorber. This effect is further accentuated by an increase in the absorption saturation energy with reverse bias, an effect that had been verified before in other semiconductor QW-based saturable absorbers. The saturation energy in QD saturable absorber lasers has also been shown to increase with reverse bias, albeit to a lesser extent [151]. If the intracavity energy of the pulses becomes comparable to the saturation energy, the discrimination between CW and pulsed operation in the saturable absorber will not be so efficient, and mode locking may even become unstable, as there is no efficient pulse shaping taking place. Such effect has been observed [81, 169] where after a certain level of reverse bias the pulses would start to slightly increase and/or the mode locking would become unstable. Finally, a very high reverse bias can break down the junction and damage the laser structure if it exceeds a critical level.

It has been widely observed by different research groups that the range of conditions (bias and temperature) for stable mode locking in semiconductor QD lasers is much larger than the range found in comparable QW lasers [189]. The region of stable mode-locked operation can also be substantially broadened by appropriate design of the QD mode-locked laser – as shown by Thompson et al. [150], a higher absorber-to-gain length ratio can expand the range of mode locking by at least twice, when moving from a 1 : 7 to a 1 : 3 ratio. As the authors point out, the longer section leads to an increased saturable absorption that can assist in mode locking over a wider range of gain currents. Design rules for QD mode-locked lasers can also be established in order to minimize the pulse duration. For instance, Thompson et al. have shown how a pulse duration reduction of 2.2 ps–800 fs can be achieved by using a 520 μm absorber instead of a 130 μm one, for the same device length of 2 mm [150], while also boosting the peak power from 20 to 260 mW. The reduction in pulse duration results therefore from both the reduced broadening effects in the now shorter gain section, while at the same time the longer absorber section strengthens the pulse shortening effects. An increase in output power was also observed, which the authors attributed to a shift in the necessary bias conditions to achieve mode locking. Due to the higher level of saturable absorption, a higher pulse energy is required to saturate the absorber, which would result in mode-locked operation at higher gain drive currents and thus with higher output power. By using a multi-section biased laser, Xin et al. [190] also demonstrated the advantage of incorporating a passive waveguide section in the laser – whereby shorter pulses could be achieved compared to the all-active laser.

There has also been much effort in designing QD-based mode-locked sources that could be deployed in the 1.55 μm optical communication band. Ultrashort pulse generation with subpicosecond duration has been achieved from single-section lasers based either on InAs quantum dots [191] or on quantum dashes [192] grown on an InP substrate. Although the authors have suggested that there are no fundamental differences between the quantum dots and the dashes in the context of mode-locked laser sources, the mechanism of the generation of ultrashort pulses in these devices – in the absence of a saturable absorber or an external RF modulation – is still unclear. This is still an ongoing area of research and more work needs to be done to understand the properties of these materials in the context of the generation of ultrashort pulses.

5.6.2
Chirp Measurement and Pulse Compression

An overview of the main results published in the past years can be found in Table 5.1. As depicted in this table, the generation of transform-limited pulses directly from passively mode-locked QD lasers has been reported by some research groups. The accepted hypothesis for the origin of this behavior stems from the opposite sign contributions of the gain and the absorber section to the overall chirp. In a monolithic two-section mode-locked semiconductor laser, where a saturable absorber and a gain section coexist, the resulting chirp is a balance between the effects caused by the absorber and the gain. In the gain section, usually a frequency up-chirp results, while the saturable absorber helps to further shape the pulse by contributing with a down-chirp [193]. With a suitable balance between both sections, the chirp can be close to zero thereby leading to transform-limited pulses. Unfortunately, this is the exception rather than the rule because this usually occurs only for a limited set of bias conditions and/or for given ratios of absorber/gain lengths. Therefore, up-chirp usually prevails for passively mode-locked lasers, leading to significant pulse broadening as the pulse propagates, due the combination of positive group velocity dispersion. The combined effect of self-phase modulation and dispersion imposes the strongest limitation on the achievable shortest duration of pulses from mode-locked semiconductor diode lasers. In order to better understand the mode-locking dynamics and to implement efficient dispersion compensation schemes, the knowledge of the spectral structure and true shape of the pulses is necessary. To achieve this understanding, it is necessary to go beyond the usual tools for pulse duration characterization of QD mode-locked lasers, which until recently have been solely based on second-harmonic autocorrelation schemes. Although autocorrelation is a useful tool that can provide a good estimate of the pulse duration, it can reveal neither the true shape of the pulse nor its phase profile. On the other hand, frequency-resolved optical gating (FROG) systems can provide full electric field characterization [194]. However, the relatively low peak power (~ 1 W or less) and picosecond pulse durations associated with mode-locked laser diodes limit the use of commercially available FROG systems, which are mostly targeted at femtosecond solid-state laser systems. The first application of this technique to characterize mode-locked QD lasers was therefore made possible by constructing a custom-made ultrasensitive FROG system [195]. This achievement resulted in many initial insights: first and foremost, it was observed that the pulses were asymmetric, which is not possible to be observed in autocorrelation traces, given that these are intrinsically symmetric. Although Xin *et al.* do acknowledge that there is a temporal ambiguity in FROG, a

Table 5.1 Performance of mode-locked quantum dot lasers using different configurations.

Number of QD layers	Setup	ML regime	τ_p (ps)	λ_0 (nm)	$\Delta\lambda$ (nm)	f_{rep} (GHz)	P_{peak} (mW)	References
2	Monolithic	Passive	17	1278	1	7.4	7	Huang et al. [118]
3	Monolithic	Hybrid/passive	<14.2	1107	0.8	10	4	Thompson et al. [174]
10	Monolithic	Passive	10	1291	0.2	18	2.5	Thompson et al. [173]
5	Monolithic	Hybrid/passive	12/7	1286	0.18	20/35	NA/6	Kuntz et al. [196]
5	Monolithic	Passive	2	1260	14	21	1100	Rafailov et al. [186, 187]
5	Monolithic	Passive	0.39	1260	8	21	3000	Rafailov et al. [186, 187]
5	Monolithic	Passive	3	1281	0.8	50	6	Kuntz et al. [168]
5	Monolithic	Passive	1.7	1277	0.86	9.7	57	Gubenko et al. [197]
NA	Monolithic	Passive	5.7	1264	4.5	5.2	290	Zhang et al. [178]
3	Monolithic	Colliding pulse	7	1107	0.32	20	13	Thompson et al. [198]
5	Monolithic (using flared waveguide)	Passive	0.79	1276	3.6	24	500	Thompson et al. [161]
10	External cavity + SOA + compressor	Passive	1.2	1268	3.1	4.95	1220	Choi et al. [79]
15	Monolithic + SOA	Hybrid	0.7	1300	8.5	20	130	Laemmlin et al. [81]
15	Monolithic + SOA	Hybrid	3.3	1300	6.5	20	26	Laemmlin et al. [81]
15	Monolithic + SOA	Hybrid	1.9	1300	2.2	40	14	Laemmlin et al. [81]
15	Monolithic + SOA	Passive	2.2	1300	4.2	80	11	Laemmlin et al. [81]
5	Monolithic	Harmonic mode locking	1.3	1280	1.4	238	10	Rae et al. [199]
6	Monolithic	Harmonic mode locking	6.4	1216	1.8	42.4	234	Xin et al. [190]

Results are presented in chronological order.
τ_p: pulse duration; λ_0: central wavelength; $\Delta\lambda$: spectral bandwidth; f_{rep}: repetition frequency; P_{peak}: peak power; NA: data not available.
Note: Except stated otherwise, the data refer to monolithic mode-locked lasers.

leading and trailing edges were assigned on the basis of the condition that the differential absorption needs to be higher than the differential gain in order to achieve stable mode locking. Using this rationale, the steeper edge was considered to be the leading edge – as assumed to be shaped by the faster absorption saturation. The spectral phase was also presented for the first time for a QD mode-locked laser, and it was shown to follow a parabolic shape, indicating that the pulses are mostly linearly chirped, at least for the bias conditions investigated in this paper – low gain current between 100–115 mA and a modest reverse bias value of −3.5 V or −4 V. Such observations suggest that there are certain conditions under which the pulses are fully recompressible, although this was not demonstrated [149, 195]. The authors also indicate that given the spectral structure obtained, the transform-limited pulse duration would be 600 fs, although the spectral bandwidth would a priori allow a pulse duration of 300 fs. This paper also presents evidence of pulse breakup for higher values of gain current (>120 mA), as clearly shown in the FROG trace and corresponding spectrogram (Figures 5.18 and 5.19). As the authors have pointed out,

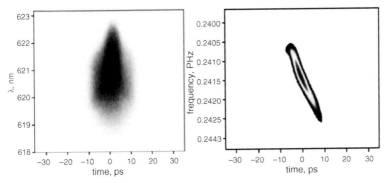

Figure 5.18 SHG FROG trace from monolithic mode-locked QD laser with gain current of 105 mA and reverse bias of 3.5 V; spectrogram of pulse retrieved from SHG FROG trace. Reproduced from Ref. [195] with permission of IET.

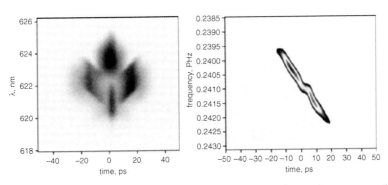

Figure 5.19 SHG FROG trace from monolithic mode-locked QD laser with gain current of 130 mA and reverse bias of 3.5 V; spectrogram of pulse retrieved from SHG FROG trace. Reprinted from Ref. [195] with permission of IET.

such pulse breakup could be an important indication that the inhomogeneous broadened spectrum fails to be fully coherently locked across its breadth, and groups of quantum dots close in energy are more easily coupled together. Such interplay between homogeneous and inhomogeneous broadening is therefore an important factor that definitely warrants further experimental and theoretical investigations. The pulse asymmetry was further investigated theoretically and modeled using a system of delay differential equations, able to predict the asymmetrized hyperbolic secant shape of the pulse, which agreed with the experimental findings [149].

FROG characterization was later used as a key tool for analyzing the chirp of the ultrashort pulses generated, for important proof-of-concept data transmission experiments [169, 170]. The authors investigated the chirp of the pulses generated directly by a passively or hybridly mode-locked 40 GHz p-doped QD laser [170]. The pulses were shown to be mostly linearly up-chirped, and as in previous work [195], the trailing edge was seen to broaden with increasing current. An important insight also revealed in this paper is the demonstration of a nearly one-to-one correlation between the pulse duration and the linear chirp. For instance, for a fixed reverse bias and increasing current, both the pulse duration and the linear chirp are shown to increase and indeed follow the same trend, with pulse durations ranging between 1.6 and 7.5 ps, for corresponding spectral chirp ranging between -0.37 and -1.2 ps nm^{-1}. The authors were able to compress the pulse duration down to 700 fs using a free space dispersion compensator based on grating and telescope optics, achieving pulse durations quite close to their transform-limited values (between 530 and 620 fs) [170], and thus only a small amount of residual chirp remained uncompensated. The possibility of pulse compression down to 700 fs was also key to successfully demonstrate 160 Gb s^{-1} communication rates using QD mode-locked lasers [170], as a period of 6.3 ps corresponds to a rate of 160 GHz, which implies that the pulse durations should be significantly smaller than this period.

Frequency-resolved Mach–Zehnder gating was also utilized by Habruseva et al. [200] to investigate the phase structure of the generated pulses. This characterization technique is, in fact, based on the same principles as FROG, but as the name implies, the gating is done via a Mach–Zehnder modulator [201]. Unfortunately, this technique is still not powerful enough to characterize pulses with a waveform instability.

5.6.3
Toward Higher Power: Tapered Lasers

In order to boost the average and peak power achievable directly from the mode-locked laser diodes, postamplification can be implemented using compact semiconductor optical amplifiers (SOAs). Furthermore, the use of postamplification allows independent control of the pulse and power characteristics. The advantages of using QD-based amplifiers have already been highlighted in Sections 3.2 and 3.3, with some examples of pulse amplification from mode-locked laser diodes.

An alternative approach to achieve pulse generation at higher output powers is by designing edge-emitting lasers with a flared or tapered geometry. Such devices are well known for their potential as compact high-power semiconductor laser sources.

Tapered lasers include a straight section – which performs the function of spatial filter in the cavity – and a tapered section with increasing width, which provides high power. Therefore, tapered lasers are a good solution to provide single spatial mode beams with high power and good beam quality. In addition to being a spatial filter, the straight section can also be used as a saturable absorber if a reverse bias voltage is applied to it. Such capability can be harnessed for generating ultrashort pulses, as demonstrated by Mar et al. [202] – by using a quantum well ($In_{0.2}Ga_{0.8}As$) tapered laser, mode locking was demonstrated in an external cavity configuration, with the generation of 3.3 ps pulses with 2 W peak power, and also evidencing a significant four-fold improvement in pulse energy over the equivalent straight ridge-waveguide laser. Due to the fact that the beam mode size in the saturable absorber section is much smaller than in the gain section, the ratio of saturation fluences of the absorber and gain sections is greatly enhanced, as these fluences scale with the area of the mode, as previously described. This enhances the pulse formation mechanisms and allows better pulse shaping and shortening, while the higher gain saturation level further relieves the effect of pulse broadening in the gain section and at the same time assists in the generation of higher power pulses.

In 2006, Thompson et al. demonstrated the generation of pulses as short as 780 fs, by using a tapered waveguide configuration in a two-section InGaAs QD laser [161]. Further optimization of the saturable absorber length resulted in the demonstration of transform-limited ultrashort pulses with pulse duration of 360 fs, average power of 15.6 mW, and peak power of 2.25 W (at 12 °C) [150], although the mechanisms of the generation of transform-limited pulses from such a device without any additional dispersion compensation are not fully understood. Very recently, we have demonstrated the highest peak power (3.6 W) ever achieved from a monolithic passively mode-locked tapered quantum dot laser based on a gain-guided configuration, with a corresponding pulse duration of 3.2 ps [203]. Picosecond pulse generation with high average power of 195 mW was also demonstrated from the same laser, which is more than one order of magnitude higher than previous results. It is anticipated that postcompression of these pulses would result in relatively high values of peak power (>20 W). We genuinely believe that the tapered laser approach is a very promising route to achieve high-peak power pulses and that many related developments will take place in near future.

5.6.4
Toward Higher Repetition Rates

To achieve higher repetition rates in mode-locked lasers, it is necessary to decrease the cavity length. This poses a significant challenge to QD lasers because of their lower gain and the operation in short cavities may shift the emission to the excited-state band [204]. To avoid this problem, a higher number of QD layers should be deployed in the active region. Using this simple approach, the highest repetition rate directly generated from a passively mode-locked QD two-section laser was 80 GHz [81], when a 15-layer structure was used. An analysis of the limits of mode locking in the ground-state and its dependence on the several device parameters can be performed, such as the one found in Ref. [205], which illustrates how the minimum gain section length that still enables ground-sate lasing can be determined for a certain modal gain, absorber section length, facet reflectivity, and internal losses.

Another method to boost the repetition rate of mode-locked lasers is to use colliding pulse mode locking. This technique is similar to passive mode locking, but the saturable absorber region is placed at the precise center of the gain section. Two counterpropagating pulses from each outer gain section therefore meet in the saturable absorber region, bleaching it much more efficiently than if just one pulse was present. This process can also result in shorter and more stable pulses. Owing to the device geometry, mode locking is achieved at the second harmonic of the fundamental (round-trip) frequency, and the pulse repetition rate is doubled. A variation of colliding pulse mode locking is harmonic mode locking, where more than two pulses circulate in the cavity, the number being equal to the harmonic.

Colliding pulse mode locking was demonstrated for QD lasers in 2005 [198], resulting in a modest repetition rate of 20 GHz. Several regimes of harmonic mode locking were investigated by Xin et al. [190] using a multicontact device and placing the absorber at several positions in the cavity, resulting in harmonic mode locking with repetition rates between 7.2 and 50.7 GHz (typical pulse durations were around 6 ps). Using a shorter device with only 1 mm cavity length (corresponding to a fundamental repetition rate of 40 GHz), Rae et al. demonstrated harmonic mode locking with repetition rates of approximately 39, 79, 118, and 237 GHz, corresponding to the first, second, third, and sixth harmonics, respectively, with typical pulse durations between 1.4 and 1.8 ps [199]. Interestingly, it was found that the average power increased linearly with increasing repetition rate – which seems to imply that the energy of the pulses remains constant, regardless of the repetition rate [150, 199], suggesting that stable mode locking seems to occur for a certain specific energy value of the pulses.

5.6.5
External Cavity QD Mode-Locked Lasers

External cavity lasers offer some advantages when compared to monolithic devices. Wavelength and repetition rate tunability are easily obtained due to the flexibility offered by the external cavity configuration. This adaptability facilitates the introduction of additional optical elements to refine the performance of the device, such as dispersion compensation schemes. In addition, owing to the reduced pulse repetition rate the requirements of carrier recovery in gain or absorption can be relaxed, while the thermal stability can also be improved due to the lower duty cycles. There is also improvement in the phase noise and timing jitter of an external cavity laser because the active waveguide occupies only a fraction of the optical cavity. Furthermore, an external cavity setup enables to extend the repetition rate of these lasers to lower values, which may prove useful for many applications where a larger interpulse period is required – for example, in some biomedical applications. There is, however, a limit to the lowest possible repetition rate achievable with semiconductor gain material and this is related to the spontaneous recombination time – which for QD materials is of the order of magnitude ~ 1 ns. As such, if the cavity length is extended to the point where a train of pulses is generated with a temporal period much longer than this recombination time, the mode-locking regime ceases to be stable because the spontaneous emission becomes a more predominant effect, as observed by a number of research groups [150, 206].

Delfyett's group was the first to develop an external cavity QD laser that exhibited ultralow noise performance under active harmonic mode locking [207], which is described in more detail in the next section. A passively mode-locked grating-coupled external cavity laser operating at 2.4 GHz was later demonstrated [154], whereby the highly chirped pulses (typical durations ~15 ps) were subsequently compressed to ~970 fs using a grating pair scheme. Furthermore, the presence of the grating in the external cavity allowed the mode-locked operation to be switched from ground-state regime to excited state. The same type of external cavity was also used to demonstrate tunable operation across the ground-state and the excited-state bands [159]. The QD device utilized by the authors incorporated 10 identical InAs QD layers, and as such the tunability exploited only the intrinsic gain and absorption inhomogeneous broadening stemming from the dispersion in QD size (as opposed to the use of chirped QD layers, which are intentionally grown for broadband operation [29]). Continuous tuning across the excited state was achieved between 1170–1220 nm approximately, while for the ground state, tuning was achieved between 1265–1295 nm [159].

An alternative approach for an all QD-based mode-locked laser in an external cavity has also been demonstrated through the combination of a gain chip incorporating 9 QD layers and a SESAM formed by 35 layers of InAs quantum dots [206, 208]. By changing the cavity length, repetition rates between 350 MHz and 1.5 GHz were demonstrated and high average output powers of up to 27 mW were reported (at 860 MHz repetition rate).

A lower repetition rate of 310 MHz was reported by Xia *et al.* [209]. At 1 GHz repetition rate, the shortest pulse duration achieved was 2.15 ps (TBWP of 2.06). By using a 150 μm thick intracavity etalon, the authors succeeded in compressing the pulse down to 930 fs, which corresponded to its transform-limited value. Most notably, and as referred to previously in the context of harmonic mode locking, the pulse energy remained constant for the different repetition rates and pulse durations.

5.7
Noise Characteristics of QD Mode-Locked Lasers

5.7.1
Timing Jitter

Timing jitter associated with the generation of ultrashort pulses is a key aspect in applications such as clock generation/extraction and optical time division multiplexing (OTDM) – where short-pulse durations and low timing jitter are necessary for low bit error rates. On the other hand, applications such as electro-optic sampling or optical analogue-to-digital conversion require not only a good level of synchronization but also some degree of averaging, in order to improve the signal-to-noise-ratio –

in fact, the presence of random fluctuations in the pulse repetition period during the averaging process could significantly degrade the temporal resolution of these techniques. Understanding and minimizing timing jitter is thus an important aspect of mode-locked laser development.

Spontaneous fluctuations in the number of photons represent one of the major sources of timing jitter. Spontaneous emission imparts random fluctuations in the index of refraction, thereby affecting the round-trip time and thus the timing of the pulses. Furthermore, the amplitude fluctuations resulting from gain and photon density variations may also lead to an increase in the timing jitter as a result of the fluctuations in the saturation levels of gain and saturable absorption that consequently will impact on the round-trip time – this is the so-called amplitude-to-timing jitter conversion process [193]. External factors can also influence the timing jitter significantly, such as fluctuations in the bias current and the noise associated with the driving electronics – it has been widely observed that the noise of RF modulating sources is passed on directly to the mode-locked laser output, and as such a driving source should be very carefully selected [210]. In the case of external cavity mode-locked lasers, thermal and mechanical instabilities can also contribute to timing jitter. It is also well known that actively and hybridly mode-locked lasers typically exhibit superior noise performance in comparison with passively mode-locked laser, which lack an active or synchronous modulation – an excellent overview based on quantum well technology can be found in Ref. [193].

QD mode-locked lasers have been gaining status as low-noise sources because of the presence of discrete levels and low internal losses, enabling low threshold current densities and reduced values of the amplified spontaneous emission coupled to the wavelength of operation, compared to other higher dimensional materials [61] – all these factors concurring to noise reduction.

The first measurement of subpicosecond timing jitter was demonstrated for the first time in 2005 [178], for a monolithic passively mode-locked 5 GHz QD laser – and right then, the QD laser showed a remarkable improvement of one order of magnitude (0.91 ps, over the offset frequency range of 30 kHz–30 MHz) compared to a typical passively mode-locked quantum well laser (12.5 ps, integrated over an offset frequency range of 150 kHz–50 MHz) [177]. A timing jitter value of 390 fs has been reported for a 20 GHz laser (integration range from 20 kHz to 50 MHz) [211], which is nearly 30 times lower than that for the similar passively mode-locked QW laser previously cited [177]. In Figure 5.20, the noise performance of 10 GHz quantum well and quantum dot passively mode-locked lasers is compared [150]. Under optimum driving conditions, a timing jitter of 122 fs was measured for the QD laser (integration range 4–80 MHz) while for the QW laser the timing jitter was 1.3 ps (in the same frequency range).

For the moment, there seems to exist, however, a certain trade-off when designing a ridge laser configuration, if the aim is to achieve a low timing jitter and the shortest pulse duration. As pointed out by Thompson and coworkers, a higher saturable absorber to total length ratio is important to assist in the generation of shorter pulses – however, as this also increases the resulting threshold current and spontaneous emission level in the mode-locked laser, the phase noise and timing jitter increase as well [150, 211]. In any case, for a given laser configuration, it is still possible to find

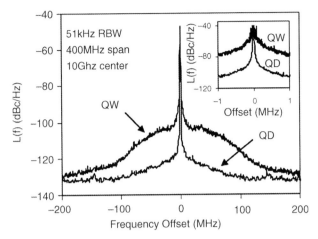

Figure 5.20 RF electrical spectrum showing the phase noise ($L(f)$) relative to the 10 GHz carrier for QW and QD mode-locked laser diodes. *Inset*: A span of 2 MHz and an RBW of 10 kHz. Reprinted from Ref. [150] © 2009 IEEE.

operating conditions that have relatively low values of timing jitter and pulse duration, as demonstrated in Ref. [212], where the authors have achieved 2 ps pulse durations for an integrated timing jitter of 500 fs (1–100 MHz), corresponding to 25 fs/cycle.

For pulse repetition rates of 40 GHz, low timing jitter values have also been reported for passively and hybridly mode-locked QD lasers, resulting in 219 fs (16–320 MHz) and 124 fs (20 kHz–32 MHz), respectively [213]. Although the passively mode-locked performance is better than what has been previously demonstrated with passively mode-locked quantum well lasers, their hybrid mode-locking performance is still inferior to the state-of-the-art hybridly mode-locked quantum well lasers at 40 GHz (73 fs, in the same integration range), which could be due to nonoptimal characteristics of the absorber, which could hinder its effective modulation [213].

Furthermore, it has also been demonstrated that not only does hybrid mode locking dramatically reduce the jitter but it also does so with a negligible impact on the pulse duration and shape, compared to passive mode locking in the same laser (it is important to note that such observations were derived from a FROG-type characterization system) [170].

Ultralow noise performance was achieved by Prof. Delfyett's group, using a QD laser in a ring cavity configuration with active harmonic mode locking to generate ultrashort pulses at a repetition rate of 12.8 GHz (via a Mach–Zender intensity modulator) [207]. In Figure 5.21, a schematic of the laser is depicted, and so are the residual phase noise and integrated timing jitter (the bump present between 10 kHz and 1 MHz results from incomplete cancellation of the synthesizer phase noise). The external cavity configuration, together with the electrical synchronization allowed by active mode locking, resulted in very low residual integrated timing jitter values of 7.5 fs (in the range between 1 Hz and 10 MHz) and was limited only by the driving synthesizer noise. This outstanding result represents the lowest timing jitter reported so far for actively mode-locked semiconductor lasers, thereby confirming the importance of QD materials in the pursuit of low-noise pulsed laser sources.

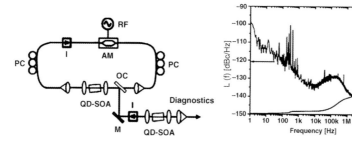

Figure 5.21 *Left*: Schematic of the unidirectional QD ring laser. QD-SOA: quantum dot semiconductor optical amplifier; I: isolator; AM: amplitude modulator; OC: 20% output coupler; PC: polarization controller; rf: rf driving signal; M: mirror. *Right*: Residual phase noise and integrated timing jitter. Reprinted with permission from Ref. [207]. Copyright 2006, American Institute of Physics.

More details are presented in Table 5.2, where the noise performance of some of the most representative low-noise QD lasers is summarized (underlying direct generation, that is, not incorporating external cavity feedback or injection locking mechanisms for noise reduction). The timing jitter is reported either in absolute terms within a certain frequency range or through the timing jitter/cycle, which has been pointed out as an alternative and perhaps more accurate method of quantifying timing jitter in passively mode-locked lasers, due to the cumulative nature of jitter in these lasers [214, 215]. Where the authors have reported the measured RF linewidth, its value is also indicated.

Finally, it is worth mentioning that from the deep involvement of the scientific community with the characterization of QD lasers, two new simplified techniques for characterizing the pulse timing jitter have emerged in the last few years, in the context of noise characterization of ultrafast QD laser diodes, based either on the analytical investigation of the measured RF linewidth [216] or on the analysis of the optical linewidth of the individual modes of the optical frequency comb generated by the mode-locked laser [218].

5.7.2
Pulse Repetition Rate Stability and Resilience to Optical Feedback

In some applications such as optical clocking and optical communications, it is crucial to deliver pulses with a predefined pulse repetition rate, within a certain frequency spread. Due to the uncertainties in the cavity length (∼1%), due to the cleaving process, the repetition rate can exhibit the same amount of deviation from the specification. In this respect, it is useful to have the possibility of fine-tuning the repetition rate via the bias conditions or temperature. However, care must be taken with this approach, as such conditions also affect greatly the optical power and the pulse duration of the pulses generated directly from the mode-locked laser. These stringent requirements led to the investigation of the variability in the pulse repetition rate with bias conditions [161, 205, 219]. For instance, Kuntz *et al.* have shown that the overall variation in the repetition rate in the whole mode-locking range

Table 5.2 Noise performance of mode-locked quantum dot lasers using different configurations.

ML regime	L_{abs}/L (%)	f_{rep} (GHz)	τ_p (ps)	Timing jitter (fs) (frequency band)	Timing jitter/cycle (fs)	RF linewidth (kHz)	References
Harmonic active (ring external cavity)	—	12.8	24–28	17.5 (1 Hz–100 MHz) 7.5 (1 Hz–10 MHz)	—	—	Choi et al. [207]
Passive	10	8.25	2/3	500/400 (1–100 MHz)	25/20	—	Todaro et al. [212]
Passive	—	17	2/1.3	—	20	—	Tourrenc et al. [214]
Passive	27	18	11	600 (2.5–500 MHz)	7.25	—	Thompson et al. [173]
Passive	30	16.1	2	1600 (20 kHz–80 MHz) 715 (100 kHz–1 GHz)	8	1.6	Kefelian et al. [216]
Passive	17	10	4.5	147 (4–80 MHz) 1000 (20 kHz–80 MHz)	—	0.5	Carpintero et al. [217]
Hybrid	17	10	4.5	197 (20 kHz–80 MHz)	—	0.04	Carpintero et al. [217]
Passive	10	40	2.5	219 (16–320 MHz) 843 (150 kHz–320 MHz)	—	—	Thompson et al. [213]
Hybrid	10	40	2.5	124 (20 kHz–320 MHz)	—	—	Thompson et al. [213]
Passive	13	5.2	5.7	910 (30 kHz–30 MHz)	—	—	Zhang et al. [178]
Passive	15	10	—	122 (4–80 MHz)	—	1.0	Thompson et al. [150]

L_{abs}/L: ratio between absorber length and total length; f_{rep}: repetition frequency; τ_p: pulse duration. Data not available, or not applicable, is represented with a dash.
Note: except stated otherwise, the entries refer to monolithic mode-locked lasers.

could be as high as 100 MHz [205]. Furthermore, it was found to decrease with increasing reverse bias voltage with a rate of 14 MHz V^{-1} and decrease with increasing current at a rate of 50 kHz A^{-1} cm^2 [205]. Similar trends were found in Ref. [161].

If the objective, on the other hand, is to stabilize the repetition rate, frequency stabilization techniques using weak optical feedback can be used. This approach has been demonstrated in a passively mode-locked quantum dot laser, through the use of a passive auxiliary optical fiber cavity [219]. In this paper, a 17-fold reduction in the shift induced by current was achieved (from -39.5 to -2.3 kHz mA^{-1}) through weak cavity coupling (-24 dB), while simultaneously the timing jitter was decreased from 1.4 to 0.9 ps. A pronounced RF linewidth reduction (from 8 KHz to 350 Hz) was also achieved by Lin *et al.*, by utilizing resonant feedback from an external fiber cavity with a length L_{ext} that was a multiple of that of the solitary laser $L_{ext}/L \sim n = 1, 2, 3, \ldots$ [220], also confirming previous theoretical predictions on the effectiveness of this approach [221]. As a consequence, the integrated rms timing jitter was also reduced from 10 to 2 ps approximately (calculated in the frequency range from 30 kHz to 30 MHz), while the 350 Hz RF linewidth was the narrowest achieved for a QD passively mode-locked laser. On the other hand, nonresonant feedback can be deleterious for mode locking as demonstrated by Grillot *et al.* [222], albeit at a relatively high feedback threshold value of -24 dB. Severe instabilities were observed, while the noise level increased in the RF electrical signal and the region of stable mode locking was reduced considerably. The authors have also shown how these effects are stronger for higher values of reverse bias applied to the saturable absorber, attributed to the lower intracavity photon density, which enhanced the sensitivity of the laser to the external feedback. In any case, the relatively high value of feedback threshold for the onset of instabilities leads to the conclusion that the operation of QD mode-locked lasers without optical isolator could be a feasible possibility in many applications. This is particularly important for intra-/interchip optical interconnects where the lasers have to be closely integrated with other devices, and where optical isolation is not feasible.

For lasers operating under hybrid or active mode-locking regimes, the locking region is also an important aspect to take into account. Schmeckebier *et al.* have demonstrated that the locking range could be significantly expanded from the previously reported 20–30 [205] to 100 MHz [170], by compressing the pulses generated directly by the mode-locked laser. Indeed, after pulse compression, the pulse duration was independent of the bias conditions, which allowed the latter to be changed in order to tune the repetition rate without altering the pulse duration. Using this approach, the authors succeeded in significantly broadening the hybrid mode-locking range, achieving an overall locking range of 101 MHz.

5.7.3
Performance Under Optical Injection

Optical injection has long been known as a valuable technique for noise reduction and synchronization between a master and a slave laser, being quite useful for the implementation of optical clock recovery. This technique relies on the injection of a certain amount of variable power from a master laser into a slave laser, within the same optical band.

A disruptive functionality has been demonstrated by Kim and Delfyett [223], who have experimentally shown for the first time the possibility to perform interband optical injection locking – that is, between a master laser mode locked via the GS transition and a slave laser mode locked via the ES and vice versa. The master laser was hybridly mode locked, while the slave laser was passively mode locked, and upon locking, a significant reduction in phase noise can be observed and the enhanced stability of the slave laser allows the observation of the optical pulse train in a digital sampling scope. The authors have observed that the locking dynamics of GS → ES and ES → GS were quite distinct. It was shown that the locking via ES → GS results in an asymmetric locking bandwidth of 500 kHz around the repetition rate of the slave laser (from −170 to + 330 kHz, at ∼1 mW injected optical power). On the other hand, in the GS → ES situation, the slave laser does not lock to the same frequency of the master laser. In this case, optical locking occurred only for repetition rates higher than the nominal slave laser repetition rate. Furthermore, a much wider locking bandwidth was observed in this scenario (6.7 MHz, at ∼1 mW injected optical power).

This is the first time that optical injection locking is achieved without the master and slave lasers sharing the same spectral band, which opens up a wealth of applications, whereby, for instance, a master laser could control a variety of slave lasers without any overlap of their spectral bandwidth [223].

Habruseva et al. have demonstrated great improvements in the mode-locking performance of a QD passively mode-locked laser through a dual-mode optical injection scheme [200]. A tunable narrow linewidth (100 kHz) master laser was used and its output modulated at a frequency half the pulse repetition rate of the slave laser, resulting in two coherent sidebands that were separated by 10 GHz.

Such output was injected into the slave laser, resulting in dramatic reduction in the spectral bandwidth (10–15 times less), without, however, impacting the pulse duration. As a result, the time–bandwidth product was significantly reduced, pulses close to Fourier transform were generated. A striking result obtained with this technique was that by tuning the master laser's wavelength, the slave laser could also be tuned, as much as 8 nm, as shown in Figure 5.22. Furthermore, phase noise was greatly reduced with injection locking, with a consequent reduction in integrated timing jitter from 11–21 ps to 1 ps–300 fs (in the range of 20 kHz–80 MHz). Typical results are shown in Figure 5.23. The linewidth of the several modes in the slave laser were also measured and, importantly, the individual modes retained the 100 kHz linewidth of the master laser. With optimization of the injection parameters and laser operating parameters, timing jitter as low as 290 fs was obtained for a pulse duration of around 4.8 ps – results which are comparable to the best results achieved with hybridly mode-locked QD lasers [217]. In fact, the authors point out that this technique could indeed be considered as a form of hybrid mode locking, but with the added functionalities of tunability and reduction in spectral bandwidth.

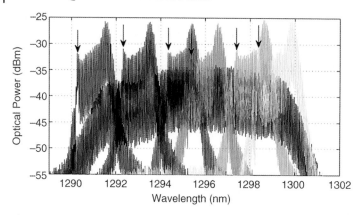

Figure 5.22 Optical spectra of the free-running (broader spectrum in black) and injection-locked QD MLL with a 17% absorber section for a number of master wavelengths (shown with arrows). Gain: 160 mA; absorber bias: −4 V. Reprinted from Ref [200] © 2010 IEEE.

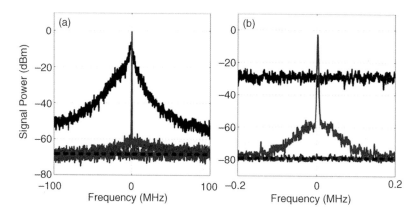

Figure 5.23 RF signals of the slave laser with a 12% absorbance section for the free-running case (black) and with dual injection (gray). Gain: 115 mA; absorber bias: −6 V. The lowest line and the black dotted line show the noise signal and noise level, respectively. (a) 200 MHz span, RBW = 100 kHz, VBW = 10 kHz; (b) 400 kHz span, RBW = 1 kHz, VBW = 100 Hz. Reprinted from Ref. [200] © 2010 IEEE.

5.8
Performance of QD Mode-Locked Lasers at Elevated Temperature

5.8.1
Stable Mode Locking at Elevated Temperature

Due to their delta function-like density of states, QDs offer great potential for designing temperature-resilient devices. If their high-speed performance is also proven to be resilient to temperature, QD lasers can become the next-generation sources for ultrafast optical telecommunications and data communications because the constraint of using thermoelectric coolers can be avoided, thus decreasing cost and complexity. In this context, we have demonstrated stable passive mode-locked operation of a two-section QD laser over an extended temperature range, at relatively high output average powers [162]. We observed very stable mode-locked operation from 20 to 70 °C, with the corresponding RF spectra exhibiting signal-to-noise ratios well over 20 dB and a −3 dB linewidth smaller than 80 kHz. The mode-locking regime became less stable only at 80 °C, where the −3 dB linewidth was 700 kHz and the signal-to-noise ratio was 15 dB. No self-pulsations were observed in the investigated temperature range. At higher temperatures, the available tuning range decreased slightly due to a shift in lasing emission to ES, with increasing reverse bias. We also observed that to provide stable mode-locking operation with increasing operating temperature, the reverse bias on the absorber section had to be reduced accordingly.

While the repetition rate was not significantly altered with increasing temperature, there was a large thermal redshift in wavelength. This is an obvious disadvantage; however, it should be noted that this structure was not optimized for high temperature operation. The results presented suggest that by carefully tailoring the semiconductor structure [224], it would be possible to obtain a mode-locked QD laser with an extremely low shift of emission wavelength with temperature.

5.8.2
Pulse Duration Trends at Higher Temperatures

In addition, to meet the requirements for high-speed communications, it is important to investigate the temperature dependence of the pulse duration. For instance, in communication systems with transmission rates of $40\,\mathrm{Gb\,s^{-1}}$ or more, the temporal interval between pulses is less than 25 ps and so the duration of the optical pulses should be well below this value at any operating temperature. We have shown that, perhaps counterintuitively, the pulse duration and the spectral width significantly decrease as the temperature is increased up to 70 °C [163]. The combination of all these effects resulted in a seven-fold decrease in the time–bandwidth product (the pulses were still highly chirped due to the strong self-phase modulation and dispersion effects in the semiconductor material).

To account for the decrease in pulse duration with temperature, a model for mode locking in quantum dot lasers was used, developed by Viktorov et al. [49]. The performed simulations reproduced successfully this trend, taking into account the temperature dependence of carrier capture and escape from the quantum dots where both increase with temperature. It was found that the role of the temperature-dependent capture and escape processes in the gain section was relatively minor in that the mode-locked pulse durations are determined principally by the capture and escape rates in the absorber section [143]. These processes in the absorber were responsible for a decrease in absorber recovery time with increasing temperature, leading to a decrease in the pulse durations. This fact was later verified using ultrafast spectroscopy to probe the absorber recovery time as a function of temperature [164, 225].

5.8.3
The Use of p-Doping in QD Mode-Locked Lasers

As suggested in Ref. [162], the performance of QD mode-locked lasers at higher temperatures could still be improved by preventing early ground-state gain saturation at higher operating temperature, by using a p-type doped laser structure [226], and/or with more QD layers [227]. In fact, p-type doping has been shown to facilitate the fabrication of CW laser devices exhibiting a higher T_0 (up to 650 K) [228], which can be further improved by increasing the number of QD layers [229]. Furthermore, p-doping has the potential of enabling an accelerated gain recovery, through a prefilling of the hole states, as previously demonstrated in high-speed laser modulation experiments [230]. This approach was subsequently taken, with the fabrication of p-doped QD two-section lasers, incorporating 15 QD layers [169]. Passive mode-locked operation of these devices was characterized with increasing temperature up to 60 °C. Due to higher available gain and the existence of p-doping, the mode-locking region involving the ground state was not limited by a transition to excited-state spectral band at higher temperatures. With increasing temperature, it was found that similar to [162], the region of stable mode locking was achieved for lower values of reverse bias applied to the saturable absorber. Furthermore, the characteristic temperature T_0 was found to be infinity for temperatures ranging from 10 to 50 °C. The possibilities demonstrated so far pave the way for the ultimate deployment of ultrastable, uncooled mode-locked diode lasers incorporating quantum dot materials.

5.9
Exploiting Different Transitions for Pulse Generation

5.9.1
Mode Locking via Ground and Excited States

It has been observed that laser emission in QD lasers can access the transitions in ground state, excited state, or both [204], as schematically represented in Figure 5.24. Subpicosecond gain recovery has been demonstrated for both GS and ES transitions in electrically pumped QD amplifiers [231]. In this reported work, the linewidth enhancement factor was shown to decrease significantly for wavelengths below the GS transition, even becoming negative at ES thereby implying a potential for chirp-free operation for the range of wavelengths involved. We have demonstrated an optically gain-switched QD laser, where pulses were generated from both GS and ES, and where the ES pulses were shorter than those generated by GS alone [232]. The potential for shorter and chirp-free pulses from ES transitions motivated us to investigate the mode-locked operation of QD lasers in this band. We demonstrated, for the first time, passive mode locking via GS (1260 nm) or ES (1190 nm) in a quantum dot laser, at repetition frequencies of 21 and 20.5 GHz, respectively [233]. The switch between these two states in the

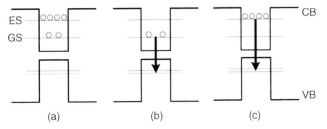

Figure 5.24 Schematic of the energy levels in a QD material (a) and radiative transitions via GS, ground state (b), and ES, excited state (c). CB: conduction band; VB: valence band.

mode-locking regime was easily achieved by changing the electrical biasing conditions, thus providing full control of the operating spectral band. It is important to stress that the average power in both operating modes was relatively high and exceeded 25 mW. In the range of bias conditions explored in this study, the shortest pulse duration measured for ES transitions was ~7 ps, where the spectral bandwidth was 5.5 nm, at an output power of 23 mW [153]. These pulse durations are similar to those generated by GS mode locking at the same power level. Although the pulses generated from both GS and ES mode locking are still far from the transform limit (with a time bandwidth product exceeding 7), they could be reduced by using external compression techniques. Finally, it is important to bear in mind that the monolithic laser investigated was not designed for ES operation, so mode locking in this band was achieved only at the expense of injecting high drive currents, which has the well-known effect of significantly broadening the pulses.

Further work in this area was done by Prof. Delfyett's group at CREOL, Florida. Instead of using a monolithic device, this group exploited the added flexibility of external cavity configurations to develop a laser where the generation at the GS and ES bands could be achieved using intracavity optical elements [154], instead of relying solely on the increase in bias conditions [153]. This configuration allowed an easier and selective feedback and lower operation current, thus preventing the possibilities of overheating the laser device and, more importantly, avoiding the temporal broadening of pulses due to high values of bias current. Using an external compression setup, it was verified that both pulses generated by GS and ES were up-chirped, and no significant differences were observed in terms of pulse duration and time–bandwidth product. After compression, pulse durations of 1.2 ps and 970 fs were achieved for ES (1193 nm) and GS (1275 nm), respectively. A similar setup was used by the same group to tune the operating wavelength across the GS spectral band (30 nm tunability) and across the ES spectral band (50 nm tunability), as has been mentioned in Section 5.5.2 [159].

More recently, we have achieved a dual-wavelength passive mode-locking regime where pulses are generated simultaneously from both ES ($\lambda = 1180$ nm) and GS ($\lambda = 1263$ nm), in a two-section GaAs-based QD laser [145, 156]. This is the widest spectral separation (83 nm) ever observed in a dual-wavelength mode-locked non-vibronic laser [234].

The QD structure used in the laser device was grown by molecular beam epitaxy on a GaAs substrate. The active region incorporated five layers of InAs QDs. A two-section QD laser diode was fabricated with a ridge waveguide 6 μm wide, a total length of 2 mm, while the saturable absorber was 300 μm long. The front and back facets were antireflection (∼3%) and high-reflection (∼95%) coated, respectively. The laser was operated at room temperature (20 °C), with its temperature thermoelectrically controlled via a Peltier cooler.

The dual-wavelength mode-locking regime was achieved under a range of bias conditions, which simultaneously satisfied the conditions for achieving mode-locking via GS and ES – for current levels in the gain section between 330 and 430 mA, and values of reverse bias between 6 and 10 V in the saturable absorber region. The excited-state levels have higher degeneracy and consequently higher saturated gain than the GS. This means that a transition from the GS to the ES can be achieved by increasing loss, which in this case can be manipulated through the increase in reverse bias applied to the saturable absorber.

The spectral separation between the two modes results in different repetition rates, due to the dispersive nature of the laser semiconductor material, which induces different cavity round-trip times for the propagation of the two modes. As such, the repetition rates of the generated pulses were 19.7 and 20 GHz for the ES ($\lambda = 1180$ nm) and GS ($\lambda = 1263$ nm), respectively (Figure 5.25).

It is our opinion that exploitation of the ES transitions – a unique feature of QD lasers – can lead to a new generation of high-speed sources, where mode locking involves electrically switchable GS or ES transitions that are spectrally distinct. It is important to add that the development of dual- and multiple wavelength ultrafast lasers is a research area that aims to address a number of important applications such as time domain spectroscopy, wavelength division multiplexing, and nonlinear optical frequency conversion. In this context, the compactness, lower cost, and direct electrical pumping associated with semiconductor lasers form a set of attractive features for reducing the footprint and complexity of the aforementioned

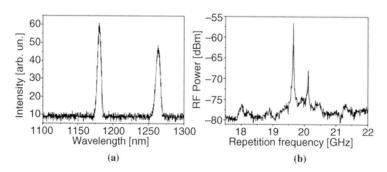

Figure 5.25 (a) Optical spectrum and (b) RF spectrum characteristic of the dual-wavelength mode-locked regime. In the RF spectrum, the observed power difference between GS and ES is a result of the different optical power levels and pulse duration for each of the pulse trains. Reproduced with permission from Ref. [145].

applications, with the potential to also open up new avenues in ultrafast optical processing and optical interconnects.

5.9.2
The Excited-State Transition as Tool for Novel Mode-Locking Regimes

QD lasers exhibit, under certain conditions, a low linewidth enhancement factor. However, the linewidth enhancement factor significantly increases as the operating current is increased, which in turn increases the frequency chirp, duration, and noise of the pulses generated by mode-locked versions of these lasers. This behavior implies that, in general, the shortest pulses generated by these mode-locked lasers are achieved while very close to threshold and using relatively high values of reverse bias [161, 190], with only a few exceptions to this trend, where subpicosecond pulses were generated at high driving currents [187, 235]. This trend limits the output power and constrains the exploitation of their typically high differential efficiency in the generation of high-power ultrashort pulses. We have exploited the CW laser emission across the excited-state transition as a novel means of circumventing this problem [158]. The coexistence of excited-state lasing with mode locking that involves the ground-state transitions enabled the generation of ultrashort and low-noise pulses under the conditions of higher levels of current injection and thus at higher optical power. This new mode-locking regime has been achieved in a monolithic two-section QD laser by significantly increasing the DC injection current supplied to the gain section and/or to the absorber section as a reverse bias. Before the onset of laser emission in the ES band, the pulse durations increased with current, as expected from greater linewidth enhancement factors [79]. However, with the appearance of simultaneous CW laser emission in the ES, the trend reverses, with pulse durations decreasing sharply from 11 ps to 6 ps (Figure 5.26). It is important to stress that the average power levels obtainable in this regime are up to three times larger than those obtained in the lower current regime, when comparing pulses of identical duration. In addition, in this new regime of operation, the noise associated with unstable mode locking involving the ground state is drastically reduced and a very sharp RF spectrum is observed with simultaneous CW lasing from excited-state populations.

A possible hypothesis for this new mode-locking regime could be based on the effect of cooling of the carriers that lie in the higher energy levels. With simultaneous laser emission from the ES, a strong depletion of the higher populated levels occurs. This carrier depletion should have an impact on the refractive index and gain of the laser, in such a manner that the linewidth enhancement factor (LEF) significantly decreases because the carrier pileup on the ES is now suppressed. It has been shown that the LEF associated with GS can even become negative with simultaneous ES emission [182]. A lower LEF along the overall cavity will result in a decrease in self-phase modulation and in the up-chirp imparted to the pulses, which together with the positive dispersion of the laser material will result in a decrease in the pulse duration. Further investigations of this regime are in progress at the time of writing, as this new regime may lead to the exploitation of the full potential of quantum dot lasers as compact sources of ultrashort, low-noise, and high-power pulses.

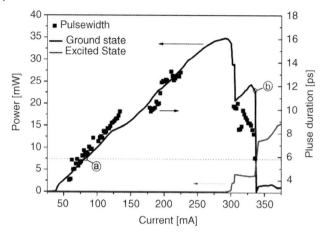

Figure 5.26 Light–current characteristics of a quantum dot mode-locked laser, at constant reverse bias, together with the measured pulse durations, while operating in a stable mode-locking regime (squares). The highlighted operating points (a) and (b) correspond to pulses of identical duration, generated in the absence or presence of simultaneous excited-state emission, respectively.

5.10
Summary and Outlook

5.10.1
QD Mode-Locked Laser Diodes: New Functionalities

Monolithic passively mode-locked quantum dot lasers can at present surpass the performance of similar quantum well lasers in terms of pulse duration [161, 187]. There are other particular features where QD lasers have already been shown to have a superior performance, notably in the case of pulse timing jitter where record low values have been reported [207, 213]. The appeal of QD lasers also resides in the novel functionalities that are distinctive of QDs. These are the exploitation of an excited-state level as a means to achieve novel mode-locking regimes; the temperature resilience offered by the quantized density of states; lower threshold and higher output power levels; and access to the enlarged spectral bandwidths associated with the inhomogeneously broadened gain features. These characteristics are not only useful from an operational point of view but also provide a more comprehensive understanding of the underlying physical mechanisms of mode locking in QD lasers. One of the views shared among many researchers in the field is that the value of the inhomogeneously broadened gain in mode-locked operation is perhaps more limited than expected. Indeed, inhomogeneous broadening seems to contribute to the higher values of frequency chirp and time–bandwidth product, instead of facilitating the generation of shorter pulses. Such behavior raises the hypothesis that the laser cavity modes that are distributed across different parts of the inhomogeneously broadened optical spectrum may fail to lock coherently. It is possible that phase locking is easier to achieve within the same population of similar quantum dots as would be expected in a part of the spectrum that is homogeneously broadened. Nevertheless, the presence of a broadband gain can still be a very useful feature, for spectral tunability, for example.

The pulse duration results from the balance between several broadening and compression effects. Self-phase modulation is the effect that contributes most to pulse broadening, while the fast recovery of the absorption in the saturable absorber is the prime shaping mechanism of the trailing edge of the pulse, resulting in shorter pulse durations with decreasing absorber recovery time. Self-phase modulation results from the interaction of the pulse with the gain and the resulting large changes in the nonlinear refractive index. This leads to a phase change across the pulse, owing to the coupling between gain and index, which is regulated via the linewidth enhancement factor. The resulting up-chirp combined with the positive dispersion of the gain material leads to substantial pulse broadening, which increases with optical power. The effect of self-phase modulation remains one of the main hurdles to the generation of shorter pulses from QD lasers, as the linewidth enhancement factor in these lasers can be very significant. Another major conclusion drawn from this work is that the role of the excited state cannot be neglected in the generation of ultrashort pulses. The population of carriers at the levels above the ground state leads to an increase in the linewidth enhancement factor [182] with a consequent negative effect on the duration of the generated pulses (via ground state). Perhaps a route to mitigate this effect could be implemented by using simultaneous laser emission involving the excited state, resulting in a strong suppression of the carrier pileup on the higher populated levels, and possible consequent decrease in the linewidth enhancement factor across the ground state [182]. An alternative technique to minimize the effect of the carrier population on the higher levels is to consider tunnel injection structures, as described in the next section.

5.10.2
Future Directions

Although there have been many advances in the control of the growth of QD laser having ultralow threshold current and temperature resilience, it is not yet understood what is the most advantageous QD structure layout to be used in the regime of mode locking. In particular, it is not clear what is the optimum level of inhomogeneous broadening that results in shorter and higher peak power pulses. Therefore, it is relevant to investigate if and how the inhomogeneously broadened spectral modes are engaged coherently in the generation of ultrashort pulses and how that effect could be used to improve the performance of the lasers toward subpicosecond pulse durations.

Further exploitation of novel QD materials based on p-doped and tunnel injection structures could also bring advantages in terms of minimizing any deleterious effect of self-phase modulation on mode-locked lasers. A theoretical and experimental comparison of undoped and p-doped lasers has shown how this technique can improve the linewidth enhancement factor. By tuning the level of doping, lasers can exhibit zero and even negative linewidth enhancement factor at low current densities [237]. Tunnel injection QD structures can also be of great interest for use in mode-locked lasers, as the injection of cold carriers may bring many benefits to the operation of mode-locked lasers. The linewidth enhancement factor has recently been calculated and has been demonstrated to be much less than that reported for other lasers [238]. Selective excitation of population in these lasers has been demonstrated, which could lead to a mitigation of the inhomogeneous broadening effects and contribute to a narrower spectrum and the production of transform-limited pulses [239].

The excited-state spectral band can also be exploited in tunable lasers using QD materials where the inhomogeneous broadening is controlled so as to maximize the overlap between ground and excited states. While in CW operation, QD lasers have been demonstrated with tunability ranges up to 202 nm with high average power (>480 mW) [240], by exploiting the gain available from the ground state and excited states. A tunable mode-locked QD laser would unlock the potential to generate ultrashort pulses from a very compact and efficient laser system having tunability over a significant spectral region.

6
Ultrashort Pulse Solid State Lasers Based on Quantum Dot Saturable Absorbers

6.1
A Brief Historical Overview of Ultrashort-Pulse Generation

The demonstration of mode locking in a range of solid-state lasers in the mid-1960s, where pulses of a few tens of picoseconds were produced, opened up the era of a new class of laser systems that are now described as ultrashort-pulse or ultrafast lasers. In the early pioneering research, the mode-locking technique was applied successfully for the generation of picosecond pulses from ruby [241], Nd:glass [242], or Nd:YAG [243] lasers that were optically pumped by flashlamps or arclamps. Mode locking was achieved by a variety of means such as the external modulation of the laser cavity frequency [241] or losses [243] (active mode locking) or by using a fast dye-based saturable absorber (passive mode locking) [242]. Impressive progress in the development of ultrashort-pulse laser technology has been achieved since then as a result of access to novel gain media, pump sources, and mode-locking devices/ techniques. Indeed, the development of high-power semiconductor lasers initiated the research work toward the development of all-solid-state [244] ultrafast laser systems toward the end of the 1980s, following a period of some 15 years when ultrafast dye lasers were the sources of choice in laser research laboratories [245, 246]. The first diode-pumped, actively mode-locked solid-state lasers such as Nd:YAG and Nd:YLF produced pulses as short as 8 ps with average powers exceeding 100 mW [247, 248]. The discovery of the Ti:sapphire (Ti:Al$_2$O$_3$) laser in 1986 [249] enabled another breakthrough in the ultrashort-pulse laser technology because with this broadband solid-state gain medium, the direct generation of sub-10fs could be envisaged. The initial attempts to generate ultrashort pulses from Ti:sapphire laser utilized mode-locking techniques such as coupled nonlinear external cavities [250], dye saturable absorbers [251], and near-infrared color filters [252]. Picosecond and femtosecond pulses were produced directly from some laser configurations, and durations as short as 200 fs were achieved using external dispersion compensation technique [250]. Further progress in discovery of novel gain media resulted in the development of a range of solid-state lasers based on Cr^{3+}-doped colquiriites (Cr^{3+}:LiCAF [253], Cr^{3+}:LiSAF [254], Cr^{4+}-doped Mg$_2$SiO$_4$ (Cr:forsterite) [255],

Ultrafast Lasers Based on Quantum Dot Structures: Physics and Devices.
Edik U. Rafailov, Maria Ana Cataluna, and Eugene A. Avrutin
Copyright © 2011 WILEY-VCH Verlag GmbH & Co. KGaA, Weinheim
ISBN: 978-3-527-40928-0

and $Y_3Al_5O_{12}$ (Cr:YAG) [256] gain media that are compatible with direct laser diode pumping and are able to produce pulses with durations <100 fs in the near-infrared spectral region [257–264]. More recently, highly efficient, compact, and low-cost femtosecond oscillators operating around 1 μm spectral region have become available due to the introduction of a range of Yb^{3+}-doped crystals [265, 266] and, simultaneously, due to rapid progress in development of InGaAs-based laser diodes operating in 0.9–1 μm region. The simple energy-level scheme of trivalent ytterbium leads to the absence of such unwanted processes as excited-state (ES) absorption, upconversion, cross-relaxation, and concentration quenching, and the small Stokes shift ($\sim 600\,cm^{-1}$) between absorption and emission reduces the thermal loading and increases the inherent laser efficiency. Ytterbium-based femtosecond lasers are able to produce pulses as short as 100 fs [267–269], deliver peak powers as high as 1.9 MW directly from the oscillator [115], and operate with optical efficiencies up to 50% [270].

These recent achievements in the development of ultrashort-pulse lasers would not have been realized without significant breakthroughs in mode-locking techniques that started in 1989. In the more recent history of ultrashort-pulse lasers, the Kerr-lens mode-locking (KLM) technique [271] and the development of novel class of semiconductor-based absorber structures (semiconductor saturable absorber mirror (SESAM) [96, 272] or saturable Bragg reflector (SBR) [273]) represent significant and major milestones. The SESAMs generally comprise one or more semiconductor saturable absorber layers (quantum wells (QWs) [274]) monolithically integrated into a mirror structure that can be employed as an additional intracavity laser element for initialization and stabilization of the mode-locking process. The performance of ultrafast lasers based on these techniques has been improved by some orders of magnitude [88, 275] and they have proved their worth in a wide range of research and industrial applications from real-time monitoring of chemical reaction dynamics [276] to applications in communications [277], metrology [278], medicine [279–280], and material processing [281].

6.2
Macroscopic Parameters of Saturable Absorbers

Given the nonresonant nature of the optical Kerr effect, the KLM technique can support the direct generation of pulses having durations around a few femtoseconds from a broadband laser [114]. This technique therefore lends itself to a wide range of gain media and the reduced intracavity losses through the absence of a physical saturable absorber can lead to a substantial enhancement in the overall laser efficiency [270]. Despite these advantages, KLM lasers have the drawbacks that they are generally not self-starting and critical laser cavity alignment is required to ensure stable mode locking. In contrast, femtosecond lasers that incorporate physical saturable absorbers (SESAMs in particular) permit greater operational tolerances and therefore offer design options for enhanced practicality. These facilitate self-starting mode-locking regimes and support the generation of ultrashort pulses from a variety of solid-state lasers that operate over a relatively wide spectral range (typically ~0.8–2.5 µm) [258, 282], deliver pulse energies of up to 2 µJ directly, and produce pulses over an attractive range of repetition frequencies from a few megahertz [283] to hundreds of gigahertz [284].

The key macroparameters of any saturable absorber, in particular a SESAM device, that determine the mode-locking laser dynamics are *saturation fluence, modulation depth, nonsaturable losses, recovery time,* and *operational spectral range*. The *saturation fluence* can be defined as a required radiation fluence on the absorber that reduces its absorption to $1/e$ of its initial value (or equivalently to increase reflectivity in case of a SESAM). Lower saturation fluence is desirable for the suppression of Q-switching instabilities [285], for low-threshold mode locking, and for stable mode locking at high pulse repetition frequencies in instances when the intracavity pulse energy is small [284]. In the case of typical SESAM devices, this value depends not only on the semiconductor material itself but also on the structure design [286]. For instance, resonant SESAM structures are characterized by lower saturation fluences ($\sim 10\,\mu J\,cm^{-2}$) compared to the antiresonant counterparts ($\sim 100\,\mu J\,cm^{-2}$). The *modulation depth* is the maximum change in transmission/reflectivity of an absorber structure. This value is particularly important for pulse duration optimization and for maintaining self-starting, stable mode locking and should be chosen appropriately for the desired gain media and laser operational parameters. For instance, large modulation depths can lead to Q-switched mode locking [285], whereas an insufficient modulation depth will not enable a self-starting mode-locked operation [287]. The typical values for the modulation depth of the SESAMs used in modern solid-state lasers are within the few percent range. The residual absorption, or *nonsaturable losses*, when the absorber is fully saturated, does not affect ultrashort-pulse formation and generation directly, but it can reduce significantly the efficiency of mode-locked laser system. Moreover, significant nonsaturable losses in high-power laser systems can lead to thermal excursions in the absorbers and subsequent breakdown effects. The absorber *recovery time* is the time period subsequent to excitation that is required for the absorber to reach the $1/e$ level of its bleached transmission. The saturable absorber can be described as fast when its recovery time is comparable to mode-locked pulse duration or slow when bleaching relaxation exceeds considerably the output pulse duration [288]. Mode locking can be maintained more easily with slow saturable absorbers because of the reduced saturation intensities that serve to facilitate a self-starting operation. However, short recovery times are required for pulse narrowing in the absence of pulse shaping mechanisms as in soliton mode locking [288]. Although much shorter pulses (compared to the absorber recovery time) can be generated from solid-state lasers that incorporate slow saturable absorbers through the soliton formation mechanism [289, 290], absorbers having a fast component ($\sim$1 ps) are more efficient for pulse shaping in the subpicosecond time domain. In bulk or quantum well semiconductor structures, carrier recombination times are usually too long (few hundreds of picoseconds) for optimized passive mode locking and to increase the depletion speed of the bands, some structural defects are introduced into the semiconductor material as traps for the excited carriers that reduce the recombination time. Incorporation of defects can be achieved by producing the semiconductor structures using low-temperature (200–400 °C) molecular beam epitaxial growth technique [291] or by applying post-growth ion implantation procedures [119]. Typical recombination times for these types of semiconductors are in range of a few picoseconds. However, such absorbers suffer from high nonsaturable losses that can reduce significantly the efficiency of ultrashort-pulse laser when it is mode locked. The *operational spectral range* of a SESAM structure is dictated by the bandgap energy of semiconductor absorber to facilitate saturable absorption in a given laser wavelength range and the reflectivity characteristic of the Bragg mirror. To date, most of the SESAM devices have been fabricated using III–V compound semiconductors. Notably, AlAs/Al$_x$Ga$_{1-x}$As-based SESAMs that have a GaAs quantum well absorber have been used for femtosecond pulse generation from both Ti:sapphire [292] and Cr:LiSAF [258] lasers that operate efficiently around 800–900 nm. For passive mode locking of solid-state lasers that operate around the 1 μm (Nd^{3+}- and Yb^{3+}-doped materials) [272, 293, 294], 1.3 μm (Cr:forsterite) [262, 295], and 1.5 μm (Cr:YAG and Er,Yb:glass) [296–298] spectral regions, In$_x$Ga$_{1-x}$As quantum well SESAMs with GaAs/AlAs-based Bragg reflectors are selected. However, it should be noted that higher indium concentrations that are required in In$_x$Ga$_{1-x}$As for the longer wavelength regimes ($x \sim 0.25$, 0.38, and 0.53 for 1, 1.3, and 1.5 μm operation, respectively) can lead to a lattice constant differential that is beyond the critical level for a coherently strained InGaAs:AlAs/GaAs Bragg mirror structure. Consequently, strain relaxation occurs with the inevitability of high nonsaturable losses and low damage thresholds. To solve this problem, SESAM structures have been designed where the InGaAs quantum well absorber has been grown on a buffer layer of InAlAs [262] or InP [296]. In addition, the alternative scheme of growing an InGaAsP or AlGaAsSb absorber on a lattice-matched InP-based mirror structure has been realized for passive mode locking in the 1.3–1.5 μm spectral region, but this offers a restricted Bragg mirror reflectivity bandwidth (<100 nm), due to the small refractive index contrast achievable with materials that are lattice-matched to InP [299, 300]. More recently, novel SESAM devices based on a Ga$_{1-x}$In$_x$N$_y$As$_{1-y}$ absorber layer have been the subject of considerable research interest [301, 302]. The ability to vary the nitrogen and indium concentrations permits the bandgap to be tailored for the 1.2–1.6 μm spectral region, through strong band bowing, while controlling the lattice constant to permit pseudomorphic growth on a GaAs substrate. Efficient picosecond [303–305] and femtosecond [306, 307] lasers have been demonstrated at around 1.3 and 1.5 μm using such low-loss GaInNAs SESAMs.

Typical bandwidths of GaAs/AlAs-based DBRs are in the 100 nm region and normally more than 25 layer pairs of semiconductor are required to achieve reflectivities exceeding 99%. For broader-band SESAM structures, material pairs having larger refractive index contrasts are required. In Ref. [308], it has been shown that only four pairs of AlGaAs/CaF$_2$ are sufficient to create a Bragg reflector with a reflectivity above 98% with a bandwidth of about 400 nm centered around 800 nm. In addition, an ultrabroadband saturable Bragg reflector covering the telecommunications region has been described in Ref. [309]. A Si/SiO$_2$-based Bragg reflector (six layer pairs) with germanium as a saturable absorber has been demonstrated to produce a reflectivity of more than 99.8% across a 1200–1900 nm spectral range.

It is evident that semiconductor-based absorber mirrors have become a key component in the development of ultrashort-pulse solid-state lasers. However, it should be noted that there are some difficulties in the implementation of such devices into mode-locked lasers where highly efficient, ultrashort-pulse generation is required in the near to midinfrared. With regard to further improvements in SESAM engineering, devices characterized by low saturation fluencies, fast recovery times (preferably, with evidence of two components: fast and slow to facilitate both low-threshold and ultrashort-pulse-enabled mode locking), low nonsaturable losses, and ability to offer more versatile characteristics for the manipulation of laser outputs would constitute an important further step toward the provision of robust femtosecond laser modules. In addition, the molecular beam epitaxial growth technique is relatively demanding and expensive and it also requires specific lattice matching conditions to be satisfied such that there are limitations in the choice of semiconductor materials that address the spectral ranges of modern lasers. It follows that the development of nonepitaxially grown saturable absorbers at low cost, with greater simplicity and versatility, would represent an alternative approach toward the provision of compact, robust, and cheaper ultrashort-pulse lasers.

In this section, we present recent achievements in the development of novel saturable absorbers that are based on semiconductor quantum dot (QD) structures and which have already been implemented in the passive mode locking of some near-infrared lasers. These are group IV–VI semiconductor nanoparticles (quantum dots) in glass matrices and self-assembled semiconductor quantum dots (group III–V) grown on semiconductor mirrors (QD SESAMs). The first group of saturable absorbers is attractive because of their ability to be fabricated using simpler and low-cost batch-melting techniques. Second, variation of the nanocrystal size leads permits spectral tuning of the excitonic absorption peak. This means that a single semiconductor material facilitates saturable absorption over a substantial spectral region simply through this variation of nanocrystal size. Quantum dot-based SESAMs grown by standard MBE techniques are characterized by a high-quality semiconductor structure and as a result low nonsaturable losses while simultaneously exhibiting subpicosecond recombination dynamics in contrast to the QW SESAM counterparts. In addition, due to the higher absorption cross sections in QDs (compared to QWs), the bleaching of such saturable absorbers can be achieved at significantly lower fluencies.

6.3
QD SESAMs for Efficient Passive Mode Locking of Solid-State Lasers Emitted around 1 μm

The successful exploitation of QD SESAMs as low-loss saturable absorbers for efficient generation of femtosecond pulses around 1 μm from Yb:KYW lasers has been reported in Ref. [310]. The QD SESAM is usually grown by MBE at around

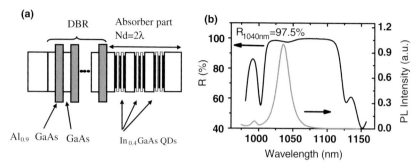

Figure 6.1 (a) Simplified schematic of a QD SESAM structure. (b) The measured reflectivity and representative photoluminescence characteristics of the QD saturable absorber.

600 °C. The distributed Bragg reflector (DBR) comprised a sequence of the 25 $\lambda/4$ GaAs and $Al_{0.9}Ga_{0.1}As$ layers chosen to provide high reflectivity at 1010–1100 nm (Figure 6.1a). An absorption section contains three QD regions separated by $\sim\lambda/2$ GaAs layers. Each QD region consists of three QD layers with dot lateral size and concentration of ~ 7 nm and $\sim 5 \times 10^{10}$ cm^{-2}, respectively, separated by 20 nm thick GaAs spacers. The Fabry–Perot cavity formed by the lower part of the GaAs–AlGaAs Bragg reflector, the absorber–spacer region, and the Fresnel reflection from the semiconductor–air interface and had total thickness of $d = 2\lambda/n$. The quantum dots were formed by the deposition of nine periods of 0.075 nm of InAs and 0.1 nm of GaAs. A test structure containing a single layer of QDs demonstrated room-temperature photoluminescence (PL) centered at 1035 nm with a full width at half maximum (FWHM) of 20 nm (Figure 6.1b). The peak of absorption of this QD structure was centered at around 1040 nm and this facilitated a maximum absorption of around 2.6% at 1042 nm.

Prior to any laser-related evaluations, the linear and nonlinear characteristics of QD SESAM have been investigated. The major motivation for utilizing these devices is that both the linear and nonlinear optical properties can be engineered over a wide range, allowing more freedom in the specific laser cavity design. The key parameters for a saturable absorber are the wavelength range and nonlinear response, including carrier dynamics and saturation fluence.

The linear optical properties of QD-based SESAMs are characterized by a broadband absorption due to the inhomogeneous broadening associated with the distribution of QD sizes. This is one of the attractive features of these devices with respect to the generation of shorter pulses from solid-state lasers compared to their QW-based counterparts. It should be mentioned that the spectral bandwidth of QD SESAMs is limited predominantly by wavelength selectivity of the DBR structure rather than parameters of the absorption medium. The modulation depth of QD SESAMs can be easily regulated by means of the number of QD layers [311]. Figure 6.2a shows that with the introduction of more QD layers in a DBR, the reflection intensity can be adjusted over a wide range from 2.5 to 30% for 9 and 80 layer SESAMs, respectively. Furthermore, the absorption band position of QD-based SESAMs can be tuned by changing the angle between the incident beam and the

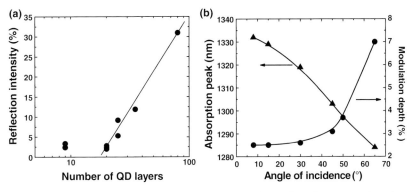

Figure 6.2 (a) Reflection intensity of QD SESAMs versus the number of incorporated QD layers. (b) Dependence of absorption band peak position and reflection intensity on angle of incidence for the nine layer QD SESAM. The solid lines are guides.

normal to the SESAM surface, as depicted in Figure 6.2b. For angles of incidence ranging from 8° to 65°, a modulation depth from 2 to 7% can be achieved with corresponding wavelength range tuning of 50 nm for a nine layer InAs SESAM (Figure 6.2b). Temperature control of the absorption band position gives a long-wavelength shift of 3 nm per 10 °C, allowing fine-tuning to achieve optimum overlap between the absorption band of the QD SESAM and the laser wavelength.

As mentioned earlier, one of the key parameters for a saturable absorber used for the generation of ultrashort pulses from a passively mode-locked laser are the saturation fluence (F_{sat}), modulation depth (ΔR), nonsaturated losses (R_{ns}), and recovery time (τ_{SA}). It was shown [40] that such multilayer QD SESAM structure is characterized by ultrafast carrier relaxation dynamics with evidence of two components: the fast component, with a typical recovery time less than 0.5 ps, and the slow component, generally exceeding 100 ps (Figure 6.3). These characteristics are useful

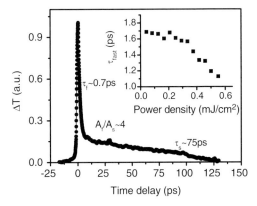

Figure 6.3 Pump–probe measurements of the carrier lifetime of a QD waveguide device. *Inset:* Power density dependence of the fast recovery time component.

from a mode-locking viewpoint because the longer time constant reduces the saturation fluence of the absorber and facilitates a self-starting operation, whereas the fast component plays a primary role in the evolution and shaping of the laser pulses. The slow component can be attributed to spontaneous carrier recombination [312], whereas the ultrafast part appearance could be due to the holes activation process [313, 314]. However, the presence of additional mechanisms affecting the bleaching recovery time, like fast Auger-assisted relaxation [48, 315], cannot be excluded as implied by the shortening of the fast relaxation component with increasing pump density (Figure 6.3, inset). To fully characterize the QD SESAM, the nonlinear reflectivity measurements have been performed as a function of the incident pulse energy fluence as shown in Figure 6.4, using a femtosecond Yb:KYW laser producing 160 fs pulses at pulse repetition frequency of 91.4 MHz and maximum average power up to 0.4 W as a pump source. Applying the model developed in Ref. [316], macroscopic parameters like nonsaturable losses modulation depth, and saturation fluence of the QD SESAM have been estimated to be $R_{ns}=0.2\%$, $\Delta R=2.35\%$, and $F_{sat}=25\,\mu J\,cm^{-2}$, respectively. In comparison, anti-resonant InGaAs QW SESAM structures are characterized by saturation fluencies of above $100\,\mu J\,cm^{-2}$. It should be noted that nonsaturable losses in a QD SESAM arise mainly from the DBR transmission and so it follows that the QD absorber region is characterized by negligible losses when fully saturated. This can be expected when semiconductor structure is grown by standard MBE ($\sim$600 °C) that leads generally to a reduced density of lattice defects and a high optical quality in the structure.

For the laser experiments on passive mode locking using QD-based SESAM, a diode-pumped Yb:KYW laser was constructed and evaluated. A highly asymmetric Z-fold laser cavity was configured with two folding mirrors having radii of curvature of 100 and 75 mm designed for high transmissivity at 980 nm ($\sim$98%) and high reflectivity at wavelengths in the range of 1025–1100 nm, an output coupler with a 3% transmission at 1040 nm, and a high-reflectivity mirror/saturable absorber. The laser gain medium was a 10 at.% Yb^{3+}-doped KYW crystal pumped by a InGaAs-tapered diode laser that produced a near-diffraction-limited 980 nm output at an average power of 3.5 W. During continuous wave (cw) operation, this Yb:KYW laser produced an output power of 2 W at 1047 nm. Mode-locked operation was obtained when one high-reflectivity cavity mirror was replaced by the QD SESAM on which the laser

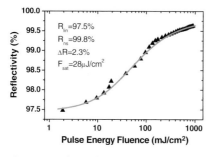

Figure 6.4 The nonlinear reflectivity change versus pulse energy density of a QD SESAM.

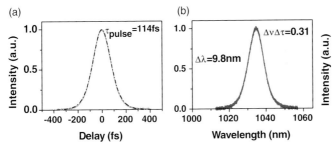

Figure 6.5 (a) Intensity autocorrelation and a fit assuming an ideal sech² pulse profile. (b) Spectrum of pulses from the Yb:KYW laser mode locked by QD SESAM.

mode was focused to a 200 μm radius spot size. Two SF10 prisms with a tip-to-tip separation of 50 cm were used to facilitate appropriate dispersion compensation. An average output power of up to 1 W was achieved in the mode-locked operation and 200 fs pulses were generated at pulse repetition frequency of 107 MHz. The shortest pulse durations produced from this Yb:KYW laser were measured to be 114 fs (assuming a sech² intensity profile) (see Figure 6.5a) at an average output power level of 0.5 W. For the observed spectral width of 9.8 nm at a center wavelength of 1035 nm, the implied duration–bandwidth product was 0.31 (Figure 6.5b).

It was demonstrated that the SESAM-like structure having a p–n junction permitted control of the saturable absorption characteristics through the application of a voltage bias [317]. This type of absorber could thus behave as an actively controlled saturable absorber mirror. For instance, by supplying a suitable bias it was possible to switch between unmode-locked and mode-locked operating regimes for the Ti:sapphire laser. Significantly, in QD-based absorber structures, this type of manipulation of macroparameters via a reverse bias can be achieved with greater efficiency than in QW-based structures because of the higher density of charge carriers accumulated at the energy of the working transition in the absence of high-energy parasitic continuum states. For example, QD-based laser diodes require a much lower laser threshold current density than the bulk or QW-based structures. For an experimental assessment of passive mode locking in a solid-state laser system driven by an actively controlled QD SESAM, the structure depicted in Figure 6.6 was designed. It had a structure similar to that used previously for the efficient mode locking of a Yb:KYW laser, but in this case following the growth process the wafer was thinned to 100 μm and the full area of the n-side was metallized with a GeAu/Ni/Au alloy. On the p-side, a ring contact extending from 160 to 760 μm diameter was formed with a ZnAu/Au alloy and the typical voltage–current characteristics for this structure are reproduced in Figure 6.7. It has been found that by applying the reverse bias in a range up to −3.5 V, the stabilization of passive mode locking of Yb:KYW laser can be achieved compared to the initial unstable multipulse operation observed with an unbiased QD SESAM. In addition, the shortening of pulse durations from ∼10 to ∼4 ps has been achieved by manipulating of reverse bias voltage [318]. These results indicate clearly the variation in macroparameters of a QD SESAM when a reverse bias voltage is applied and which in turn leads to changes in the mode-locking

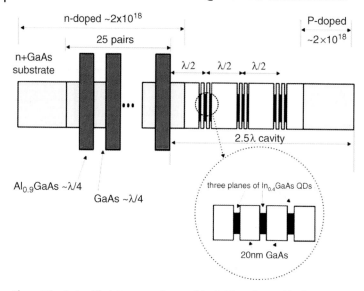

Figure 6.6 A simplified structure scheme of the QD-based saturable absorber device incorporating a p–n junction at 1030 nm.

dynamics of a solid-state laser. Similar effect has been observed in two-section QD semiconductor laser where stable mode locking with pulse durations as short as 400 fs was observed only at a particular reverse bias voltage on the QD absorber component [187]. It is known that in whole semiconductor laser system, stable ultrashort-pulse generation can be achieved when the absorber recovers faster than the gain medium. It can thus be assumed that in applying a reverse voltage, the absorber bleaching recovery dynamics are changed. A direct measurement of the absorption dynamics in an InAs p–i–n ridge waveguide QD modulator under reverse bias has been demonstrated recently [99]. A significant decrease of the absorption recovery time from 62 ps to 700 fs was observed as the applied reverse bias was increased from 0 to -10 V at room temperature (see Figure 6.8). The obtained

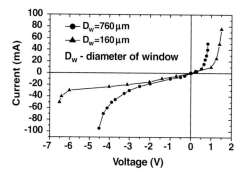

Figure 6.7 Typical voltage–current characteristics for the QD SESAM structure with p–n junction having two different window diameters.

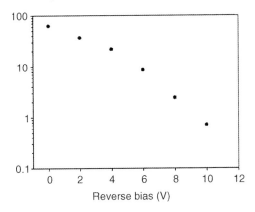

Figure 6.8 Absorption recovery times from exponential fitting as a function of reverse bias voltage.

recovery time behavior was attributed to thermionic emission at low applied fields, while for higher applied fields, tunneling was considered to be the dominant escape mechanism for the carriers.

These observations show that a suitably designed QD SESAM can serve as a low-loss and a low-saturation fluence passive modulator for the efficient generation of femtosecond pulses from solid-state lasers that operate in the near-infrared spectral range. Moreover, such devices offer more accessible manipulation of their macro-parameters for optimized and controlled mode locking.

6.4
QD SESAMs for Efficient Passive Mode Locking of Solid-State Lasers Emitted around 1.3 μm

The passive mode locking of solid-state lasers operating around 1.3 μm spectral region is of continuing interest for laboratory-related investigations in data communications and biophotonics. Typically, GaAs-based Bragg mirrors with InGaAs quantum wells [262, 295] or InGaAsP and AlGaAsSb quantum wells on an InP Bragg mirror [299] have been used for ultrashort-pulse generation at around 1300 nm. However, these saturable absorbers have acknowledged shortcomings that include high nonsaturable losses together with low damage thresholds or narrow reflectivity bandwidths for the Bragg mirrors. Another route adopted for the production of femtosecond pulses in this spectral region involves the use of a GaAs-based Bragg mirror having GaInNAs quantum wells [306, 307] where the inclusion of nitrogen in the quantum well structure serves to reduce the strain and shifts the absorption edge into the 1200–1400 nm region.

In this section, we describe an alternative approach to the generation of femtosecond pulses from solid-state lasers operating around 1.3 μm that involves the exploitation of low-loss QD-based saturable absorbers. Femtosecond Cr:forsterite laser systems that have been passively mode locked using resonant QD SESAMs [319]

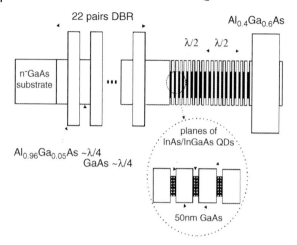

Figure 6.9 A simplified structure scheme of the QD-based saturable absorber device at 1300 nm.

that had suitably matched ground-state (GS) or a first excited-state optical transitions will be considered.

The quantum dot absorber structures used for these experimental assessments were grown by molecular beam epitaxy and comprise a GaAs buffer layer deposited on a GaAs(100) substrate, a distributed Bragg reflector with reflectivity from 1250 to 1400 nm, a first 0.1 μm thick GaAs matrix layer, plane layers of InAs/InGaAs QDs separated by 50 nm thick GaAs spacers, a second 0.1 μm thick GaAs matrix layer, a 0.2 μm thick $Al_{0.4}Ga_{0.6}As$ barrier layer, and a 10 nm thick GaAs cap layer, the number of planes of InAs/InGaAs QD differs from 20 to 35 layers for different QD SESAMs. The DBR consisted of a 22-period quarter-wave $Al_{0.95}Ga_{0.05}As$/GaAs stack. The QD plane is formed by deposition of 2.3 monolayers (ML) of InAs followed by a 5 nm thick $In_{0.16}Ga_{0.84}As$ covering layer (Figure 6.9). A schematic of the laser cavity and pumping geometry used for the Cr:forsterite laser is shown in Figure 6.10. The Nd:YVO$_4$ pump laser produced up to 10 W of linearly polarized light at 1064 nm that was focused into a 32 μm spot size inside the 12 mm-long Brewster-cut crystal. The asymmetric laser

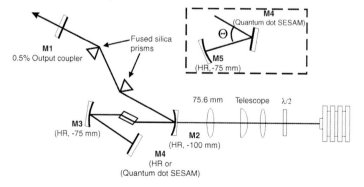

Figure 6.10 Schematic of a Cr:forsterite laser. *Top right*: The angle of incidence on a QD SESAM, θ/2, can be changed.

cavity included two curved folding mirrors that had radii of curvature of 100 and 75 mm for the long and short arms, respectively. A wedged 0.5% output coupler located at the end of the long arm of the cavity and a high-reflector mirror or the QD SESAM structure terminated the short arm of the cavity. Two intracavity-fused silica prisms were used for compensation of positive group velocity dispersion. As shown in Figure 6.10, two configurations were used to terminate the short arm of the cavity. A four-mirror cavity was utilized for perpendicular incorporation of the QD SESAM in the incoming beam. To achieve relative flexibility in the choice of incidence angle ($\Theta/2$) on the QD SESAM, the laser cavity short arm was reconfigured as shown in Figure 6.10 (top right). Continuous wave operation of the Cr:forsterite laser was realized with a lasing threshold of around 1.4 W of incident pump power and corresponding maximum average output powers of 95 and 251 mW using output couplers of 0.5 and 2%, respectively, for a pump power of 6 W at 1280 nm.

The QD SESAM having 20 layers of InAs QDs with the absorption peak at 1330 nm was used to initiate mode locking. Initially, this device was introduced into the laser cavity at normal incidence. The shortest pulses achieved in this configuration had durations around 58 ps at 1278 nm. This relatively long-pulse duration can be explained by a negligible modulation depth at this wavelength. A resonant-like structure of the QD SESAM allowed the resonance conditions to be changed by employing the SESAM at different angles of incidence (Figure 6.11). It can be seen that in varying the angle of incidence from 8° to 65°, the resonance peak of absorption can be shifted from 1330 to 1284 nm. The laser cavity was then altered to incorporate the QD device at an angle as shown in Figure 6.10 (top right). For this configuration, a second 75 mm radius of curvature folding mirror was used to terminate the cavity, thereby allowing the spot size incident upon the SESAM to be kept at the appropriate size (~30 μm in radius) while allowing access to a range of angles. Figure 6.12 shows the effect that this range of angles has on the pulse duration and the output power. It can be seen that as the angle is increased, the pulse duration is reduced dramatically. The shortest pulses were produced at an incidence angle of 45° and the intensity autocorrelation measurements implied a pulse duration of 158 fs. The corresponding spectral bandwidth was measured to be of 10.5 nm giving the duration–bandwidth

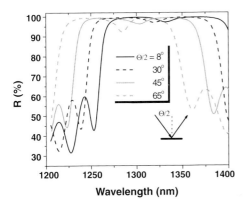

Figure 6.11 QD SESAM reflectivity dependence on angles of incidence.

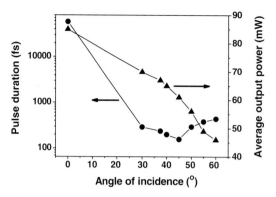

Figure 6.12 Mode-locked pulse durations and average output powers from a Cr:forsterite laser as a function of angle between normal of the QD SESAM and laser cavity axis.

product of 0.32. An output power of 64 mW was measured at a pulse repetition rate of 194 MHz for an output coupling of 0.5%. For angles different from this optimum orientation, the pulse durations ranged from 440 fs for $\theta/2 = 60°$ to 240 fs for 30° while still remaining transform limited where some prism insertion adjustment was made to optimize the compensation of positive dispersion. (Angles smaller than 30° could not be accessed due to constraints on the positioning of the mirror mounts.) For angles exceeding 60°, the losses in the cavity increased to the point that mode locking was no longer possible, and for angles of $\theta/2 > 70°$ even a continuous wave operation could no longer be sustained. The main reason for the variation in pulse duration with tilting of QD SESAM device is attributable to the modulation characteristics of the absorber. As the angle is increased, the QD absorption band translates toward and then through the lasing band around 1280 nm, thereby allowing larger modulation depths to be accessed. At $\theta/2 = 45°$, where the shortest pulses were achieved, a modulation depth of 0.5% can be estimated. Here it should be noted that the maximum overlap between the absorption peak of the QDs and the laser spectrum occurs at $\theta/2 = 60°$ where a modulation depth up to 2% can be achieved. However, for this larger angle, the output power was relatively low due to the losses introduced by the DBR structure (poor reflectivity) and therefore the fluence on the QD device was insufficient to fully saturate the device. In addition, around resonant peak the QD structure is characterized by an enormous amount of group delay dispersion, making the generation of ultrashort pulses through the soliton mode-locking regime relatively difficult.

The ability to choose from a range of pulse durations at particular laser wavelength is desirable for those applications where optimization of peak power is required. QD-based SESAMs can thus be tailored to have a resonant-like structure for such laser systems.

Another advantage of QD is the possibility to exploit the ground- and excited-state transitions to develop saturable absorber mirrors. In the ground-state absorption transitions described above, quantum dot SESAMs have been utilized for ultrashort-pulse generation around 1.3 μm. Exploitation of the higher degeneracy QD excited states allows such devices to be characterized by higher absorption coefficients compared to GS counterparts [320]. This, in turn, could lead to reduced number of QD layers to reach a desirable modulation depth and lower saturation fluences can be expected compared to GS devices. Moreover, shorter absorption recovery time in QD SESAM working at ES transitions can be anticipated compared to GS devices due to fast relaxation of carriers to lower dot states. Previously, the first ES QD SESAM was demonstrated for passive mode locking of a Nd:YVO$_4$ laser at 1064 nm, where 84 ps pulses were generated with an output power of 3 W [321]. Here, we include some experimental results on femtosecond pulse generation from a solid-state Cr^{4+}:forsterite laser system that incorporated QD SESAM devices operating either on ground-state or a first excited-state (ES1) transitions.

The SESAM structures, used for these experimental assessments, comprised a 25-period AlAs/GaAs distributed Bragg reflector with QDs embedded in a GaAs cavity deposited on the top. The DBR had a design wavelength λ_0 of 1290 nm, which corresponded to the center of the high-reflectivity stopband in the frequency domain. The high-reflectivity region covered the 1230–1350 nm spectral band. Each sample contained nine dot layers, incorporated into a $9\lambda_0/4$ low-finesse cavity leading to resonance at λ_0. The dot layers were ordered in groups of three, each separated by 40 nm GaAs spacer layers, and these groups were designed to be centered at the antinodes of the field pattern. The index profile and calculated square electric field pattern at λ_0 is illustrated in Figure 6.13a [322]. The field was calculated using a

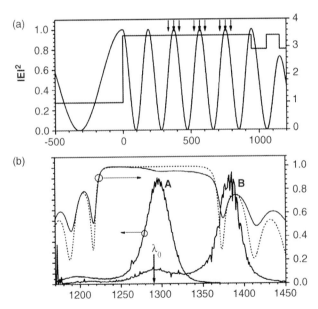

Figure 6.13 (a) The refractive index profile for the low-finesse, resonant SESAM with a $9\lambda/4$ cavity and the calculated field pattern at λ_0. The arrows denote the position of the QD layers. (b) Room-temperature PL spectra obtained from the GS and ES samples grown without the DBR. The dotted line is the calculated reflectivity obtained from the ES sample and the solid line is its measured equivalent.

standard transfer matrix technique. The devices were grown by solid-source molecular beam epitaxy. The dots in the ground-state sample (A) were single-layer dots with a density of 2.2×10^{10} cm^{-2} per layer, designed to have the GS transition resonant with λ_0 at 295 K. They were grown by deposition of 2.4 ML of InAs at a growth rate of 0.014 ML s^{-1} and a substrate temperature of 485 °C. The QDs were then capped by 4 nm In$_{0.1}$Ga$_{0.9}$As at 470 °C, followed by 11 nm GaAs also grown at 470 °C, with further GaAs grown at 580 °C. The dot layers in the ES1 sample (B) had bilayers [323] consisting of a seed layer of QDs and a second QD layer separated by 10 nm of GaAs. The seed layer was grown by deposition of 2.4 ML InAs at 0.014 ML s^{-1} at a substrate temperature of 492 °C. The second QD layer was grown by deposition of 3.3 ML InAs, also at 0.014 ML s^{-1} and at a substrate temperature of 467 °C. It was capped with 15 nm GaAs grown at 467 °C and further GaAs grown at 580 °C. The density of QDs in the second layer was 1.6×10^{10} cm^{-2}. The overall thickness of the GaAs cavity was the same as the GS sample, so the seed layer and 10 nm GaAs spacer layer were included in the GaAs buffer between each second layer of the bilayers. The corresponding state energies of the seed layer dots located at considerably shorter wavelengths than 1290 nm and therefore did not participate in the absorption. Room-temperature photoluminescence (PL) from two samples of dots grown of the same recipe is shown in Figure 6.13b [322] and clearly demonstrates the position of the dot states relative to the SESAM design wavelength.

For the laser experiments, the Cr:forsterite laser setup described in this section has been used, the only difference being that laser cavity mode was focused to a spot radius of 200 μm on the QD SESAM that was terminating the short arm at normal incidence. Both devices were capped with 190 nm MgO layers, which allow broadband elimination of GDD on reflection while maintaining a low saturation fluence and a high modulation depth. With the ground-state SESAM in place, pulse durations as short as 91 fs were measured assuming sech^2 pulse shape. For the observed spectral width of 20 nm at a center wavelength of 1280 nm, the deduced time–bandwidth product was 0.32 confirming the generation of near-transform-limited pulses. The saturation fluence of the SESAM structure was estimated to be as low as 9 μJ cm^{-2} by measuring the threshold for mode locking. The excited-state version of this device gave similar results, where pulse durations as short as 86 fs were measured (see Ref. [322]). The saturation fluence of 25 μJ cm^{-2} was marginally higher than that of the ground-state device. A maximum 55 mW of output power was generated for both devices.

Later on, absorption recovery dynamics of GS and ES QD SESAMs were measured using time-resolved photoluminescence and heterodyne four-wave mixing experimental methods [324]. It was found that both devices are characterized by a multi-exponential recovery dynamics with characteristic time constants of 0.8, 70, and 765 ps for the GS transition and 0.4, 26, and 348 ps for the ES band.

The operation of a p–n junction QD SESAM and its effects on the mode-locked pulse duration of a Cr:fosterite laser was also investigated in the 1.3 μm spectral range [325]. Stable mode locking was achieved when the p–n junction QD SESAM was both unbiased and biased. A 2.7 times reduction in pulse duration, producing near-transform-limited pulses of duration 6.4 ps, was observed when a DC bias of −4.5 V

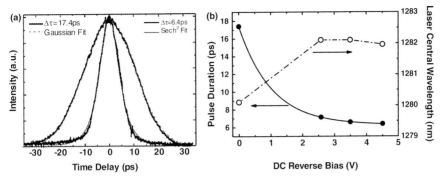

Figure 6.14 (a) Intensity autocorrelation trace of the pulses obtained with 0 and −4.5 V applied on the p–n QD SESAM. (b) Mode-locked pulse width shortening (filled dots) and laser central wavelength shift (hollow dots) obtained with an increase in applied reverse bias on the p–n junction QD SESAM. The black line is an exponential fit and dashed line is for eye guide.

was applied. Figure 6.14a represents a typical autocorrelation trace acquired from the passively mode-locked Cr:forsterite laser in this regime. The output pulse duration was deduced to be 17.4 ps at FWHM, assuming a Gaussian intensity profile. No dispersion compensation was used. A DC reverse bias on the p–n junction QD SESAM was applied for voltages from 0 to −4.5 V in the same experimental setup and for the same pumping conditions. The effect this reverse bias had on the mode-locked behavior of the laser was investigated. Stable mode-locked operation of the Cr:forsterite laser was attained across the whole bias range with pronounced pulse shortening observed with increasing applied reverse voltage (see Figure 6.14b).

With these initial results, a direct comparison of mode-locked parameters from solid-state laser system could be made for the QD SESAM devices in which GS and ES transitions were being utilized for ultrashort-pulse generation. Although this gives an initial experimental step in gaining some further understanding of the dynamics and mechanisms that underlie ground- and excited-state quantum dot SESAMs, the authors appreciate that more detailed studies on the effect of state degeneracy and recovery time of ES1 devices will be required to exercise more fully the probable degrees of performance tailoring that this category of devices will offer in the future.

6.5
QD SESAMs for the Passive Mode Locking of Fiber Lasers

Recent progress in the passive mode locking of fiber-based laser systems has been significantly enhanced by the advances in semiconductor technology. Ultrashort-pulse generation from fiber lasers has been demonstrated recently in the spectral range from 0.9 to 1.6 µm, earlier occupied predominantly by the ultrashort solid-state-based laser counterparts [326]. It was shown that SESAMs with fast recovery time are more favorable for self-starting capability and subsequent pulse shaping in fiber laser systems where high-intensity amplified spontaneous emission takes place [327]. In this context QD-based absorber structures look as a promising approach to initiate and stabilize mode locking in a fiber laser. Such approach has been undertaken for the first time in Ref. [328] where Yb–fiber laser has been mode locked using QD SESAM.

The InGaAs QD absorber mirror that was used had been grown epitaxially on a GaAs substrate. The mirror consisted of 33 pairs of $Al_{0.9}Ga_{0.1}As/GaAs$ quarter-wave layers forming a Bragg reflector with a center wavelength of 1060 nm and a bandwidth of 100 nm. The absorber section of the structure included 8 multiple stacks of 10 InGaAs QD layers with 10 nm thick spacer layers of GaAs. The adjacent multiple stacks were separated by 23.7 nm of high-temperature grown GaAs to eliminate the effects of indium segregation and 13.7 nm of low-temperature grown GaAs to decrease the carrier lifetime. The modulation depth, saturation fluence, and nonsaturable loss of the QD SESAM were estimated to be 17.6%, 14.8 µJ cm^{-2}, and 7.6%, respectively, based on the data from the nonlinear reflectivity measurements performed at the resonant wavelength of 1042 nm (similar to Figure 6.4). It should be noted that the highest modulation depth of the nonlinear response was expected near 1042 nm because this corresponds to the spectral "hole" in the unsaturated reflectivity characteristics. Consequently, the modulation depth decreased gradually with detuning from the resonant wavelength. The performance of the QD absorber mirror was investigated within the cavity configuration, as shown in Figure 6.15a. The laser cavity consisted of a 75 cm long piece of Yb-doped fiber having an absorption of

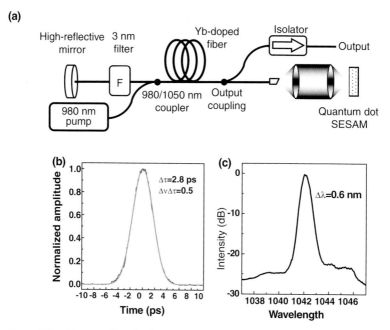

Figure 6.15 (a) Setup of the fiber laser used to test the QD absorber. The overall dispersion of the cavity was estimated to be -0.1 ps nm^{-2}. (b) Optical spectrum and (c) intensity autocorrelation of the output from the mode-locked Yb-doped fiber laser.

434 dB m^{-1} at 980 nm pumped through a 980/1050 nm wavelength-selective coupler, an output coupler provided by a tap coupler, a broadband high-reflector mirror, and terminated by the 80 layer QD SESAM. A tunable intracavity 3 nm bandwidth filter was used to optimize the mode-locked operation. Optimum focusing of the laser beam onto the absorber mirror resulted in self-starting mode locking at 1042 nm and it was clearly evident that this was initiated and stabilized by the saturation dynamics of the QD component. Stable mode locking was obtained at 1040 nm with a low-threshold pump power of 30 mW at 980 nm for an output coupling of 7%. A sequence of pulses at a 50 MHz repetition rate with an average output power of 5 mW was obtained for a pump power of 100 mW and with output coupling of 30%. Pulse durations as short as 2.8 ps (Figure 6.15b) were measured and the time–bandwidth product of 0.5 that was estimated from the observed spectral width of 0.6 nm at a center wavelength of 1040 nm (Figure 6.15c) was close to the transform limit for the Gaussian pulse shapes assumed. The low value of time–bandwidth product despite the high normal dispersion in the laser cavity indicated a strong pulse shaping influence due to the QD SESAM that had a high modulation contrast and a short recovery time. Indeed, with this high modulation depth in the nonlinear reflectivity, stable mode locking was achieved without any intracavity dispersion compensation.

These results thus demonstrate the enormous potential of QD-based SESAMs as strong pulse shapers in fiber laser systems. In turn, this will be able to counteract the effects of the high levels of amplified spontaneous emission and cavity dispersion that are typical for these lasers.

6.6
Mode-Locked Semiconductor Disk Lasers Incorporating QD SESAMs

Diode-pumped semiconductor disk lasers (SDLs), also referred to as vertical external cavity surface emitting lasers (VECSELs), are showing considerable promise as new types of ultrashort-pulse solid-state laser sources operating with multigigahertz repetition rates [329]. Ultrashort-pulse generation from compact semiconductor lasers is of wide applications-related interest especially for high-capacity telecommunication and data communication systems, photonic switching devices, optical interconnections, clocks for very large-scale integrated microprocessors, and high-speed electro-optic sampling systems. During the past few years, integrated absorber concepts have been discussed extensively for high repetition rate passively mode-locked VECSELs because of their near-diffraction-limited output beams, large gain cross sections, and high integrability. There are, however, several issues to be resolved on the way toward wafer-scale integration. To date, most passively mode-locked VECSELs have been based on QW structures for both gain and SESAM parts. To achieve stable mode locking in a whole semiconductor laser system, the SESAM needs to saturate earlier and recover faster than the gain [275], but as both gain and saturable absorption have similar saturation fluences, the mode area on the SESAM has to be much smaller than that on the gain structure. In practice, the high gain media/absorber mode areas ratio in short cavities, with correspondingly high pulse repetition rates, can limit achievable compactness and the ultimate stability of such a laser. The tight focusing on the SESAM also reveals a problem relating to the thermal management. The decreasing peak power with increased repetition rate would impose even tighter constraints on the focusing of the beam on the SESAM and this in turn raises the temperature significantly, thereby increasing the SESAM losses and limiting the power level obtainable from the laser. It became apparent that QD-based structures could offer a solution to these problems [112]. Indeed, the high repetition rate passive mode locking was achieved from VECSEL using a low saturation fluence QD SESAM [330]. Nearly transform-limited 3.3 ps pulses were produced with a repetition rate up to 50 GHz at around 960 nm. QD SESAM had a resonant design and was characterized by a saturation fluence as low as $1.7\,\mu J\,cm^{-2}$ and modulation depth of $\sim$3%. The mode radii on both the gain structure and the SESAM were approximately the same ($\sim$60 μm). Since a stable mode locking can be achieved from a VECSEL that incorporates a QD SESAM with similar mode radii on both gain and absorber, further integration of absorber structure into VECSEL gain region would be possible. The monolithic gain medium/saturable absorber concept was realized with a mode-locked integrated external cavity surface emitting laser (MIXSEL) [331]. The epitaxial layer structure of this laser has both the gain and the absorber layers integrated, which can thus be optimized separately by simply using different growth parameters. The MIXSEL thus represents a promising development that could offer an attractive route toward a future high-volume, wafer-scale fabrication of compact, ultrafast lasers. In particular, the MIXEL where saturable absorber consists of a single self-assembled InAs quantum dot layer at low growth temperatures (430 °C), while the gain section was formed by seven layers of InGaAs quantum wells grown at 520 °C, delivered 40 mW of average power in 35 ps pulses at repetition rate of 2.8 GHz [331].

Detailed investigation of the nonlinear optical properties of QD SESAMs that have been grown at different conditions with varying dot density for optimized mode locking of VECSELs has been recently reported in Ref. [332]. Particularly, it was shown that for the fast relaxation component enhancement, the QDs should be grown close to the formation threshold ($\sim$1.6 In monolayer coverage) and designed for the operation at around QDs photoluminescence peak. In addition, it was shown that postgrowth annealing leads to the substantial reduction in saturation fluence value (up to a factor of 9) in the samples having a small In monolayer coverage together with further fast component enhancement. These observations have had the major impact on optimization of QD SESAMs for efficient ultrashort-pulse generation from highly compact semiconductor laser systems.

It is evident that a great progress has been achieved in the development of passively mode-locked semiconductor disk lasers operating in the few picoseconds regime at multigigahertz repetition rates. However, since semiconductor structures are typically characterized by a broad gain, a key challenge is generation of transform-limited femtosecond pulses from such laser systems. A femtosecond VECSEL has been demonstrated for the fist time by Garnache *et al.* producing sub-500 fs soliton-like pulses [333]. More recently, 290 fs pulses from a semiconductor disk laser [334] and generation of 260 fs pulses using an optical Stark effect in the QW SESAM have been demonstrated [335].

The first demonstration of a sequence of subpicosecond pulses generated from a QW VECSEL gain structure using a QD SESAM is presented in Ref. [336]. The QD SESAM, grown by MBE, consisted of a highly reflecting 25-pair DBR centered at a wavelength of 1070 nm and a 2λ GaAs microcavity with a design wavelength of 1050 nm containing three groups of three layers of InAs/GaAs quantum dots grown by the Stranski–Krastanov (SK) technique and positioned at the antinodes of the electric field. The quantum dot density was $\sim 5 \times 10^{10}\,\mathrm{cm}^{-2}$. The asymmetric laser cavity was formed by the QD SESAM acting as an end mirror, a 38 mm radius high reflector and the gain structure acting as folding mirrors, and a 75 mm radius of curvature 0.6% transmission output coupler. The laser had a cavity frequency of 895.5 MHz. The gain structure was optically pumped by a 1.3 W, 808 nm fiber-coupled diode laser. The laser mode radii on the gain medium and SESAM were 60 and 22.5 µm, respectively, giving an area ratio of 7. The gain element and SESAM were mounted on temperature-controlled copper mounts to allow temperature tuning. The laser output was a circularly symmetric TEM_{00} beam with average output power during mode-locked operation up to 60 mW before thermal rollover was observed. The shortest pulse duration was generated when the QD SESAM was at a temperature of 20 °C, the QW gain structure at 0.9 °C, and the pump power at 0.93 W. The output power under these conditions was measured to be 45.5 mW, and the pulse fluence on the SESAM was $530\,\mathrm{\mu J\,cm^{-2}}$. The pulse had a near-Gaussian profile and the measured autocorrelation yielded a pulse duration of 870 fs with the measured spectral width of 2 nm centered at 1027.5 nm (Figure 6.16). This gave a time–bandwidth product of 0.49. The slight chirp seen with the minimum pulse duration is thought to be due to self-phase modulation in the SESAM.

We believe that further optimization of the QD SESAM design is feasible to obtain high modulation depth, with a short microcavity structure, to facilitate the generation of subpicosecond pulses over a broader wavelength range. This would provide

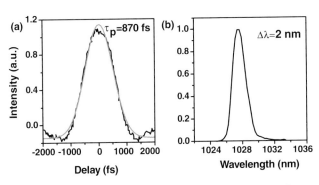

Figure 6.16 (a) Intensity autocorrelation of the pulse train (gray line) and the Gaussian fit (dark line), showing a 1.14 × Fourier-limited pulse with duration of 870 fs. (b) Optical spectrum of the semiconductor disk laser output.

operational scope for wavelength tuning of the mode-locked pulses as well as access to shorter pulse durations.

6.7
Optically Pumped Quantum Dot VECSELs

Optically pumped semiconductor disk lasers (OP-SDLs) [337], also known as vertical external cavity surface emitting lasers [338], firmly entered the research and market of photonics mainly due to their ability to combine high output power and good beam quality [339], while allowing the advantages of intracavity techniques such as frequency mixing [340] and mode locking [341, 342] to be exploited. Also, SDL technique was demonstrated to cover spectral range from deep UV using the fourth harmonic generation [343] to mid-IR incorporating IV–VI materials [344], thus giving access to a number of different applications. Recently, the SDL architecture has been combined with quantum dot-based gain materials to explore their advantages provided by three-dimensional carrier confinement, like temperature insensitivity and broad gain bandwidth [13]. Examples of SDLs based on both InAs/GaAs submonolayer (SML) and InGaAs Stranski–Krastanov grown quantum dots have

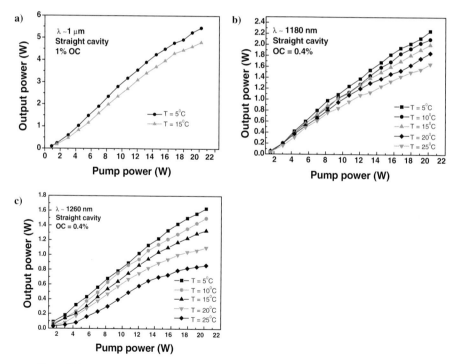

Figure 6.17 Output power versus pump power characteristics of QD SDL at 1040 nm (a), 1180 nm (b), and 1260 nm (c).

been successfully demonstrated [345, 346]. Power scaling of QD-based SDLs up to 5.5 W cw output power level has been demonstrated recently at 1040 nm (see Figure 6.17a). Our recent results demonstrated that SK QD SDLs can also efficiently generate at around 1200 and 1270 nm wavelengths, which expands the spectral coverage of such types of lasers. An intracavity diamond heat spreader was used to power scale the devices that helped to reach output powers up to 2.25 W for 1180 nm [347] and up to 1.6 W at 1270 nm [348], as depicted in Figure 6.17b and c.

7
Saturable Absorbers Based on QD-Doped Glasses

7.1
II–VI Semiconductor Nanocrystals in Glass

In the early 1980s, it was shown that semiconductor microcrystals (quantum dots (QDs)) with radii much larger than the semiconductor lattice and comparable to the electron–hole (exciton) Bohr radius have characteristic electronic properties that are different from those in bulk structure [349–351]. The three-dimensional confinement leads to a discrete structure of electron–hole energy states (atomic-like energy structure) where a blueshift of the exciton lines arises in comparison to the bandgap in bulk semiconductors [322]. This energy difference is proportional to the $\hbar/2K\bar{a}^2$ value, where $\bar{a}$ is the average nanocrystal radius and K is determined to be $m_e^{-1} + m_h^{-1}$ in the case of strong confinement or m_e and $m_e + m_h$ in the case of intermediate or weak confinements (m_e and m_h are electron and hole effective masses, respectively). The different types of quantum confinements are determined by the difference between semiconductor QD size ($\bar{a}$) and the semiconductor material electron and hole Bohr radii (a_e and a_h) [349]. In the case where $\bar{a} < a_e, a_h$, a strong confinement is realized and both the valence and conductive states are discrete. If $a_h < \bar{a} < a_e$, the case of intermediate confinement applies, when only electrons are confined. Finally, the weak case represents the situation when $\bar{a} > a_e, a_h$, where neither electrons nor holes have discrete states but excitons are confined. It follows that by fabricating nanocrystals of different sizes, the absorption edge of such structures can be tuned over wide wavelength ranges. In some pioneering research on the optical properties of quantum dot structures, particular attention has been laid on silicate glasses doped with II–VI semiconductor nanocrystals (CdS_xSe_{1-x}, $CdSe_xTe_{1-x}$) [23, 352–355]. Initially, these structures attracted attention as sharp cutoff optical filters and were marketed by Corning Glass Works for operation in yellow to red spectral regions [352]. They were produced typically by adding a few weight percent of cadmium, selenium, sulfur, tellurium, and their compounds to the glass melt at temperatures in the range of 1300–1400 °C. Subsequent processing included high optical quality glass structure formation at 450–550 °C (annealing step) and heat treatment to initiate the process of semiconductor nanocrystal formation [353]. This last step is especially relevant to the control of the physical and optical properties of quantum

Ultrafast Lasers Based on Quantum Dot Structures: Physics and Devices.
Edik U. Rafailov, Maria Ana Cataluna, and Eugene A. Avrutin
Copyright © 2011 WILEY-VCH Verlag GmbH & Co. KGaA, Weinheim
ISBN: 978-3-527-40928-0

dots within glass matrices. It has been shown that, in general, longer periods of heat treatment at higher temperatures lead to the formation of larger nanocrystals with higher densities. For instance, CdSe nanocrystals with diameters of ~2.5–3.5 nm, ~4.4 nm, and ~8 nm can be obtained at sample treatment temperatures of 600, 650, and 700 °C, respectively, for 0.5 h-long period. In this case, the location of first excitonic transition was shifted from around 580 to 650 nm and high-energy structures have become prominent. On the other hand, increasing treatment time from 0.5 to 4 h ($T = 593$ °C) was sufficient to shift the first absorption peak from around 500 to 560 nm (implying the formation of larger nanocrystals) and the absorption coefficient of the sample was increased from 30 to 70 cm^{-1}. More pronounced quantum confinement effects on absorption and photoluminescence were observed at particular treatment conditions when quantum dot diameters in the 2.5–4 nm range were realized. (The Bohr radii of the electron and hole in bulk CdSe are approximately 3 and ~0.5 nm, respectively; so there is a case of intermediate confinement.)

It was predicted that three-dimensional localization of free carriers in semiconductor material not only leads to modification of the linear optical properties but would also enhance nonlinear characteristics [357]. Indeed, the density of states of QD can be represented as a set of delta function peaks centered on atom-like energy levels in contrast to the continuum that applies to bulk semiconductors. The density of the charge carriers accumulated at the energy of a specific transition is thus significantly higher due to the absence of any high-energy parasitic states of the continuum. Absorption saturation effects have been observed in CdSe$_x$S$_{1-x}$-doped glasses and saturation fluences have been measured to be in the 1–10 mJ cm^{-2} range [358, 359]. However, it was found that samples containing smaller particles, where quantum effects are more prominent, are characterized by higher saturation fluences [359]. It was explained by the fact that absorption nonlinearities depend not only on oscillator strength but also on the carrier recombination time. If it is assumed that smaller nanocrystals have faster relaxation times, then this in turn implies higher pump energy requirements for band filling. Indeed, as reported by the authors of Ref. [360] for Corning filters under picosecond excitation, exciton lifetimes were measured to be 18, 85, and 109 ps for quantum dot sizes of 9.4, 10.6, and 13.3 nm, respectively. In addition, such dependencies have been observed for measurements undertaken in the femtosecond regime [361, 362]. Bleaching decays vary from 85 to 270 fs in CdSe nanocrystals of 2 and 4 nm diameters [362]. The linear dependence of relaxation time on particle size can be related to carrier trapping on QD surface defects in which case the trapping rate should be proportional to QD surface-to-volume ratio $S_d/V_d \sim 1/R$, where R is the nanocrystal radius. However, for CdSe QDs of ~3–7 nm radii, the bleaching relaxation rate ($1/\tau$) was found to be proportional to R^{-6} [35]. A reasonable explanation of this dependence can be suggested on the basis that confinement-induced squeezing of the electron wave function leads to a better overlap with the wave function of deep trap states through which the carrier relaxation takes place. Using 60 fs excitation pulses, recombination times of around 200 fs have been measured in commercial Schott glass filters accompanied by a second slower relaxation component having a typical time constant of around 80 ps [363]. In addition, it was found that relaxation times strongly depend on excitation energies, which confirms a carrier density-dependent band filling process. Subsequently, a polarization decay time of 25 fs was measured in CdSe borosilicate glass through hole-burning assessments [364].

Thus, the research carried out on the optical properties of II–VI semiconductor QDs in glasses has indicated that such materials can be suitable for applications in ultrafast optoelectronics. In particular, optical saturable absorbers (SAs) based on QD-doped glasses can facilitate ultrafast modulation of laser output through the passive mode-locking technique. In 1990, using an infrared color filter as a saturable absorber, pulses as short as 2.7 ps have been generated by a Ti:sapphire laser in wavelength range from 785 to 855 nm [365]. Later work showed that silica-based glasses doped with CdTe could be used to initiate the generation of femtosecond pulses ($\tau_{pulse} = 52$ fs) from a Ti:sapphire laser operating around 850 nm through the Kerr-lens mode-locking technique [366]. These saturable absorbers had characteristic saturation fluences of ~ 1 mJ cm^{-2} and relaxation times in a range of several picoseconds to ~ 400 fs depending on exposure time [366].

Given the comprehensive studies of the physical and optical properties of II–VI semiconductor quantum dots in glasses and their formation processes, it is clear that there is a strong limitation to their practical use due to "narrow" operational spectral ranges (~ 500–700 nm) dictated by bulk bandgaps of these materials ($\lambda_g \sim 0.8$ μm). To achieve quantum confinement effects at longer wavelengths, appropriate semiconductor materials having low-energy bandgaps thus represent more favorable options.

7.2
IV–VI Semiconductor QD-Doped Glasses for Ultrashort-Pulse Generation from Solid-State Lasers

Bandgaps of semiconductors from the IV–VI group, namely, the lead salts (PbS, PbSe, and PbTe), are in the range of ~ 0.3–0.4 eV. This means that this type of semiconductor structures can exhibit quantum confinement effects in the technologically important wavelength regions of 1–3 μm depending on nanocrystallite size. In addition, in all these materials, electron and hole Bohr radii are ~ 10 nm. It follows that the strong quantum confinement case can be realized in lead salt QDs having radii of ≤ 10 nm. This is opposite to the situation with II–VI semiconductor QDs, where the small Bohr radius for the holes of around 0.5 nm prevents the possibility of efficient quantum confinement. Because the electron can be confined strongly ($a_e \sim 3$ nm), this leads to significant asymmetry in the density of the charge carriers, which in turn changes the efficiency of exciton–phonon coupling and the corresponding particle lineshape. In addition, given that strong quantum confinement of charge carriers can be achieved in relatively large QDs in lead salts, the surface effects can be minimized due to large volume-to-surface ratio that enables investigation of the intrinsic QD properties.

The initial attempts to produce lead salt QDs included such methods and materials as chemical reactions in porous [367] and organic [368] glasses. Despite the demonstration of nanocrystallite formations of 2–20 nm diameter, the discrete absorption bands corresponding to particular exciton transitions were not observed because of the inherent dispersion of particle sizes. Later, using the method of direct crystallization of semiconductor nanoparticles from the inorganic glass phase, similar to those applied for CdS and CdSe QDs grown earlier, nanocrystals of both PbS and PbSe in silicate glass have been obtained in which well-pronounced absorption peaks were evident in 1–2.5 μm spectral region [369]. Applying double-stage temperature treatment processes, QD sizes have been controlled in the 8–13 nm range. A representative example of absorption spectra of different size PbS QDs embedded into a glass structure is reproduced in Ref. [322]. Subsequent successful growth of PbTe QDs in silicate glass has been demonstrated that have particular interest for data communication-related applications around the 1.3 and 1.5 μm wavelength regions [370]. To increase the concentration of semiconductor nanocrystals in glass, the growth of PbSe QDs in phosphate glass was introduced [371, 372]. The solubility of PbSe in melt batches was as high as 1.5 wt.% that order of magnitude larger compared to the solubility of II–VI semiconductors in silicate glasses.

Surveying the research work performed on the development of lead salt QD-doped glasses and investigation of their optical properties [372–379], it can be concluded that this technology is now well established for a range of high optical quality silicate and phosphate glasses doped with PbS, PbSe, and PbTe semiconductor QDs having sizes from few nanometers (strong confinement) up to 20 nm (weak confinement) with corresponding radius distributions of $\Delta R/R \sim 5\text{–}7\%$. In the following sections, we describe several examples of lead salt QD-doped glasses used as saturable absorbers for passive mode locking of near-infrared solid-state lasers. Particular attention will be paid to the nonlinear parameters of QD-based saturable absorbers that determine the key output characteristics of mode-locked solid-state lasers.

7.3
QD-Doped Glass Saturable Absorbers for Passive Mode Locking around 1.3 μm

The first successful use of lead salt QD-doped glasses for particular practical application was reported in Ref. [380]. Using a thin plate (150 μm) of PbS-doped silicate glass ($R = 2.7$ nm), acting as an intracavity saturable absorber, continuous wave (cw) generation of picosecond pulses was achieved in a mode-locked Cr: forsterite laser operating in 1200–1300 nm spectral range. Near-transform-limited 4.6 ps pulses were produced at a repetition frequency of 110 MHz and the average power was 75 mW. The QD-based saturable absorber was characterized by a low-signal absorption of 1.34 cm^{-1}, nonsaturable losses of 0.03 cm^{-1}, and a saturation intensity of 0.18 MW cm^{-2} (33 ps pulses were used for nonlinear transmission measurements). The measured saturation fluence of the absorber was in a range of values reported earlier for II–VI semiconductor QDs [358, 381]. The temporal dynamics of this QD saturable absorber were evaluated using a pump–probe technique with operational wavelengths that corresponded to the lowest electron–hole transition around 1.3 μm [103]. It was found that the relaxation kinetics strongly depend on the pump–pulse energies. Similar to the behavior previously reported for CdSe and CdS QDs [35, 382], the bleaching recovery times reduced with increase of pump energy for both the slow and fast components. Such behavior can be explained within the saturation of carrier trapping at QD surface defects [383] or based on Auger-like recombination mechanisms (i.e., high pump power leads to increased density of free and trapped carriers and thus to smaller tunneling distance for secondary recombination) [382]. However, further increase of pump energy leads to induced photodarkening of the QD sample and at energies of around 10 mJ cm^{-2}, QD-doped glass did not act as a saturable absorber. In addition to these observations, several new effects in QD relaxation dynamics were identified that favor efficient passive mode locking in solid-state laser systems. Specifically, it was found that following excitation at a pump wavelength of 1.3 μm, QD-doped glass having its first excitonic transition at around 2 μm ($R = 5$ nm) had faster relaxation dynamics (~1 ps) and lower saturation fluences.

Detailed study of the bleaching decay of PbS QDs in phosphate glass and their use as a saturable absorber for passive mode locking of a flashlamp-pumped Nd:YAP laser at 1.3 μm have been reported in Ref. [384]. Samples with QD radii of 2.4–2.9 nm have been used with corresponding first excitonic transition in the range of 1.2–1.4 μm [322]. Two-component carrier relaxation dynamics were identified according to the following approximation: $\Delta OD = A_f \exp(-t/\tau_f) + A_s \exp(-t/\tau_s)$, where A_s, A_f, τ_f, and τ_s are the amplitudes (A) and relaxation time constants (τ) of the fast (f) and slow (s) bleaching components. The time constants were found to be in the picosecond and nanosecond ranges at pump energies up to 14 mJ cm^{-2} (excitation pulse duration is 15 ps) at around 1.08 μm. The fast component of relaxation was found to be proportional to R^9 [322]. This was attributed to strong quantum confinement similar to that reported for CdSe QDs [362]. However, such carrier relaxation dependence on PbS dot radii quite differs from observations reported for PbSe QDs earlier [102], where the bleaching relaxation rate $1/\tau$ was found to be ~$1/R^3$ during pump–probe experiments undertaken at around 0.77 μm (excitation pulse duration of 0.2 ps) and probe wavelengths coincident with first excitonic transition of QD samples (from ~1.6 to ~3 μm). To explain such a cubic dependence, the authors suggested the presence of additional volume V_s surrounding the quantum dot, which contains nonradiative centers and where carriers move following optical excitation. In this case, the decay rate is proportional to V_s/V_D ratio, and if it is assumed that V_s is independent of dot radius, then $1/R^3$ dependence may be expected. Apart from the presence of fast and slow components in the bleaching relaxation dynamics of QD-based saturable absorbers, which play significant role in ultrashort-pulse formation, the ratio of their decay amplitudes (A_f and A_s) is another important parameter for optimizing low-threshold and self-starting mode-locking operations in a solid-state laser. Depending on the A_f/A_s ratio, the saturable absorber can be considered to be either slow or fast. It has been shown that in case of PbS QDs in phosphate glass, this ratio varies according to the wavelength difference between the first excitonic transition and probe beam. It can be seen that when the probe is centered on the peak of first excitonic transition, only a fast component is present in the carrier relaxation dynamics [322]. The shift of probe wavelength to lower or higher energy sides of the excitonic band leads to higher A_s/A_f values, which arises mainly due to the variation in the fast component amplitude A_f, while the slow component remains essentially the same. This can be understood if it is assumed that the slow relaxation involves the deep electron traps [383].

Using PbS QD-doped glasses with dot radii of 2.55 and 2.9 nm (see Ref. [322]), Q-switched mode locking has been realized in a flashlamp-pumped Nd:YAP laser operating at 1.3 μm. Single mode-locked pulses as short as 150 ps have been produced with corresponding Q-switched pulse energy of up to 250 μJ. The saturation fluences of these investigated QD-based absorbers were measured to be in the range of 0.8–6 μJ cm^{-2}.

7.4
Cr:YAG Laser Passively Mode Locked with a QD-Doped Glass Saturable Absorber

After the demonstration of the successful deployment of lead salt QD-doped glass saturable absorbers for ultrashort-pulse generation around 1.3 μm and detailed studies of their nonlinear characteristics, it became evident that there are no serious restrictions in the use of such devices for passive mode locking in other spectral regions. It simply requires appropriate absorber design mainly relating to the location of the first excitonic transition.

In Ref. [385], passive mode locking of a Cr:YAG continuous wave laser at around 1.5 μm has been reported using PbS QD-doped silicate glass as a saturable absorber. A double-stage process of heat treatment allowed obtaining a required volume of PbS crystallite fraction with a predetermined dot size. The first stage in this process is a prolonged low-temperature treatment for PbS nucleation (480 °C, 72 h). This was followed by a nanocrystallite growth phase that was performed at 520 °C for a period of 25 h. The correlation between time–temperature heat treatment regimes and the location of the first excitonic resonance of the PbS QDs was found by the analysis of small-angle X-ray scattering (SAXS) data (scattering by monodispersed spherical particles assumed). Reference [322] shows the room-temperature absorption spectrum of PbS QDs-doped glass sample that was assessed for passive mode locking around 1.5 μm. The first excitonic peak was observed at 1.4 μm and the average dot diameter was estimated to be 5.4 nm.

The relaxation time characteristics for the PbS-doped glass sample are indicated in Ref. [322]. To obtain this measurement, 15 ps pulses from a passively mode-locked Nd:YAlO$_3$ laser ($\lambda = 1.08$ μm) were used as the optical excitation to the second excitonic transition of PbS QDs. The third Raman Stokes ($\lambda = 1.524$ μm) from a KGd(WO$_4$)$_2$ crystal, excited by an intensity component of the laser pulse, was used as a probe pulse. The pump beam had a diameter of 3 mm giving the pump intensity on the sample of $\sim$1 GW cm^{-2} ($\sim$15 mJ cm^{-2}). A double-exponential fit to the experimental data (see Ref. [322]) while taking pump and probe pulse durations into account implies a value of 25 ± 3 ps for the fast relaxation component. The ratio

A_s/A_f was found to be 0.05 that indicates that the slow bleaching relaxation component is not appreciable for this PbS-doped glass. Absorption saturation measurements of the PbS-doped glass were performed at 1.34 μm with energy fluences on the sample of up to 10 J cm^{-2}. The intensity-dependent transmission (squares) of the PbS-doped glass at 1.34 μm is shown in Ref. [322]. The intensity-dependent absorption α is modeled within the frame of a two-level scheme for a fast saturable absorber. For a pump pulse that is much longer than the bleaching relaxation time of the saturable absorber having a nonsaturable absorption coefficient of α_{ns},

$$\alpha = \frac{\alpha_0}{1 + I/I_{sat}} + \alpha_{ns},$$

where α_0 is the saturable absorption coefficient, I is the input intensity, and I_{sat} is the saturation intensity. The saturable absorber parameters were thus estimated to be $\alpha_0/\alpha_{ns} = 2.4$ and $I_{sat} = 8$ MW cm^{-2}. The nonsaturable absorption includes the excited-state absorption (ESA) from the lowest energy excitonic state. Using the above values, it was possible to estimate ground-state absorption cross section as well as saturation fluence for the PbS-doped glass. Assuming that the relaxation time at the wavelength 1.34 μm has the same value as at 1.52 μm ($\tau = 25$ ps), the ground-state absorption cross section at 1.34 μm can be estimated from $\sigma_{gsa} = h\nu/I_{sat}\tau$ to be $\approx 7 \times 10^{-16}$ cm^2. From the absorption spectrum (given in Ref. [322]), the ground-state absorption cross section at 1.51 μm can be calculated to be $\sigma_{gsa} \approx 3 \times 10^{-16}$ cm^2. Therefore, the saturation fluence at the wavelength of 1.51 μm is evaluated as $F_{sat} = h\nu/\sigma_{gsa} \approx 500$ μJ cm^{-2}.

For incorporation as a saturable absorber in the laser cavity, the PbS-doped glass plate was polished to a thickness of 90 μm and antireflection coated for wavelengths centered at 1500 nm. The small-signal absorption of the sample ranged from 3 to 0.3% over the 1450–1550 nm region. Self-starting mode-locked operation of the Cr:YAG laser was obtained when the glass sample was inserted in contact with the high-reflector (HR) mirror in the short arm of the cavity where the beam waist was 20 μm (see schematic of laser cavity [322]). A 235 MHz sequence of pulses with an average output power of 35 mW was obtained. Assuming sech^2 intensity profiles, the pulse duration was deduced from autocorrelation measurements to be 10 ps [322]. For the observed spectral width of 0.3 nm at a center wavelength of 1509.5 nm, the implied duration–bandwidth product was 0.4 [322]. Stable mode-locked operation was observed when the PbS-doped glass plate was placed within 1 mm of the high reflectivity mirror such that the laser mode radius varied from 20 to 31 μm corresponding to intensities on the saturable absorber of 0.2–0.1 GW cm^{-2}. Mode locking was maintained throughout the day-long operational period implying that photodarkening in the QD-doped glass was not present. Tunability of the Cr:YAG laser during this mode-locked operation was achieved from 1460 to 1550 nm where tunability to the shorter wavelength region was limited by an increase in the absorption of the saturable absorber. An absence of mode locking at wavelengths longer than 1550 nm can be accounted for by the decrease in the modulation depth of the saturable absorber.

7.5
PbS QD-Doped Glass Saturable Absorbers for Passive Mode Locking around 1 µm and Their Nonlinear Characteristics

The nonlinear response characteristics of QD-doped glasses not only depend on quantum confinement conditions determined by the ratio of dot and exciton radii but are also governed by the light intensity being applied during nonlinear measurements. It has been clearly identified in most instances that an increase of intensity leads to a shortening of the fast relaxation component in the bleaching kinetics [103, 363, 382, 386]. This shortening of the bleaching relaxation time in QDs is not the only consequence of the increased pump intensity because a second effect is the change in the relative impact of the slow and fast component amplitudes A_s/A_f [382]. Therefore, before describing the use of PbS QD-doped glasses as a saturable absorber for efficient passive mode locking in Yb:KYW laser around 1040 nm, it is useful to discuss their nonlinear characteristics under different pump conditions. Another interesting aspect is that PbS QDs are usually synthesized in silicate glass. However, a phosphate glass matrix has some advantages for PbS QDs preparation compared to silicate through access to a higher concentration of semiconductor compounds due to the greater solubility. In addition, a narrower distribution can be achieved for the QD size in the course of the phase decomposition of the semiconductor solid solution in phosphate glass matrices due to the higher oversaturation of the solid solution. Moreover, the lower characteristic temperature and the lower viscosity of phosphate glasses afford more modest annealing temperatures necessary to grow QDs in such matrices and the annealing time is much shorter than for a silicate glass (tens of minutes instead of hours) [378]. Using such a PbS QDs-doped phosphate glass as a saturable absorber, a neodymium laser has been passively mode locked very successfully at 1340 nm [384]. However, the comparison of the nonlinear optical properties of PbS QDs in different glass matrices had not been made. For such assessments, we prepared silicate and phosphate glasses containing approximately 0.4 and 1 wt% of PbS QDs, respectively, having radii around 2 nm with corresponding first exciton transition at around 1 µm [104]. For the nonlinear optical characterizations, the silicate glass samples were polished to a thickness of 2 mm and the phosphate glass was polished to a thickness of 0.5 mm. All samples exhibited an internal small-signal transmission of approximately 50% around 1040 nm. The nonlinear response characteristics of PbS QDs have been studied by a standard pump–probe technique, where a passively mode-locked Yb:KYW laser that produced 144 fs pulses at 1048 nm with an average output power of up to 400 mW at a pulse repetition frequency of 86 MHz was used as the pump source. The pump intensity on the sample was varied in the range of 0.9–15 GW cm^{-2} (0.13–2.2 mJ cm^{-2}). The bleaching value is defined as a differential absorption $\Delta OD = -\log(T/T_0)$, where T and T_0 are transmissions of the sample with and without pump beam, respectively. Increase of pump fluence leads to the shortening of the fast time constants τ_f for all the PbS QDs-doped glasses studied and also to the shortening of the slow time constants τ_s. A summary of these dependencies is presented in Ref. [322] where it can be seen for $R = 1.9$ nm QDs that an increase of the pump fluence at the factor of

≈15 (from 4 to 68 µJ cm^{-2}) leads to a shortening of a similar magnitude in the fast time constant τ_f (from 27 to 2 ps). The same slope (approximately -1) was observed for the 2.15 nm PbS QDs in silicate glass: a shortening of τ_f from 27 to 5 ps with an increase of the pump fluence from 14.5 to 68 µJ cm^{-2} (approximately ×5). Double-exponential features for the bleaching relaxation appear for PbS QDs in silicate and phosphate glasses at 30 and 130 µJ cm^{-2}, respectively. The fast and slow relaxation times in phosphate glass are longer in comparison to that of silicate matrix. The ratio between the amplitudes of the slow and the fast components, A_s/A_f, increases with pump fluence for all samples, as it is evident in Ref. [322]. One additional point to note is that the ratio A_s/A_f shows lower values in phosphate glass than in the silicate counterpart offering lower nonsaturable (residual) absorption due to ESA in PbS QDs at the higher levels of pump power. The observations of fast component relaxation time shortening and the slow component appearing with increase in pump fluences can be explained in the following terms. Under conditions of high pump intensities ($I > 10$ GW cm^{-2}), a QD can exhibit multiphoton absorption and more than one electron–hole pair can be generated. Multiparticle interactions in such QDs can readily lead to trapping of carriers in deep defect states, which exist at the dot surfaces or interfacial layers to produce characteristically longer lifetimes. The Auger process [382] is one of the possible mechanisms for these multiparticle interactions. Such interactions at high pump levels lead to the appearance of the second component in the observed bleaching relaxation due to the trapping of carriers at deep defect states [383]. This also accounts for a shortening in the fast decay time due to the additional mechanism of excited carrier recombination.

It is interesting to compare the fast time constants τ_f for QDs of different sizes taken at different pump intensities. Reference [322] presents such a comparison for three pump levels for QDs synthesized in the silicate glass matrices together with data extracted from Ref. [386]. It was found that variation of τ_f with dot size at pump fluence of ≈30 µJ cm^{-2} (triangles) can be approximated with an R^3 dependence (solid line). At a pump fluence of ≈13 µJ cm^{-2} (circles), the approximation is much worse (dashed line); and at lower pump level ≈4 µJ cm^{-2} (squares), the approximation does not fit. The data obtained at high pump intensities are not in contradiction to the R^3 dependence, which implies that there is the possibility of carrier transitions out of the PbS QD to the volume V_S under intense pumping in addition to other relaxation processes such as direct electron–hole recombination [102].

Intensity-dependent transmission measurements demonstrated the bleaching effect for all PbS QD-doped glasses. The best fits to experimental data gave the values of saturation fluence of 15 and 12 µJ cm^{-2} and nonsaturable absorption coefficients of 2.5 and 3 cm^{-1} for glass samples with QDs of 1.9 and 2.15 nm radii, respectively. The difference in nonsaturable absorption coefficients for these glass samples cannot be associated with imperfections only (according to absorption spectra), it can rather be also attributed primarily to the excited-state absorption in the QDs. Since the pump wavelength for these samples lies on the lower and higher energy sides of the excitonic absorption band, different ESA features can be associated with the diverse ESA cross sections for QDs of different sizes or their variation along the absorption band. It has already been reported [386] that ESA cross sections become higher for larger PbS QDs if measured at wavelengths corresponding to the maxima of excitonic absorption bands. ESA spectra for PbS QDs have not been studied yet, but an indication of an increase of the ESA at the higher energy side of an excitonic peak compared to the lower one can be found in Ref. [384]. We therefore believe that both these factors have impact and such a behavior of the ESA enables to make some recommendations in respect of further developments. The decrease of nonsaturable losses in the SA can be achieved by using glasses with smaller QDs and tuning the laser emission wavelength to the lower energy side of excitonic absorption band.

As a follow-up to this work, a saturable absorber sample for mode-locking experiments with a Yb:KYW laser was prepared by the diffusion bonding technique from the PbS QDs ($R = 1.9$ nm)-doped glass characterized above and a pure glass substrate. The PbS-doped glass was wedged and polished to achieve a gradual change of the internal transmission from 97 to 100%. A plane–parallel disk of this saturable absorber was also prepared with a high reflectivity coating at around 1050 nm on one face and an antireflection coating on the other face (intracavity surface). To achieve saturation of absorption, the intracavity radiation was focused into the absorber to a spot of 55 µm radius. Pulses with durations of 2.6 ps (assuming a sech^2 intensity profile) were produced at an average output power of 250 mW and with the corresponding spectral width of 0.44 nm (centered at 1034.5 nm) [322], the deduced time–bandwidth product is 0.32. The pulse train, recorded with a fast photodiode and a 300 MHz digital oscilloscope, showed stable cw mode-locked operation with less than 2% pulse energy fluctuation [322]. The pulse repetition frequency was measured to be 99 MHz. The threshold for the stable mode-locked operation was achieved when the laser produced 190 mW of output power and this corresponded to a pulse energy fluence of 670 µJ cm^{-2} incident on the absorber. At this fluence, the SA is bleached to the maximum achievable level that corresponds to a modulation depth of $\approx 0.4\%$. At the maximum available pump power of 950 mW, an output power of 250 mW was achieved that corresponds to an optical-to-optical conversion efficiency of 26% (diode pump power to mode-locked average output power). In a cw nonmode-locked operation with a similar cavity design, but with a high reflector instead of the saturable absorber, the laser power was 370 mW that implies a cw-to-mode-locked conversion efficiency of 68%.

8
Emerging Applications of Ultrafast Quantum Dot Lasers

Mode-locked semiconductor lasers are well suited for a wide range of applications, including optical fiber communications, optical clock distribution, clock recovery, radio over fiber signal generation, and optical sampling of high-speed signals and metrology [11, 387–390]. In this section, we will explore what are the most particularly promising application areas for QD-based ultrafast sources.

8.1
Optical Communications

QD mode-locked lasers and semiconductor optical amplifiers are poised to make a large impact on the next generation of optical networks and communication systems. The current InAs QD technology is well suited for optical communications using the O-band (1260–1360 nm) as defined by the International Telecommunication Union (ITU-T). The O-band coincides with the spectral window of lowest dispersion in optical fibers and is of particular interest for metropolitan networks. Quantum dot lasers can now be routinely grown on GaAs substrates, which is a major advantage over quantum well technology because quantum well lasers emitting at the same spectral range have to be grown on InP, with the associated poor performance due to high nonradiative Auger recombination. The telecom optical C-band of 1530–1565 nm (erbium window) has also been addressed, with much research done on this topic in recent times [391], particularly with materials based on quantum dashes grown on InP.

For optical time-division multiplexing systems (OTDMs), the possibility to generate pulses at very high repetition rate and with record low jitter is highly desirable, where a mode-locked QD laser can be used either as a pulse source or in a clock recovery circuit. In OTDM, different slower data signals are interleaved to form a single faster signal. If a shorter pulse duration is used, the switching window can be narrowed, enabling higher bit rates. Very recently, Schmeckebier *et al.* demonstrated the enormous potential of QD mode-locked lasers for optical time-division multiplexing up to 160 Gbit s^{-1}, after temporally interleaving the split output of a 40 GHz hybridly mode-locked QD laser [170]. It is important to emphasize that a 160 GHz rate

Ultrafast Lasers Based on Quantum Dot Structures: Physics and Devices.
Edik U. Rafailov, Maria Ana Cataluna, and Eugene A. Avrutin
Copyright © 2011 WILEY-VCH Verlag GmbH & Co. KGaA, Weinheim
ISBN: 978-3-527-40928-0

corresponds to a temporal periodicity of ~6.3 ps, which implies that the pulse duration has to be significantly shorter than this period – and indeed, the authors have succeeded in compressing the pulses down to 700 fs, prior to multiplexing. Return-to-zero eye diagrams for transmission rates of 40 and 80 Gbit s^{-1} were also presented in this paper, depicting a clear open eye. An important issue that requires attention in this type of application is the variation of the pulse repetition rate with the different bias conditions [205]. As the cavity length determines the pulse repetition rate, and the length can be met within a deviation of ~1% due to uncertainties in the cleaving process, the pulse repetition rate is affected by the same degree of deviation. For future implementation, the inclusion of a refraction index tuning section would perhaps be useful to compensate for this deviation and meet the specifications of the end user [205].

On the other hand, the broad spectral bandwidth offered by QD structures can be deployed in wavelength-division multiplexing systems (WDMs). In this transmission format, each signal is assigned a different wavelength, enabling the independent propagation down the fiber. By engineering the inhomogeneous broadening caused by the size distribution of QDs, a very broad laser emission spectrum can be obtained with nearly uniform intensity distribution [13]. Using slicing techniques, this spectrum can be converted into an array of different equally spaced wavelengths, which is a cost-effective technique to use WDM, without resorting to the use of several laser sources. In fact, a uniform 93-channel multiwavelength QD laser has been demonstrated recently [392] over a wavelength range from 1638 to 1646 nm. The 93 channels were generated directly from the single monolithic chip, without external components or slicing setups, as the channel/mode spacing was simply determined by the length of the cavity (4.5 mm). The stability of the laser modes was attributed to the inhomogeneous broadened nature of the gain.

With the additional features of low threshold and resilience to temperature, QD lasers have become suitable candidates for the next generation of telecommunication sources [170], either fiber based or free space. In fact, the use of QD lasers in space applications is very promising, as they also exhibit enhanced radiation hardness, when compared to QW materials [393, 394]. Due to the cluster-like nature of QD gain material and enhanced carrier confinement, the impact of high-energy radiation on active layers and the consequent generation of defects is less likely to decrease the efficiency of the material, as the carriers are spatially confined. This greatly enhances the reliability of the lasers used in spaceborne applications, such as intersatellite communications.

Optical clocking is also a very important application that has driven, and continues to drive, the research in mode-locked laser diodes [384]. All-optical clock recovery using QD lasers has been intensively investigated in the past few years, but almost exclusively in the InP-based QD materials. Excellent phase noise characteristics and low timing jitter have been demonstrated in all-optical clock recovery at 40 and 160 Gbit s^{-1} using InAs/InP QD lasers [395, 396].

Finally, very recent results have been reported showing the promise of QD mode-locked lasers for the generation of microwave signals, directly from the intrawaveguide saturable absorber [397], schematically represented in Figure 8.1. In this paper,

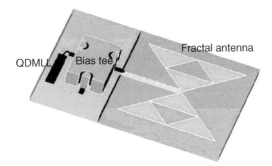

Figure 8.1 Optical integration of the QDMLL with a reconfigurable bowtie antenna. First, high-frequency electrical pulse signals are generated from the saturable absorber of the QDMLL. These signals are then routed by a bias tee and a coplanar waveguide to a reconfigurable bowtie antenna. This integrated unit can then be used as a cellular block in more complex arrays that are controlled, for example, by field programmable gate arrays. Reprinted from Ref. [397] © 2009 IEEE.

Lin et al. demonstrate a differential efficiency of 33% in optical-to-RF power conversion, while the best extraction efficiency of the saturable absorber is shown to be about 86% (for a 10 GHz signal). These results could pave the way for a new range of applications where monolithic QD passively mode-locked lasers could be used as compact RF sources for wireless communications [398].

8.2
Datacoms

Mode-locked semiconductor lasers can also be used as optical interconnects for clock distribution, either on-chip or at an interchip/interboard level [5]. In recent years, there has been much interest in lower repetition rate mode-locked lasers (5 GHz and less) that have potential as optical clocks for computer motherboards, a technology that is being developed to avoid interaction with electrical signal lines. In order to achieve such low repetition rates, it is necessary to fabricate lasers with very long cavities, around 8 mm or more. Technologically, it is less of a challenge to fabricate these long lasers from QD materials as it is with QW lasers, because they exhibit comparatively lower internal optical losses (and thus lower threshold). This relaxes the need of insertion of a low-loss passive waveguide to minimize threshold current and pulse duration, as it is common practice with long-cavity QW lasers [399]. Long-cavity mode-locked QD lasers have been reported since 2004 [178, 400], and their fabrication has reached a level of maturity such that they are now commercially available through Innolume GmBH. This company has recently announced its commitment to the application of QD lasers in silicon-based photonics, as the emission range of 1064–1310 nm coincides with the transparency window of silicon.

A quantum dot crosspoint switch integrated with a monolithic mode-locked QD laser has been demonstrated by Wang et al. [401], where successful routing of the

10 GHz picosecond pulse train was performed, with very little distortion of the pulse – these are promising results for the applications in high-speed switching and routing.

8.3
Biophotonics and Medical Applications

Telecoms and datacoms are two particular niches of applications that semiconductor ultrafast laser diodes have been traditionally designed to address [11, 12]. The ultimate goal is to access applications that have mainly been in the domain of solid-state lasers. Such is the case of biophotonics and medical applications in particular, where compact, rugged, and turnkey sources are crucial for the deployment of sophisticated and noninvasive optical diagnostics and therapeutics. In this respect, compact and simple semiconductor ultrafast sources based on QD materials can offer a number of advantages.

Optical coherence tomography (OCT) – a technique that enables imaging with resolutions up to the micrometer level – is one of the medical diagnostics that may benefit in the near future from developments in QD ultrafast lasers. The resolution achieved by this technique is determined by the wavelength and spectral bandwidth of the optical source and it is desirable to achieve as high a bandwidth as possible. The optical source used in OCT should also have a short coherence length. All these requirements can be satisfied by mode-locked lasers, which have been the best performing sources deployed for OCT, in particular Ti:sapphire lasers (800 nm) and Cr:forsterite and fiber lasers (1300 nm). However, in order to turn OCT an interesting and practical tool, it is crucial to decrease the footprint and complexity of the laser system. Superluminescent diodes have been used to this end, but power levels can be very low.

An alternative could be to use QD-based lasers. The spectral range that is most routinely accessed with QD lasers (around 1.3 μm) can penetrate deeper into biological tissue [402, 403], as it suffers less scattering and absorption than at 800 nm. In fact, the use of this longer wavelength has allowed imaging depths of 3 mm in nontransparent tissues [404].

In order to become competitive sources, QD lasers will have to generate a spectral bandwidth at least comparable to what has been obtained directly by Cr:forsterite and fiber lasers (~100 nm). The possibility of engineering the inhomogeneous broadening in QD laser devices could afford some potential in the fabrication of inexpensive and efficient alternative sources for OCT, as revealed by the recent demonstration of a 75 nm broad spectrum from a cw QD laser [29].

8.4
Outlook

It can be concluded from the content of this book that there is an ongoing evolution in the development of quantum dot optoelectronic devices from laboratory configurations to an increasingly practical and potentially deployable system status. This is especially notable if the progression toward quantum dot structures can be sustained and confirmed because this category of devices would address a number of key specifications for the types of sources needed in ultrafast optoelectronic and photonic technologies. Further progress will rely on both the new insights from the fundamental science and the technical innovations, and so we can predict that research and development activities in this field are set to remain at the forefront of international endeavor for the foreseeable future.

References

1 Zewail, A.H. (2000) Femtochemistry: atomic-scale dynamics of the chemical bond using ultrafast lasers (Nobel lecture). *Angew. Chem., Int. Ed.*, **39**, 2587–2631.
2 Wada, O. et al. (1999) Femtosecond semiconductor-based optoelectronic devices for optical communications and signal-processing systems, in *Femtosecond Technology: from Basic Research to Application Technology* (eds T. Kamiya et al.), Springer, Berlin, pp. 59–174.
3 Knox, W.H. (2000) Ultrafast technology in telecommunications. *IEEE J. Sel. Top. Quantum Electron.*, **6**, 1273–1278.
4 Morioka, T. (2007) Ultrafast optical technologies for large-capacity TDM/WDM photonic networks. *J. Opt. Fiber Commun. Rep.*, **4**, 14–40.
5 Keeler, G.A. et al. (2003) The benefits of ultrashort optical pulses in optically interconnected systems. *IEEE J. Sel. Top. Quantum Electron.*, **9**, 477–485.
6 Valdmanis, J.A. and Mourou, G. (1986) Subpicosecond electrooptic sampling: principles and applications. *IEEE J. Quantum Electron.*, **22**, 69–78.
7 Juodawlkis, P.W. et al. (2001) Optically sampled analog-to-digital converters. *IEEE Trans. Microwave Theory Tech.*, **49**, 1840–1853.
8 Delfyett, P.J. et al. (2002) Signal processing at the speed of lightwaves. *IEEE Circuits Devices*, **18**, 28–35.
9 The Nobel Prize in Chemistry 1999. Nobelprize.org, accessed 5 Aug. 2010, http://nobelprize.org/nobel_prizes/chemistry/laureates/1999
10 The Nobel Prize in Physics 2005. Nobelprize.org, accessed 5 Aug. 2010, http://nobelprize.org/nobel_prizes/physics/laureates/2005
11 Vasil'ev, P. (1995) *Ultrafast Diode Lasers: Fundamentals and Applications*, Artech House, Boston, MA.
12 Avrutin, E.A. et al. (2000) Monolithic and multi-gigahertz mode-locked semiconductor lasers: constructions, experiments, models and applications. *IEE Proc. Optoelectron.*, **147**, 251–278.
13 Rafailov, E.U. et al. (2007) Mode-locked quantum-dot lasers. *Nat. Photon.*, **1**, 395–401.
14 Dupuis, R.D. (1987) An introduction to the development of the semiconductor laser. *IEEE J. Quantum Electron.*, **23**, 651–657.
15 Alferov, Z.I. (2001) The double heterostructure concept and its applications in physics, electronics, and technology (Nobel lecture). *Rev. Mod. Phys.*, **73**, 767.
16 Saleh, B.E.A. and Teich, M.C. (1991) *Fundamentals of Photonics*, John Wiley & Sons, Inc., New York.
17 Dingle, R. and Henry, C.H. (1976) Quantum effects in heterostructure lasers. US Patent 3,982,207.
18 Arakawa, Y. and Sakaki, H. (1982) Multidimensional quantum well laser and temperature dependence of its threshold current. *Appl. Phys. Lett.*, **40**, 939–941.

19 Goldstein, L. et al. (1985) Growth by molecular beam epitaxy and characterization of InAs/GaAs strained-layer superlattices. *Appl. Phys. Lett.*, **47**, 1099–1101.

20 Ustinov, V.M. et al. (2003) *Quantum Dot Lasers*, Oxford University Press, New York.

21 Smowton, P.M. et al. (2001) Optical mode loss and gain of multiple-layer quantum-dot lasers. *Appl. Phys. Lett.*, **78**, 2629–2631.

22 Bhattacharyya, D. et al. (1999) Spectral and dynamic properties of InAs–GaAs self-organized quantum-dot lasers. *IEEE J. Sel. Top. Quantum Electron.*, **5**, 648–657.

23 Alivisatos, A.P. (1996) Semiconductor clusters, nanocrystals, and quantum dots. *Science*, **271**, 933–937.

24 Gaponenko, S.V. (1998) *Optical Properties of Semiconductor Nanocrystals*, Cambridge University Press, Cambridge, UK.

25 Yoffe, A.D. (2001) Semiconductor quantum dots and related systems: electronic, optical, luminescence and related properties of low dimensional systems. *Adv. Phys.*, **50**, 1–208.

26 Dery, H. et al. (2004) On the nature of quantum dash structures. *J. Appl. Phys.*, **95**, 6103–6111.

27 Kovsh, A.R. et al. (2004) Long-wavelength (1.3–1.5 μm) quantum dot lasers based on GaAs, Physics and Simulation of Optoelectronic Devices XII, Bellingham, pp. 31–45.

28 Liu, H.Y. et al. (2005) High-performance three-layer 1.3-μm InAs–GaAs quantum-dot lasers with very low continuous-wave room-temperature threshold currents. *IEEE Photon. Technol. Lett.*, **17**, 1139–1141.

29 Kovsh, A. et al. (2007) Quantum dot laser with 75 nm broad spectrum of emission. *Opt. Lett.*, **32**, 793–795.

30 Asryan, L.V. and Suris, R.A. (1996) Inhomogeneous line broadening and the threshold current density of a semiconductor quantum dot laser. *Semicond. Sci. Technol.*, **11**, 554–567.

31 Ray, S.K. et al. (2006) Broad-band superluminescent light-emitting diodes incorporating quantum dots in compositionally modulated quantum wells. *IEEE Photon. Technol. Lett.*, **18**, 58–60.

32 Rossetti, M. et al. (2006) High-power quantum-dot superluminescent diodes with p-doped active region. *IEEE Photon. Technol. Lett.*, **18**, 1946–1948.

33 Qasaimeh, O. (2003) Effect of inhomogeneous line broadening on gain and differential gain of quantum dot lasers. *IEEE Trans. Electron. Devices*, **50**, 1575–1581.

34 Dery, H. and Eisenstein, G. (2005) The impact of energy band diagram and inhomogeneous broadening on the optical differential gain in nanostructure lasers. *IEEE J. Quantum Electron.*, **41**, 26–35.

35 Klimov, V.I. and McBranch, D.W. (1998) Femtosecond 1p-to-1s electron relaxation in strongly confined semiconductor nanocrystals. *Phys. Rev. Lett.*, **80**, 4028–4031.

36 Benisty, H. et al. (1991) Intrinsic mechanism for the poor luminescence properties of quantum-box systems. *Phys. Rev. B*, **44**, 10945–10948.

37 Mukai, K. et al. (1996) Phonon bottleneck in self-formed $In_xGa_{1-x}As$/GaAs quantum dots by electroluminescence and time-resolved photoluminescence. *Phys. Rev. B*, **54**, R5243–R5246.

38 Borri, P. et al. (2006) Ultrafast carrier dynamics in InGaAs quantum dot materials and devices. *J. Opt. A: Pure Appl. Opt.*, **8**, S33–S36.

39 Borri, P. et al. (2000) Spectral hole-burning and carrier-heating dynamics in InGaAs quantum-dot amplifiers. *IEEE J. Sel. Top. Quantum Electron.*, **6**, 544–551.

40 Rafailov, E.U. et al. (2004) Fast quantum-dot saturable absorber for passive mode-locking of solid-state lasers. *IEEE Photon. Technol. Lett.*, **16**, 2439–2441.

41 Klimov, V.I. et al. (2000) Mechanisms for intraband energy relaxation in semiconductor quantum dots: the role of electron–hole interactions. *Phys. Rev. B*, **61**, R13349.

42 Schaller, R.D. et al. (2005) Breaking the phonon bottleneck in semiconductor nanocrystals via multiphonon emission induced by intrinsic nonadiabatic interactions. *Phys. Rev. Lett.*, **95**, 196401.

43 Klimov, V.I. (2006) Mechanisms for photogeneration and recombination of multiexcitons in semiconductor nanocrystals: implications for lasing and solar energy conversion. *J. Phys. Chem. B*, **110**, 16827–16845.

44 Reschner, D.W. et al. (2009) Pulse amplification and spatio-spectral hole-burning in inhomogeneously broadened quantum-dot semiconductor optical amplifiers. *IEEE J. Quantum Electron.*, **45**, 21–33.

45 Stier, O. et al. (1999) Electronic and optical properties of strained quantum dots modeled by 8-band k center dot p theory. *Phys. Rev. B*, **59**, 5688–5701.

46 Grundmann, M. (1999) The present status of quantum dot lasers. *Physica E: Low-Dimensional Syst. Nanostruct.*, **5**, 167–184.

47 Crowley, M.T. et al. (2009) Interconnection between ground state and excited state gain in InAs/GaAs quantum dot semiconductor optical amplifiers. *Phys. Status Solidi B*, **246**, 868–871.

48 Berg, T.W. et al. (2001) Ultrafast gain recovery and modulation limitations in self-assembled quantum-dot devices. *IEEE Photon. Technol. Lett.*, **3**, 541–543.

49 Viktorov, E.A. et al. (2006) Model for mode locking in quantum dot lasers. *Appl. Phys. Lett.*, **88**, 201102.

50 Uskov, A.V. et al. (2004) Theory of pulse-train amplification without patterning effects in quantum-dot semiconductor optical amplifiers. *IEEE J. Quantum Electron.*, **40**, 306–320.

51 Blood, P. (2009) Gain and recombination in quantum dot lasers. *IEEE J. Sel. Top. Quantum Electron.*, **15**, 808–818.

52 Xing, C. and Avrutin, E.A. (2005) Multimode spectra and active mode locking potential of quantum dot lasers. *J. Appl. Phys.*, **97**, 104301.

53 Berg, T.W. and Mork, J. (2004) Saturation and noise properties of quantum-dot optical amplifiers. *IEEE J. Quantum Electron.*, **40**, 1527–1539.

54 Grillot, F. et al. (2009) Spectral analysis of 1.55-μm InAs–InP(113)B quantum-dot lasers based on a multipopulation rate equations model. *IEEE J. Quantum Electron.*, **45**, 873–879.

55 Jiang, H. and Singh, J. (1999) Nonequilibrium distribution in quantum dots lasers and influence on laser spectral output. *J. Appl. Phys.*, **85**, 7438–7442.

56 Erneux, T. et al. (2009) The fast recovery dynamics of a quantum dot semiconductor optical amplifier. *Appl. Phys. Lett.*, **94**, 113501.

57 Markus, A. and Fiore, A. (2004) Modeling carrier dynamics in quantum-dot lasers. *Phys. Status Solidi A*, **201**, 338–344.

58 Ludge, K. et al. (2009) Publisher's Note: Turn-on dynamics and modulation response in semiconductor quantum dot lasers [Phys. Rev. B 78, 035316 (2008)]. *Phys. Rev. B*, **79**, 079903.

59 Lüdge, K. and Schöll, E., Quantum dot lasers – desynchronised nonlinear dynamics of electrons and holes, *IEEE J. Quantum Electron.*, **45**, 1396–1403.

60 Asryan, L.V. and Suris, R.A. (1997) Charge neutrality violation in quantum-dot lasers. *IEEE J. Sel. Top. Quantum Electron.*, **3**, 148–157.

61 Berg, T.W. and Mork, J. (2003) Quantum dot amplifiers with high output power and low noise. *Appl. Phys. Lett.*, **82**, 3083–3085.

62 Chen, Y.-H. and Mørk, J. (2010) Microwave signal processing based on ultrafast dynamics in quantum dot waveguides. International Conference on Transparent Optical Networks, Munich, p. WeB3.3.

63 Malic, E. et al. (2007) Coulomb damped relaxation oscillations in semiconductor quantum dot lasers. *IEEE J. Sel. Top. Quantum Electron.*, **13**, 1242–1248.

64 Grundmann, M. et al. (1997) Carrier dynamics in quantum dots: modeling with master equations for the transitions between micro-states. *Phys. Status Solidi B*, **203**, 121–132.

65 Lang, R. (2001) Problems in recent analysis of injected carrier dynamics in semiconductor quantum dots. *Appl. Phys. Lett.*, **79**, 3912–3913.

66 Zhukov, A.E. et al. (1999) Gain characteristics of quantum dot injection lasers. *Semicond. Sci. Technol.*, **14**, 118–123.

67 Sugawara, M. et al. (2002) Quantum-dot semiconductor optical amplifiers for

high-bit-rate signal processing up to 160 Gb s^{-1} and a new scheme of 3R regenerators. *Meas. Sci. Technol.*, **13**, 1683–1691.

68 Akiyama, T., Hatori, N., Nakata, Y., Ebe, H., and Sugawara, M. (2003) Pattern-effect-free amplification and cross-gain modulation achieved by using ultrafast gain nonlinearity in quantum-dot semiconductor optical amplifiers. *Phys. Status Solidi B*, **238**, 301–304.

69 Spyropoulou, M. et al. (2008) A high-speed multiwavelength clock recovery scheme for optical packets. *IEEE Photon. Technol. Lett.*, **20**, 2147–2149.

70 Qasaimeh, O. (2004) Ultrafast dynamic properties of quantum dot semiconductor optical amplifiers including line broadening and carrier heating. *IEE Proc. Optoelectron.*, **151**, 205–210.

71 Coldren, L.A. and Corzine, S.W. (1995) *Diode Lasers and Photonic Integrated Circuits*, John Wiley & Sons, Inc., New York.

72 Ju, H. et al. (2006) Effects of two-photon absorption on carrier dynamics in quantum-dot optical amplifiers. *Appl. Phys. B*, **82**, 615–620.

73 Agrawal, G.P. (1997) *Fiber-Optic Communication Systems*, John Wiley & Sons, Inc., New York.

74 Agrawal, G.P. and Olsson, N.A. (1989) Self-phase modulation and spectral broadening of optical pulses in semiconductor laser amplifiers. *IEEE J. Quantum Electron.*, **25**, 2297–2306.

75 Mork, J. and Mecozzi, A. (1997) Theory of nondegenerate four-wave mixing between pulses in a semiconductor waveguide. *IEEE J. Quantum Electron.*, **33**, 545–555.

76 Mecozzi, A. and Mork, J. (1997) Saturation effects in nondegenerate four-wave mixing between short optical pulses in semiconductor laser amplifiers. *IEEE J. Sel. Top. Quantum Electron.*, **3**, 1190–1207.

77 Bakonyi, Z. et al. (2003) High-gain quantum-dot semiconductor optical amplifier for 1300 nm. *IEEE J. Quantum Electron.*, **39**, 1409–1414.

78 Rafailov, E.U. et al. (2003) Amplification of femtosecond pulses over by 18 dB in a quantum-dot semiconductor optical amplifier. *IEEE Photon. Technol. Lett.*, **15**, 1023–1025.

79 Choi, M.T. et al. (2005) Ultrashort, high-power pulse generation from a master oscillator power amplifier based on external cavity mode locking of a quantum-dot two-section diode laser. *Appl. Phys. Lett.*, **87**, 221107.

80 Tucker, R.S. et al. (1988) Optical time-division multiplexing for very high bit-rate transmission. *J. Lightwave Technol.*, **6**, 1737–1749.

81 Laemmlin, M. et al. (2006) Distortion-free optical amplification of 20–80 GHz modelocked laser pulses at 1.3 μm using quantum dots. *Electron. Lett.*, **42**, 697–699.

82 Gehrig, E. and Hess, O. (2004) Minimized deterioration of ultrashort pulses in quantum dot optical amplifiers. Semiconductor Lasers and Laser Dynamics, pp. 477–485.

83 Manning, R.J. et al. (1997) Semiconductor laser amplifiers for ultrafast all-optical signal processing. *J. Opt. Soc. Am. B*, **14**, 3204–3216.

84 Akiyama, T. et al. (2002) Pattern-effect-free semiconductor optical amplifier achieved using quantum dots. *Electron. Lett.*, **38**, 1139–1140.

85 Song, X.B. et al. (2007) Integrated packet-level all-optical clock recovery scheme based on a resonator and a quantum dot optical amplifier, in *Proceedings of the 9th International Conference on Transparent Optical Networks*, IEEE, New York.

86 Kehayas, E. et al. (2005) 40 Gb/s all-optical packet clock recovery with ultrafast lock-in time and low inter-packet guardbands. *Opt. Express*, **13**, 475–480.

87 Ben Ezra, Y. et al. (2009) Ultrafast all-optical processor based on quantum-dot semiconductor optical amplifiers. *IEEE J. Quantum Electron.*, **45**, 34–41.

88 Keller, U. (2003) Recent developments in compact ultrafast lasers. *Nature*, **424**, 831–838.

89 Piwonski, J.P.T., Madden, G., Houlihan, J., Huyet, G., Viktorov, E.A., Erneux, T., and Mandel, P. (2010) The nonlinear absorption and phase recovery of

quantum dot based reverse-biased waveguide electro-absorbers. International Conference on Transparent Optical Networks (ICTON-2010), Munich, p. ThB4.4.

90 Ryvkin, B.S. et al. (2009) Use of the Franz–Keldysh effect in self-biased p-i-n-heterostructures for saturable absorber mirrors. *Semicond. Sci. Technol.*, **24**, 025001.

91 Elsäßer, W. and Breuer, M.H.S. (2010) State-switched modelocking of two-segment quantum dot laser via self-electro-optical quantum dot absorber. *Electron. Lett.*, **46**, 2.

92 Nikolaev, V.V. and Avrutin, E.A. (2003) Photocarrier escape time in quantum-well light-absorbing devices: effects of electric field and well parameters. *IEEE J. Quantum Electron.*, **39**, 1653–1660.

93 Avrutin, E. et al. (2009) External electrical and optical effects in the operation of monolithic mode-locked laser diodes and the potential of nanostructure technologies in reducing these effects, in *Proceedings of the 11th International Conference on Transparent Optical Networks*, IEEE, New York.

94 Karin, J.R. et al. (1994) Ultrafast dynamics in field-enhanced saturable absorbers. *Appl. Phys. Lett.*, **64**, 676–678.

95 Hofmann, M. et al. (1997) Temporal and spectral dynamics in multiquantum-well semiconductor saturable absorbers. *IEEE Photon. Technol. Lett.*, **9**, 622–624.

96 Keller, U. et al. (1996) Semiconductor saturable absorber mirrors (SESAM's) for femtosecond to nanosecond pulse generation in solid-state lasers. *IEEE J. Sel. Top. Quantum Electron.*, **2**, 435–453.

97 Unlu, M.S. and Strite, S. (1995) Resonant-cavity enhanced photonic devices. *J. Appl. Phys.*, **78**, 607–639.

98 Viktorov, E.A. et al. (2009) Recovery time scales in a reversed-biased quantum dot absorber. *Appl. Phys. Lett.*, **94**, 263502.

99 Malins, D.B. et al. (2006) Ultrafast electroabsorption dynamics in an InAs quantum dot saturable absorber at 1.3 µm. *Appl. Phys. Lett.*, **89**, 171111.

100 Piwonski, T. et al. (2009) Intradot dynamics of InAs quantum dot based electroabsorbers. *Appl. Phys. Lett.*, **94**, 123504.

101 Nikolaev, V.V. and Avrutin, E.A. (2004) Quantum-well design for monolithic optical devices with gain and saturable absorber sections. *IEEE Photon. Technol. Lett.*, **16**, 24–26.

102 Okuno, T. et al. (2000) Size-dependent picosecond energy relaxation in PbSe quantum dots. *Appl. Phys. Lett.*, **77**, 504–506.

103 Wundke, K. et al. (2000) PbS quantum-dot-doped glasses for ultrashort-pulse generation. *Appl. Phys. Lett.*, **76**, 10–12.

104 Malyarevich, A.M. et al. (2006) Nonlinear spectroscopy of PbS quantum-dot-doped glasses as saturable absorbers for the mode locking of solid-state lasers. *J. Appl. Phys.*, **100**, 23108.

105 Zhang, J.Z. (2000) Interfacial charge carrier dynamics of colloidal semiconductor nanoparticles. *J. Phys. Chem. B*, **104**, 7239–7253.

106 Padilha, L.A. et al. (2005) Ultrafast optical switching with CdTe nanocrystals in a glass matrix. *Appl. Phys. Lett.*, **86**, 161111.

107 Flytzanis, C. (2005) Nonlinear optics in mesoscopic composite materials. *J. Phys. B: At. Mol. Opt.*, **38**, S661–S679.

108 Klimov, V.I. et al. (1996) Optical nonlinearities and carrier trapping dynamics in CdS and Cu_xS nanocrystals. *Superlatt. Microstruct.*, **20**, 395–404.

109 Klimov, V.I. and Karavanskii, V.A. (1996) Mechanisms for optical nonlinearities and ultrafast carrier dynamics in Cu_xS nanocrystals. *Phys. Rev. B*, **54**, 8087–8094.

110 Wise, F.W. (2000) Lead salt quantum dots: the limit of strong quantum confinement. *Acc. Chem. Res.*, **33**, 773–780.

111 Huang, X.D. et al. (2001) Bistable operation of a two-section 1.3 µm InAs quantum dot laser: absorption saturation and the quantum confined Stark effect. *IEEE J. Quantum Electron.*, **37**, 414–417.

112 Lorenser, D. et al. (2004) Towards wafer-scale integration of high repetition rate passively mode-locked surface-emitting semiconductor lasers. *Appl. Phys. B*, **79**, 927–932.

113 Brown, C.T.A. et al. (2004) Compact laser-diode-based femtosecond sources. *New J. Phys.*, **6**, 175.

114 Ell, R. et al. (2001) Generation of 5-fs pulses and octave-spanning spectra directly from a Ti:sapphire laser. *Opt. Lett.*, **26**, 373–375.

115 Innerhofer, E. et al. (2003) 60-W average power in 810-fs pulses from a thin-disk Yb:YAG laser. *Opt. Lett.*, **28**, 367–369.

116 Sucha, G. (2003) Overview of industrial and medical applications of ultrafast lasers, in *Ultrafast Lasers: Technology and Applications* (eds M.E. Fermann et al.), Marcel Dekker, Inc., New York, pp. 323–358.

117 Kim, K. et al. (2005) 1.4 kW high peak power generation from an all semiconductor mode-locked master oscillator power amplifier system based on eXtreme Chirped Pulse Amplification (X-CPA). *Opt. Express*, **13**, 4600–4606.

118 Huang, X.D. et al. (2001) Passive mode-locking in 1.3 μm two-section InAs quantum dot lasers. *Appl. Phys. Lett.*, **78**, 2825–2827.

119 Delpon, E.L. et al. (1998) Ultrafast excitonic saturable absorption in ion-implanted InGaAs/InAlAs multiple quantum wells. *Appl. Phys. Lett.*, **72**, 759–761.

120 Haus, H.A. and Silberberg, Y. (1985) Theory of mode-locking of a laser diode with a multiple-quantum-well structure. *J. Opt. Soc. Am. B*, **2**, 1237–1243.

121 Leegwater, J.A. (1996) Theory of mode-locked semiconductor lasers. *IEEE J. Quantum Electron.*, **32**, 1782–1790.

122 New, G.H.C. (1974) Pulse evolution in mode-locked quasicontinuous lasers. *IEEE J. Quantum Electron.*, **10**, 115–124.

123 Vladimirov, A.G. et al. (2004) Delay differential equations for mode-locked semiconductor lasers. *Opt. Lett.*, **29**, 1221–1223.

124 Vladimirov, A.G. and Turaev, D. (2005) Model for passive mode locking in semiconductor lasers. *Phys. Rev. A*, **72**, 033808.

125 Haus, H.A. (1975) Theory of mode-locking with a slow saturable absorber. *IEEE J. Quantum Electron.*, **11**, 736–746.

126 Koumans, R. and van Roijen, R. (1996) Theory for passive mode-locking in semiconductor laser structures including the effects of self-phase modulation, dispersion, and pulse collisions. *IEEE J. Quantum Electron.*, **32**, 478–492.

127 Arahira, S. and Ogawa, Y. (1997) Repetition-frequency tuning of monolithic passively mode-locked semiconductor lasers with integrated extended cavities. *IEEE J. Quantum Electron.*, **33**, 255–264.

128 Dubbeldam, J.L.A. et al. (1997) Theory of mode-locked semiconductor lasers with finite absorber relaxation times. *Appl. Phys. Lett.*, **70**, 1938–1940.

129 Khalfin, V.B. et al. (1995) A theoretical model of synchronization of a mode-locked semiconductor laser with an external pulse stream. *IEEE J. Sel. Top. Quantum Electron.*, **1**, 523–527.

130 Mulet, J. and Balle, S. (2005) Mode-locking dynamics in electrically driven vertical-external-cavity surface-emitting lasers. *IEEE J. Quantum Electron.*, **41**, 1148–1156.

131 Martins, J.F. et al. (1995) Monolithic multiple colliding pulse mode-locked quantum-well lasers: experiment and theory. *IEEE J. Sel. Top. Quantum Electron.*, **1**, 539–551.

132 Bischoff, S. et al. (1997) Monolithic colliding pulse mode-locked semiconductor lasers. *Quantum Semiclass. Opt.*, **9**, 655–674.

133 Heck, M.J.R. et al. (2006) Simulation and design of integrated femtosecond passively mode-locked semiconductor ring lasers including integrated passive pulse shaping components. *IEEE J. Sel. Top. Quantum Electron.*, **12**, 265–276.

134 Bandelow, U. et al. (2001) Impact of gain dispersion on the spatio-temporal dynamics of multisection lasers. *IEEE J. Quantum Electron.*, **37**, 183–188.

135 Avrutin, E.A. et al. (1996) Analysis of dynamics of monolithic passively mode-locked laser diodes under external periodic excitation. *IEE Proc. Optoelectron.*, **143**, 81–88.

136 Asryan, L.V. and Suris, R.A. (2000) Longitudinal spatial hole burning in a quantum-dot laser. *IEEE J. Quantum Electron.*, **36**, 1151–1160.

137 Avrutin, E.A. et al. (2005) Monolithic mode-locked semiconductor lasers, in *Optoelectronic Devices: Advanced*

Simulation and Analysis (ed. J. Piprek), Springer, New York, pp. 185–215.
138. Avrutin, E.A. *et al.* (2003) Dynamic modal analysis of monolithic mode-locked semiconductor lasers. *IEEE J. Sel. Top. Quantum Electron.*, **9**, 844–856.
139. Kuznetsov, M. (1985) Pulsations of semiconductor lasers with a proton bombarded segment: well-developed pulsations. *IEEE J. Quantum Electron.*, **21**, 587–592.
140. Avrutin, E.A. *et al.* (1991) Effect of nonlinear amplification on characteristics of quality modulation regime in semiconducting laser with fast saturated absorbers. *Pisma Zh. Tekh. Fiz.*, **17**, 49–54.
141. Salvatore, R.A. *et al.* (1996) Supermodes of high-repetition-rate passively mode-locked semiconductor lasers. *IEEE J. Quantum Electron.*, **32**, 941–952.
142. Viktorov, E.A. *et al.* (2007) Stability of the mode-locked regime in quantum dot lasers. *Appl. Phys. Lett.*, **91**, 231116.
143. Cataluna, M.A. *et al.* (2007) Temperature dependence of pulse duration in a mode-locked quantum-dot laser. *Appl. Phys. Lett.*, **90**, 101102.
144. Fiol, G. *et al.* (2010) Hybrid mode-locking in a 40 GHz monolithic quantum dot laser. *Appl. Phys. Lett.*, **96**, 011104.
145. Cataluna, M.A. *et al.* (2010) Dual-wavelength mode-locked quantum-dot laser, via ground and excited state transitions: experimental and theoretical investigation. *Opt. Express*, **18**, 12832–12838.
146. Breuer, S. *et al.* (2010) Two-state passive mode-locking of quantum dot semiconductor lasers: classical state scenario and novel reverse state dynamics. International Conference on Transparent Optical Networks, Munich, p. WeD4.4.
147. Rossetti, M. and Montrosset, I. (2010) Time-domain travelling-wave model for quantum dot passively mode-locked lasers. *IEEE J. Quantum Electron.*, **46**, 11.
148. Radziunas, M. *et al.* (2010) Traveling wave modeling, simulation, and analysis of quantum-dot mode-locked semiconductor lasers. Semiconductor Lasers and Laser Dynamics IV, Brussels, p. 77200X.
149. Usechak, N.G. *et al.* (2009) Modeling and direct electric-field measurements of passively mode-locked quantum-dot lasers. *IEEE J. Sel. Top. Quantum Electron.*, **15**, 653–660.
150. Thompson, M.G. *et al.* (2009) InGaAs quantum-dot mode-locked laser diodes. *IEEE J. Sel. Top. Quantum Electron.*, **15**, 661–672.
151. Thompson, M.G. *et al.* (2004) Properties of InGaAs quantum dot saturable absorbers in monolithic mode-locked lasers. IEEE 19th International Semiconductor Laser Conference, Matsue Shi, Japan, pp. 53–54.
152. Delfyett, P.J. *et al.* (1998) Intracavity spectral shaping in external cavity mode-locked semiconductor diode lasers. *IEEE J. Sel. Top. Quantum Electron.*, **4**, 216–223.
153. Cataluna, M.A. *et al.* (2006) Stable mode locking via ground- or excited-state transitions in a two-section quantum-dot laser. *Appl. Phys. Lett.*, **89**, 81124.
154. Kim, J. *et al.* (2006) Pulse generation and compression via ground and excited states from a grating coupled passively mode-locked quantum dot two-section diode laser. *Appl. Phys. Lett.*, **89**, 261106.
155. Breuer, S. *et al.* (2010) State-switched modelocking of two-segment quantum dot laser via self-electro-optical quantum dot absorber. *Electron. Lett.*, **46**, 161–162.
156. Cataluna, M.A. *et al.* (2009) Dual-wavelength mode-locked GaAs-based quantum-dot laser. CLEO/Europe and EQEC 2009 Conference Digest, Germany, paper CB4_6.
157. Breuer, S. *et al.* (2010) Reverse ground-state excited-state transition dynamics in two-section quantum dot semiconductor lasers: mode-locking and state-switching. *Proc. SPIE*, **7720**, 772011.
158. Cataluna, M.A. *et al.* (2006) New mode locking regime in a quantum-dot laser: enhancement by simultaneous CW excited-state emission. Conference on Lasers and Electro-Optics/Quantum Electronics and Laser Science Conference, Long Beach, CA.
159. Kim, J. *et al.* (2006) Wavelength tunable mode-locked quantum-dot laser. Enabling Photonics Technologies for Defense, Security, and Aerospace

Applications II, Kissimmee, FL, pp. M2430–M3430.
160 Delfyett, P.J. et al. (2006) Recent advances in stabilized ultrafast modelocked semiconductor diode lasers for high speed information based applications. 19th Annual Meeting of the IEEE Lasers and Electro-Optics Society, Montreal, Canada, p. 791.
161 Thompson, M.G. et al. (2006) Subpicosecond high-power mode locking using flared waveguide monolithic quantum-dot lasers. *Appl. Phys. Lett.*, **88**, 133119.
162 Cataluna, M.A. et al. (2006) Stable mode-locked operation up to 80 °C from an InGaAs quantum-dot laser. *IEEE Photon. Technol. Lett.*, **18**, 1500–1502.
163 Cataluna, M.A. et al. (2006) Temperature dependence of pulse duration in a mode-locked quantum-dot laser: experiment and theory. 19th Annual Meeting of the IEEE Lasers and Electro-Optics Society, Montreal, Canada, pp. 798–799.
164 Cataluna, M.A. et al. (2010) Temperature dependence of electroabsorption dynamics in an InAs quantum-dot saturable absorber at 1.3 μm and its impact on mode-locked quantum-dot lasers, *Appl. Phys. Lett.*, **97**, 121110.
165 Moore, S.A. et al. (2006) Reduced surface sidewall recombination and diffusion in quantum-dot lasers. *IEEE Photon. Technol. Lett.*, **18**, 1861–1863.
166 Ouyang, D. et al. (2003) High performance narrow stripe quantum-dot lasers with etched waveguide. *Semicond. Sci. Technol.*, **18**, L53–L54.
167 Ouyang, D. et al. (2004) Impact of the mesa etching profiles on the spectral hole burning effects in quantum dot lasers. *Semicond. Sci. Technol.*, **19**, L43–L47.
168 Kuntz, M. et al. (2004) Direct modulation and mode locking of 1.3 μm quantum dot lasers. *New J. Phys.*, **6**, 181.
169 Fiol, G. et al. (2009) Quantum-dot semiconductor mode-locked lasers and amplifiers at 40 GHz. *IEEE J. Quantum Electron.*, **45**, 1429–1435.
170 Schmeckebier, H. et al. (2010) Complete pulse characterization of quantum-dot mode-locked lasers suitable for optical communication up to 160 Gbit/s. *Opt. Express*, **18**, 3415–3425.
171 Ribbat, C. et al. (2003) Complete suppression of filamentation and superior beam quality in quantum-dot lasers. *Appl. Phys. Lett.*, **82**, 952–954.
172 Gehrig, E. et al. (2004) Dynamic filamentation and beam quality of quantum-dot lasers. *Appl. Phys. Lett.*, **84**, 1650–1652.
173 Thompson, M.G. et al. (2004) Transform-limited optical pulses from 18 GHz monolithic modelocked quantum dot lasers operating at 1.3 μm. *Electron. Lett.*, **40**, 346–347.
174 Thompson, M.G. et al. (2003) 10 GHz hybrid modelocking of monolithic InGaAs quantum dot lasers. *Electron. Lett.*, **39**, 1121–1122.
175 Grundmann, M. et al. (2000) Progress in quantum dot lasers: 1100 nm, 1300 nm, and high power applications. *Jpn. J. Appl. Phys. 1*, **39**, 2341–2343.
176 Elliott, S.N. et al. (2008) Higher catastrophic optical mirror damage power density level at facet from quantum dot material, in *Proceedings of the 21st International Semiconductor Laser Conference*, IEEE, New York.
177 Derickson, D.J. et al. (1991) Comparison of timing jitter in external and monolithic cavity mode-locked semiconductor lasers. *Appl. Phys. Lett.*, **59**, 3372–3374.
178 Zhang, L. et al. (2005) Low timing jitter, 5 GHz optical pulses from monolithic two-section passively mode-locked 1250/1310 nm quantum dot lasers for high-speed optical interconnects. Optical Fiber Communication Conference, Anaheim, CA.
179 Newell, T.C. et al. (1999) Gain and linewidth enhancement factor in InAs quantum-dot laser diodes. *IEEE Photon. Technol. Lett.*, **11**, 1527–1529.
180 Martinez, A. et al. (2005) Static and dynamic measurements of the alpha-factor of five-quantum-dot-layer single-mode lasers emitting at 1.3 μm on GaAs. *Appl. Phys. Lett.*, **86**, 211115.
181 Su, H. and Lester, L.F. (2005) Dynamic properties of quantum dot distributed feedback lasers: high speed, linewidth and chirp. *J. Phys. D: Appl. Phys.*, **38**, 2112–2118.
182 Dagens, B. et al. (2005) Giant linewidth enhancement factor and purely frequency

modulated emission from quantum dot laser. *Electron. Lett.*, **41**, 323–324.

183 Kim, J. and Delfyett, P.J. (2009) Above threshold spectral dependence of linewidth enhancement factor, optical duration and linear chirp of quantum dot lasers. *Opt. Express*, **17**, 22566–22570.

184 Vazquez, J.M. et al. (2006) Linewidth enhancement factor of quantum-dot optical amplifiers. *IEEE J. Quantum Electron.*, **42**, 986–993.

185 Schneider, S. et al. (2004) Linewidth enhancement factor in InGaAs quantum-dot amplifiers. *IEEE J. Quantum Electron.*, **40**, 1423–1429.

186 Rafailov, E.U. et al. (2004) High-power ultrashort pulses output from a mode-locked two-section quantum-dot laser. Conference on Lasers and Electro-Optics/International Quantum Electronics Conference, San Francisco, CA, pp. 1031–1032.

187 Rafailov, E.U. et al. (2005) High-power picosecond and femtosecond pulse generation from a two-section mode-locked quantum-dot laser. *Appl. Phys. Lett.*, **87**, 81107.

188 Huang, X.D. et al. (2001) Bistable operation of a two-section 1.3-μm InAs quantum dot laser: absorption saturation and the quantum confined Stark effect. *IEEE J. Quantum Electron.*, **37**, 414–417.

189 Viktorov, E.A. et al. (2007) Stability of the mode-locked regime in quantum dot lasers. *Appl. Phys. Lett.*, **91**, 231116.

190 Xin, Y.C. et al. (2007) Reconfigurable quantum dot monolithic multi-section passive mode-locked lasers. *Opt. Express*, **15**, 7623–7633.

191 Renaudier, J. et al. (2005) 45 GHz self-pulsation with narrow linewidth in quantum dot Fabry–Perot semiconductor lasers at 1.5 μm. *Electron. Lett.*, **41**, 1007–1008.

192 Gosset, C. et al. (2006) Subpicosecond pulse generation at 134 GHz using a quantum-dash-based Fabry–Perot laser emitting at 1.56 μm. *Appl. Phys. Lett.*, **88**, 241105.

193 Derickson, D.J. et al. (1992) Short pulse generation using multisegment mode-locked semiconductor lasers. *IEEE J. Quantum Electron.*, **28**, 2186–2202.

194 Kane, D.J. and Trebino, R. (1993) Characterization of arbitrary femtosecond pulses using frequency-resolved optical gating. *IEEE J. Quantum Electron.*, **29**, 571–579.

195 Xin, Y.C. et al. (2008) Frequency-resolved optical gating characterisation of passively modelocked quantum-dot laser. *Electron. Lett.*, **44**, 1255–1256.

196 Kuntz, M. et al. (2004) 35 GHz mode-locking of 1.3 μm quantum-dot lasers, *Appl. Phys. Lett.*, **85**, 843.

197 Gubenko, A.E. et al. (2004) Mode-locking at 9.7 GHz repetition rate with 1.7 ps pulse duration in two-section QD lasers, in *19th International Semiconductor Laser Conference*, Matsue Shi, Japan, 51–52.

198 Thompson, M.G. et al. (2005) Colliding-pulse modelocked quantum dot lasers. *Electron. Lett.*, **41**, 248–250.

199 Rae, A.R. et al. (2006) Harmonic mode-locking of a quantum-dot laser diode. 19th Annual Meeting of the IEEE Lasers and Electro-Optics Society, Montreal, Canada, pp. 874–875.

200 Habruseva, T. et al. (2010) Quantum-dot mode-locked lasers with dual-mode optical injection. *IEEE Photon. Technol. Lett.*, **22**, 359–361.

201 Dorrer, C. and Kang, I. (2002) Simultaneous temporal characterization of telecommunication optical pulses and modulators by use of spectrograms. *Opt. Lett.*, **27**, 1315–1317.

202 Mar, A. et al. (1995) High-power mode-locked semiconductor lasers using flared waveguides. *Appl. Phys. Lett.*, **66**, 3558–3560.

203 Nikitichev, D.I. et al. (2010) High-power passively mode-locked tapered InAs/GaAs quantum-dot lasers, *Appl. Phys. B*, **101**, 587–591.

204 Markus, A. et al. (2003) Simultaneous two-state lasing in quantum-dot lasers. *Appl. Phys. Lett.*, **82**, 1818–1820.

205 Kuntz, M. et al. (2007) High-speed mode-locked quantum-dot lasers and optical amplifiers. *Proc. IEEE*, **95**, 1767–1778.

206 McRobbie, A.D. et al. (2007) High power all-quantum-dot-based external cavity modelocked laser. *Electron. Lett.*, **43**, 812–813.

207 Choi, M.-T. et al. (2006) Ultralow noise optical pulse generation in an actively

mode-locked quantum-dot semiconductor laser. *Appl. Phys. Lett.*, **88**, 131106.
208 McRobbie, A.D. et al. (2006) High power all-quantum-dot based external cavity mode-locked laser. 19th Annual Meeting of the IEEE Lasers and Electro-Optics Society, Montreal, Canada, pp. 796–797.
209 Xia, M. et al. (2008) External-cavity mode-locked quantum-dot lasers for low repetition rate, sub-picosecond pulse generation. Conference on Lasers and Electro-Optics & Quantum Electronics and Laser Science Conference, vols.1–9, pp. 1137–1138.
210 Burns, D. et al. (1990) Noise characterization of a mode-locked InGaAsP semiconductor diode laser. *IEEE J. Quantum Electron.*, **26**, 1860–1863.
211 Rae, A.R. et al. (2006) Absorber length optimisation for sub-picosecond pulse generation in passively mode-locked 1.3 μm quantum-dot laser diodes. Semiconductor Lasers and Laser Dynamics II, pp. F1841–F2841.
212 Todaro, M.T. et al. (2006) Simultaneous achievement of narrow pulse width and low pulse-to-pulse timing jitter in 1.3 μm passively mode-locked quantum-dot lasers. *Opt. Lett.*, **31**, 3107–3109.
213 Thompson, M.G. et al. (2006) Monolithic hybrid and passive mode-locked 40 GHz quantum dot laser diodes. 32nd European Conference on Optical Communication, Nice, France.
214 Tourrenc, J.P. et al. (2006) Cross-correlation timing jitter measurement of high power passively mode-locked two-section quantum-dot lasers. *IEEE Photon. Technol. Lett.*, **18**, 2317–2319.
215 Jiang, L.A. et al. (2002) Measuring timing jitter with optical cross correlations. *IEEE J. Quantum Electron.*, **38**, 1047–1052.
216 Kefelian, F. et al. (2008) RF linewidth in monolithic passively mode-locked semiconductor laser. *IEEE Photon. Technol. Lett.*, **20**, 1405–1407.
217 Carpintero, G. et al. (2009) Low noise performance of passively mode-locked 10-GHz quantum-dot laser diode. *IEEE Photon. Technol. Lett.*, **21**, 389–391.
218 Habruseva, T. et al. (2009) Optical linewidth of a passively mode-locked semiconductor laser. *Opt. Lett.*, **34**, 3307–3309.
219 Breuer, S. et al. (2010) Investigations of repetition rate stability of a mode-locked quantum dot semiconductor laser in an auxiliary optical fiber cavity. *IEEE J. Quantum Electron.*, **46**, 150–157.
220 Lin, C.Y. et al. (2010) rf linewidth reduction in a quantum dot passively mode-locked laser subject to external optical feedback. *Appl. Phys. Lett.*, **96**, 51118
221 Avrutin, E.A. and Russell, B.M. (2009) Dynamics and spectra of monolithic mode-locked laser diodes under external optical feedback. *IEEE J. Quantum Electron.*, **45**, 1456–1464.
222 Grillot, F. et al. (2009) Optical feedback instabilities in a monolithic InAs/GaAs quantum dot passively mode-locked laser. *Appl. Phys. Lett.*, **94**, 153503.
223 Kim, J. and Delfyett, P.J. (2008) Interband optical pulse injection locking of quantum dot mode-locked semiconductor laser. *Opt. Express*, **16**, 11153–11161.
224 Klopf, F. et al. (2002) Correlation between the gain profile and the temperature-induced shift in wavelength of quantum-dot lasers. *Appl. Phys. Lett.*, **81**, 217–219.
225 Malins, D.B. et al. (2007) Temperature dependence of electroabsorption dynamics in an InAs quantum dot saturable absorber at 1.3 μm. European Conference on Lasers and Electro-Optics, Munich, Germany.
226 Shchekin, O.B. and Deppe, D.G. (2002) 1.3 μm InAs quantum dot laser with $T_0 = 161$ K from 0 to 80 °C. *Appl. Phys. Lett.*, **80**, 3277–3279.
227 Zhukov, A.E. et al. (1999) Long-wavelength lasing from multiply stacked InAs/InGaAs quantum dots on GaAs substrates. *Appl. Phys. Lett.*, **75**, 1926–1928.
228 Mikhrin, S.S. et al. (2005) High power temperature-insensitive 1.3 μm InAs/InGaAs/GaAs quantum dot lasers. *Semicond. Sci. Technol.*, **20**, 340–342.
229 Shchekin, O.B. and Deppe, D.G. (2002) Low-threshold high-T_0 1.3-μm InAs quantum-dot lasers due to p-type

modulation doping of the active region. *IEEE Photon. Technol. Lett.*, **14**, 1231–1233.

230 Fathpour, S. *et al.* (2005) High-speed quantum dot lasers. *J. Phys. D: Appl. Phys.*, **38**, 2103–2111.

231 Schneider, S. *et al.* (2005) Excited-state gain dynamics in InGaAs quantum-dot amplifiers. *IEEE Photon. Technol. Lett.*, **17**, 2014–2016.

232 Rafailov, E.U. *et al.* (2006) Investigation of transition dynamics in a quantum-dot laser optically pumped by femtosecond pulses. *Appl. Phys. Lett.*, **88**, 41101.

233 Cataluna, M.A. *et al.* (2005) Ground and excited-state modelocking in a two-section quantum-dot laser. 18th Annual Meeting of the IEEE Lasers and Electro-Optics Society, Sydney, Australia, pp. 870–871.

234 Leitenstorfer, A. *et al.* (1995) Widely tunable two-color mode-locked Ti:sapphire laser with pulse jitter of less than 2 fs. *Opt. Lett.*, **20**, 916–918.

235 Thompson, M.G. *et al.* (2008) Regimes of mode-locking in tapered quantum dot laser diodes, in *Proceedings of the 21st International Semiconductor Laser Conference*, IEEE, New York.

236 Kim, J. and Chuang, S.L. (2006) Theoretical and experimental study of optical gain, refractive index change, and linewidth enhancement factor of p-doped quantum-dot lasers, *IEEE J. Quantum Electron.*, **42**, 942–952.

237 Alexander, R.R. *et al.* (2007) Zero and controllable linewidth enhancement factor in p-doped 1.3 μm quantum-dot lasers, *Jap. J. Appl. Phys.*, **46**, 2421–2423.

238 Mi, Z. and Bhattacharya, P. (2007) Analysis of the linewidth enhancement factor of long-wavelength tunnel-injection quantum-dot lasers, *IEEE J. Quantim Electron.*, **43**, 363–369.

239 George, A.A. *et al.* (2007) Long wavelength quantum-dot lasers selectivity populated using tunnel injection, *Semiconductor Sci. Tech.*, **22**, 557–560.

240 Fedorova, K.A. *et al.* (2010) Broadly-tunable high-power InAs/GaAs quantum-dot external-cavity diode lasers. *Opt. Express*, **18**, 19438–19443.

241 Deutsch, T. (1965) Mode-locking effects in an internally modulated ruby laser. *Appl. Phys. Lett.*, **7**, 80–82.

242 DeMaria, A.J. *et al.* (1966) Self mode-locking of lasers with saturable absorbers. *Appl. Phys. Lett.*, **8**, 174–176.

243 DiDomenico, J.M. *et al.* (1966) Generation of ultrashort optical pulses by mode locking in YAG:Nd laser. *Appl. Phys. Lett.*, **8**, 180–183.

244 Byer, R.L. (1988) Diode-laser-pumped solid-state lasers. *Science*, **239**, 742–747.

245 Ippen, E.P. *et al.* (1972) Passive mode locking of the cw dye laser. *Appl. Phys. Lett.*, **21**, 348–350.

246 Valdmanis, J.A. *et al.* (1985) Generation of optical pulses as short as 27 femtoseconds directly from a laser balancing self-phase modulation, group-velocity dispersion, saturable absorption, and saturable gain. *Opt. Lett.*, **10**, 131–133.

247 Maker, G.T. and Ferguson, A.I. (1989) Mode-locking and Q-switching of a diode-laser pumped neodymium-doped yttrium lithium fluoride laser. *Appl. Phys. Lett.*, **54**, 403–405.

248 Weingarten, K.J. *et al.* (1990) 2-Gigahertz repetition-rate, diode-pumped, mode-locked Nd:YLF laser. *Opt. Lett.*, **15**, 962–964.

249 Moulton, P.F. (1986) Spectroscopic and laser characteristics of Ti:Al$_2$O$_3$. *J. Opt. Soc. Am. B*, **3**, 125–133.

250 Goodberlet, J. *et al.* (1989) Femtosecond passively mode-locked Ti:Al$_2$O$_3$ laser with a nonlinear external cavity. *Opt. Lett.*, **14**, 1125–1127.

251 Sarukura, N. *et al.* (1990) CW passive mode locking of a Ti:sapphire laser. *Appl. Phys. Lett.*, **56**, 814–815.

252 Sarukura, N. *et al.* (1990) All solid-state CW passively mode-locked Ti:sapphire laser using a colored glass filter. *Appl. Phys. Lett.*, **57**, 229–230.

253 Payne, S.A. *et al.* (1988) LiCaAlF$_6$-Cr^{3+}: a promising new solid-state laser material. *IEEE J. Quantum Electron.*, **24**, 2243–2252.

254 Payne, S.A. *et al.* (1989) Laser performance of LiSrAlF$_6$:Cr^{3+}. *J. Appl. Phys.*, **66**, 1051–1056.

255 Petricevic, V. *et al.* (1988) Laser action in chromium-doped forsterite. *Appl. Phys. Lett.*, **52**, 1040–1042.

256 Angert, N.B. et al. (1988) The laser action in impurity color centers in yttrium–aluminum garnet crystals in the wavelength range of 1.35–1.45 μm. *Kvant. Elektron.*, **15**, 113–115.

257 Sorokina, I.T. et al. (1997) 14-fs pulse generation in Kerr-lens mode-locked prismless Cr:LiSGaF and Cr:LiSAF lasers: observation of pulse self-frequency shift. *Opt. Lett.*, **22**, 1716–1718.

258 Kopf, D. et al. (1997) Broadly tunable femtosecond Cr:LiSAF laser. *Opt. Lett.*, **22**, 621–623.

259 Wagenblast, P. et al. (2003) Diode-pumped 10-fs Cr^{3+}:LiCAF laser. *Opt. Lett.*, **28**, 1713–1715.

260 Demirbas, U. et al. (2008) Highly efficient, low-cost femtosecond Cr^{3+}: LiCAF laser pumped by single-mode diodes. *Opt. Lett.*, **33**, 590–592.

261 Yanovsky, V. et al. (1993) Generation of 25-fs pulses from a self-mode-locked Cr: forsterite laser with optimized group-delay dispersion. *Opt. Lett.*, **18**, 1541–1543.

262 Zhang, Z.G. et al. (1997) Self-starting mode-locked femtosecond forsterite laser with a semiconductor saturable-absorber mirror. *Opt. Lett.*, **22**, 1006–1008.

263 Ripin, D.J. et al. (2002) Generation of 20-fs pulses by a prismless Cr^{4+}:YAG laser. *Opt. Lett.*, **27**, 61–63.

264 Naumov, S. et al. (2004) Directly diode-pumped Kerr-lens mode-locked Cr^{4+}: YAG laser. *Opt. Lett.*, **29**, 1276–1278.

265 Deloach, L.D. et al. (1993) Evaluation of absorption and emission properties of Yb^{3+} doped crystals for laser applications. *IEEE J. Quantum Electron.*, **29**, 1179–1191.

266 Brenier, A. (2001) A new evaluation of Yb^{3+}-doped crystals for laser applications. *J. Lumin.*, **92**, 199–204.

267 Lagatsky, A.A. et al. (2005) Yb^{3+}-doped YVO_4 crystal for efficient Kerr-lens mode locking in solid-state lasers. *Opt. Lett.*, **30**, 3234–3236.

268 Rivier, S. et al. (2006) Passively mode-locked Yb:LuVO$_4$ oscillator. *Opt. Express*, **14**, 11668–11671.

269 Boudeile, J. et al. (2007) Continuous-wave and femtosecond laser operation of Yb: CaGdAlO$_4$ under high-power diode pumping. *Opt. Lett.*, **32**, 1962–1964.

270 Lagatsky, A.A. et al. (2004) Highly efficient and low threshold diode-pumped Kerr-lens mode-locked Yb:KYW laser. *Opt. Express*, **12**, 3928–3933.

271 Spence, D.E. et al. (1991) 60-fsec pulse generation from a self-mode-locked Ti: sapphire laser. *Opt. Lett.*, **16**, 42–44.

272 Keller, U. et al. (1992) Solid-state low-loss intracavity saturable absorber for Nd:YLF lasers: an antiresonant semiconductor Fabry–Perot saturable absorber. *Opt. Lett.*, **17**, 505–507.

273 Tsuda, S. et al. (1996) Mode-locking ultrafast solid-state lasers with saturable Bragg reflectors. *IEEE J. Sel. Top. Quantum Electron.*, **2**, 454–464.

274 Islam, M.N. et al. (1989) Color center lasers passively mode-locked by quantum wells. *IEEE J. Quantum Electron.*, **25**, 2454–2463.

275 Keller, U. (2004) Ultrafast solid-state lasers, in *Progress in Optics*, vol. 46, Elsevier Science, Amsterdam, pp. 1–115.

276 Zewail, A.H. (2000) Femtochemistry: atomic-scale dynamics of the chemical bond. *J. Phys. Chem. A*, **104**, 5660–5694.

277 Desouza, E.A. et al. (1995) Wavelength-division multiplexing with femtosecond pulses. *Opt. Lett.*, **20**, 1166–1168.

278 Udem, T. et al. (2002) Optical frequency metrology. *Nature*, **416**, 233–237.

279 Loesel, F.H. et al. (1998) Non-thermal ablation of neural tissue with femtosecond laser pulses. *Appl. Phys. B*, **66**, 121–128.

280 Juhasz, T. et al. (1999) Corneal refractive surgery with femtosecond lasers. *IEEE J. Sel. Top. Quantum Electron.*, **5**, 902–910.

281 Nolte, S. et al. (1997) Ablation of metals by ultrashort laser pulses. *J. Opt. Soc. Am. B*, **14**, 2716–2722.

282 Sorokina, I.T. et al. (2006) A SESAM passively mode-locked Cr:ZnSe laser, in *Advanced Solid-State Photonics 2006 Technical Digest*, The Optical Society of America, Washington, DC.

283 Papadopoulos, D.N. et al. (2003) Passively mode-locked diode-pumped Nd:YVO$_4$ oscillator operating at an ultralow repetition rate. *Opt. Lett.*, **28**, 1838–1840.

284 Krainer, L. et al. (2002) Compact Nd:YVO$_4$ lasers with pulse repetition rates up to 160 GHz. *IEEE J. Quantum Electron.*, **38**, 1331–1338.

285 Honninger, C. et al. (1999) Q-switching stability limits of continuous-wave passive mode locking. *J. Opt. Soc. Am. B*, **16**, 46–56.

286 Spuhler, G.J. et al. (2005) Semiconductor saturable absorber mirror structures with low saturation fluence. *Appl. Phys. B*, **81**, 27–32.

287 Krausz, F. et al. (1992) Femtosecond solid-state lasers. *IEEE J. Quantum Electron.*, **28**, 2097–2122.

288 Kartner, F.X. et al. (1998) Mode-locking with slow and fast saturable absorbers: what's the difference? *IEEE J. Sel. Top. Quantum Electron.*, **4**, 159–168.

289 Kartner, F.X. and Keller, U. (1995) Stabilization of soliton-like pulses with a slow saturable absorber. *Opt. Lett.*, **20**, 16–18.

290 Jung, I.D. et al. (1995) Experimental verification of soliton mode locking using only a slow saturable absorber. *Opt. Lett.*, **20**, 1892–1894.

291 Gupta, S. et al. (1992) Ultrafast carrier dynamics in III–V semiconductors grown by molecular-beam epitaxy at very low substrate temperatures. *IEEE J. Quantum Electron.*, **28**, 2464–2472.

292 Jung, I.D. et al. (1995) Scaling of the antiresonant Fabry–Perot saturable absorber design toward a thin saturable absorber. *Opt. Lett.*, **20**, 1559–1561.

293 Au, J.A.D. et al. (1997) 60-fs pulses from a diode-pumped Nd:glass laser. *Opt. Lett.*, **22**, 307–309.

294 Brunner, F. et al. (2000) Diode-pumped femtosecond Yb:KGd(WO$_4$)$_2$ laser with 1.1-W average power. *Opt. Lett.*, **25**, 1119–1121.

295 Guerreiro, P.T. et al. (1997) Self-starting mode-locked Cr:forsterite laser with semiconductor saturable Bragg reflector. *Opt. Commun.*, **136**, 27–30.

296 Collings, B.C. et al. (1996) Saturable Bragg reflector self-starting passive mode locking of a Cr^{4+}:YAG laser pumped with a diode-pumped Nd:YVO$_4$ laser. *Opt. Lett.*, **21**, 1171–1173.

297 Spalter, S. et al. (1997) Self-starting soliton-modelocked femtosecond Cr^{4+}:YAG laser using an antiresonant Fabry–Perot saturable absorber. *Appl. Phys. B*, **65**, 335–338.

298 Spuhler, G. et al. (1999) Passively modelocked diode-pumped erbium–ytterbium glass laser using a semiconductor saturable absorber mirror. *Electron. Lett.*, **35**, 567–569.

299 Adachi, S. (1987) Band gaps and refractive indexes of AlGaAsSb, GaInAsSb, and InPAsSb: key properties for a variety of the 2–4-μm optoelectronic device applications. *J. Appl. Phys.*, **61**, 4869–4876.

300 Grange, R. et al. (2006) Antimonide semiconductor saturable absorber for passive mode locking of a 1.5-μm Er:Yb: glass laser at 10 GHz. *IEEE Photon. Technol. Lett.*, **18**, 805–807.

301 Kondow, M. et al. (1996) GaInNAs: a novel material for long-wavelength-range laser diodes with excellent high-temperature performance. *Jpn. J. Appl. Phys. 1*, **35**, 1273–1275.

302 Calvez, S. et al. (2008) GaInNAs(Sb) surface normal devices. *Phys. Status Solidi A*, **205**, 85–92.

303 Sun, H.D. et al. (2002) Low-loss 1.3-μm GaInNAs saturable Bragg reflector for high-power picosecond neodymium lasers. *Opt. Lett.*, **27**, 2124–2126.

304 Liverini, V. et al. (2004) Low-loss GaInNAs saturable absorber mode locking a 1.3-μm solid-state laser. *Appl. Phys. Lett.*, **84**, 4002–4004.

305 Rutz, A. et al. (2005) 1.5 μm GaInNAs semiconductor saturable absorber for passively modelocked solid-state lasers. *Electron. Lett.*, **41**, 321–323.

306 McWilliam, A. et al. (2005) Low-loss GaInNAs saturable Bragg reflector for mode-locking of a femtosecond Cr^{4+}:forsterite laser. *IEEE Photon. Technol. Lett.*, **17**, 2292–2294.

307 Metzger, N.K. et al. (2008) Femtosecond pulse generation around 1500 nm using a GaInNAsSb SESAM. *Opt. Express*, **16**, 18739–18744.

308 Schon, S. et al. (2000) Ultrabroadband AlGaAs/CaF$_2$ semiconductor saturable absorber mirrors. *Appl. Phys. Lett.*, **77**, 782–784.

309 Grawert, F.J. et al. (2005) 220-fs erbium–ytterbium:glass laser mode locked by a broadband low-loss silicon/germanium saturable absorber. *Opt. Lett.*, **30**, 329–331.

310 Lagatsky, A.A. et al. (2007) Low-loss quantum-dot-based saturable absorber for efficient femtosecond pulse generation. *Appl. Phys. Lett.*, **91**, 231111.

311 Rafailov, E.U. et al. (2008) Compact and efficient mode-locked lasers based on QD-SESAMs, in *Solid State Lasers and Amplifiers III*, vol. 6998 (eds J.A. Terry et al.), SPIE (International Society for Optical Engineering), Bellingham, WA, pp. B9980–B10980.

312 Yang, W. et al. (1997) Effect of carrier emission and retrapping on luminescence time decays in InAs/GaAs quantum dots. *Phys. Rev. B*, **56**, 13314–13320.

313 Borri, P. et al. (2000) Ultrafast gain dynamics in InAs–InGaAs quantum-dot amplifiers. *IEEE Photon. Technol. Lett.*, **12**, 594–596.

314 Quochi, F. et al. (2002) Ultrafast carrier dynamics of resonantly excited 1.3-μm InAs/GaAs self-assembled quantum dots. *Physica B: Condens. Matter*, **314**, 263–267.

315 Ohnesorge, B. et al. (1996) Rapid carrier relaxation in self-assembled $In_xGa_{1-x}As$/GaAs quantum dots. *Phys. Rev. B*, **54**, 11532–11538.

316 Haiml, M. et al. (2004) Optical characterization of semiconductor saturable absorbers. *Appl. Phys. B*, **79**, 331–339.

317 Stormont, B. et al. (2004) Extended-cavity surface-emitting diode laser as active mirror controlling modelocked Ti:sapphire laser. *Electron. Lett.*, **40**, 732–734.

318 Lagatsky, A.A. et al. (2005) Quantum-dot-based saturable absorber with p–n junction for mode-locking of solid-state lasers. *IEEE Photon. Technol. Lett.*, **17**, 294–296.

319 McWilliam, A. et al. (2006) Quantum-dot-based saturable absorber for femtosecond mode-locked operation of a solid-state laser. *Opt. Lett.*, **31**, 1444–1446.

320 Warburton, R.J. et al. (1998) Coulomb interactions in small charge-tunable quantum dots: a simple model. *Phys. Rev. B*, **58**, 16221–16231.

321 Lumb, M.P. et al. (2009) Post-growth tailoring of quantum-dot saturable absorber mirrors by chemical etching. *Appl. Phys. B*, **94**, 393–398.

322 Lagatsky, A.A. et al. (2010) Ultrashort-pulse lasers passively mode locked by quantum-dot-based saturable absorbers. *Prog. Quantum Electron.*, **34**, 1–45.

323 Le Ru, E.C. et al. (2003) Strain-engineered InAs/GaAs quantum dots for long-wavelength emission. *Phys. Rev. B*, **67**, 165303.

324 Lumb, M.P. et al. (2009) Ultrafast absorption recovery dynamics of 1300 nm quantum dot saturable absorber mirrors. *Appl. Phys. Lett.*, **95**, 041101.

325 Zolotovskaya, S.A. et al. (2009) Electronically controlled pulse duration passively mode-locked Cr:forsterite laser. *IEEE Photon. Technol. Lett.*, **21**, 1124–1126.

326 Okhotnikov, O. et al. (2004) Ultra-fast fibre laser systems based on SESAM technology: new horizons and applications. *New J. Phys.*, **6**, 177.

327 Herda, R. and Okhotnikov, O.G. (2005) Effect of amplified spontaneous emission and absorber mirror recovery time on the dynamics of mode-locked fiber lasers. *Appl. Phys. Lett.*, **86**, 11113.

328 Herda, R. et al. (2006) Semiconductor quantum-dot saturable absorber mode-locked fiber laser. *IEEE Photon. Technol. Lett.*, **18**, 157–159.

329 Keller, U. and Tropper, A.C. (2006) Passively modelocked surface-emitting semiconductor lasers. *Phys. Rep.*, **429**, 67–120.

330 Lorenser, D. et al. (2006) 50-GHz passively mode-locked surface-emitting semiconductor laser with 100-mW average output power. *IEEE J. Quantum Electron.*, **42**, 838–847.

331 Maas, D. et al. (2007) Vertical integration of ultrafast semiconductor lasers. *Appl. Phys. B*, **88**, 493–497.

332 Maas, D.H.J.C. et al. (2008) Growth parameter optimization for fast quantum dot SESAMs, *Opt. Express.*, **16**, 18646–18656.

333 Garnache, A. et al. (2002) Sub-500-fs soliton-like pulse in a passively mode-locked broadband surface-emitting laser with 100 mW average power. *Appl. Phys. Lett.*, **80**, 3892–3894.

334 Klopp, P. et al. (2008) 290-fs pulses from a semiconductor disk laser. *Opt. Express*, **16**, 5770–5775.

335 Wilcox, K.G. et al. (2008) Ultrafast optical Stark mode-locked semiconductor laser. *Opt. Lett.*, **33**, 2797–2799.

336 Wilcox, K.G. et al. (2009) 870-fs Passively mode-locked quantum-dot SESAM semiconductor disk laser. Advanced Solid-State Photonics.

337 Basov, N.G. et al. (1966) Semiconductor lasers with radiating mirrors. *IEEE J. Quantum Electron.*, **2**, 154.

338 Kuznetsov, M. et al. (1999) Design and characteristics of high-power (>0.5-W CW) diode-pumped vertical-external-cavity surface-emitting semiconductor lasers with circular TEM_{00} beams. *IEEE J. Sel. Top. Quantum Electron.*, **5**, 561–573.

339 Rudin, B. et al. (2008) Highly efficient optically pumped vertical-emitting semiconductor laser with more than 20 W average output power in a fundamental transverse mode. *Opt. Lett.*, **33**, 2719–2721.

340 Hastie, J.E. et al. (2006) Tunable ultraviolet output from an intracavity frequency-doubled red vertical-external-cavity surface-emitting laser. *Appl. Phys. Lett.*, **89**, 061114.

341 Keller, U. and Tropper, A.C. (2006) Passively modelocked surface-emitting semiconductor lasers. *Phys. Rep.*, **429**, 67–120.

342 Klopp, P. et al. (2009) Mode-locked InGaAs–AlGaAs disk laser generating sub-200-fs pulses, pulse picking and amplification by a tapered diode amplifier. *Opt. Express*, **17**, 10820–10834.

343 Kaneda, Y. et al. (2008) Continuous-wave all-solid-state 244 nm deep-ultraviolet laser source by fourth-harmonic generation of an optically pumped semiconductor laser using $CsLiB_6O_{10}$ in an external resonator. *Opt. Lett.*, **33**, 1705–1707.

344 Rahim, M. et al. (2008) Optically pumped 5 μm IV–VI VECSEL with Al-heat spreader. *Opt. Lett.*, **33**, 3010–3012.

345 Germann, T.D. et al. (2008) Quantum-dot semiconductor disk lasers. *J. Cryst. Growth*, **310**, 5182–5186.

346 Butkus, M. et al. (2009) High-power quantum-dot-based semiconductor disk laser. *Opt. Lett.*, **34**, 1672–1674.

347 Rautiainen, J. et al. (2010) 1170 nm optically-pumped quantum dot disk laser. CLEO'10, San Francisco, CA.

348 Butkus, M. et al. (2010) 1270 nm optically-pumped quantum dot disk laser. Laser Optics, St. Petersburg, Russia.

349 Efros, A.L. (1982) Interband absorption of light in a semiconductor sphere. *Sov. Phys. Semicond.*, **16**, 772–775.

350 Brus, L.E. (1984) Electron–electron and electron–hole interactions in small semiconductor crystallites: the size dependence of the lowest excited electronic state. *J. Chem. Phys.*, **80**, 4403–4409.

351 Ekimov, A.I. et al. (1985) Quantum size effect in semiconductor microcrystals. *Solid State Commun.*, **56**, 921–924.

352 Murray, C.B. et al. (1993) Synthesis and characterization of nearly monodisperse CDE (E = S, SE, TE) semiconductor nanocrystallites. *J. Am. Chem. Soc.*, **115**, 8706–8715.

353 Borrelli, N.F. et al. (1987) Quantum confinement effects of semiconducting microcrystallites in glass. *J. Appl. Phys.*, **61**, 5399–5409.

354 Brus, L. (1991) Quantum crystallites and nonlinear optics. *Appl. Phys. A*, **53**, 465–474.

355 Vossmeyer, T. et al. (1994) CdS nanoclusters: synthesis, characterization, size-dependent oscillator strength, temperature shift of the excitonic transition energy, and reversible absorbency shift. *J. Phys. Chem.*, **98**, 7665–7673.

356 Warnock, J. and Awschalom, D.D. (1985) Quantum size effects in simple colored glass. *Phys. Rev. B*, **32**, 5529–5531.

357 Schmittrink, S. et al. (1987) Theory of the linear and nonlinear optical properties of semiconductor microcrystallites. *Phys. Rev. B*, **35**, 8113–8125.

358 Roussignol, P. et al. (1985) Optical phase conjugation in semiconductor-doped glasses. *Opt. Commun.*, **55**, 143–148.

359 Hall, D.W. and Borrelli, N.F. (1988) Absorption saturation in commercial and quantum-confined $CdSe_xS_{1-x}$-doped glasses. *J. Opt. Soc. Am. B*, **5**, 1650–1654.

360 Warnock, J. and Awschalom, D.D. (1986) Picosecond studies of electron

confinement in simple colored glasses. *Appl. Phys. Lett.*, **48**, 425–427.

361 Peyghambarian, N. *et al.* (1989) Optical nonlinearities and femtosecond dynamics of quantum confined CdSe microcrystallites, in *Optical Switching in Low-Dimensional Systems* (eds H. Haug and L. Banyai), Academic Press, New York, pp. 191–201.

362 Mittleman, D.M. *et al.* (1994) Quantum-size dependence of femtosecond electronic dephasing and vibrational dynamics in CdSe nanocrystals. *Phys. Rev. B*, **49**, 14435–14447.

363 Nuss, M.C. *et al.* (1986) Femtosecond carrier relaxation in semiconductor-doped glasses. *Appl. Phys. Lett.*, **49**, 1717–1719.

364 Peyghambarian, N. *et al.* (1989) Femtosecond optical nonlinearities of CdSe quantum dots. *IEEE J. Quantum Electron.*, **25**, 2516–2522.

365 Sarukura, N. *et al.* (1990) All solid-state cw passively mode-locked Ti:sapphire laser using a colored glass filter. *Appl. Phys. Lett.*, **57**, 229–230.

366 Bilinsky, I.P. *et al.* (1999) Self-starting mode locking and Kerr-lens mode locking of a Ti:Al$_2$O$_3$ laser by use of semiconductor-doped glass structures. *J. Opt. Soc. Am. B*, **16**, 546–549.

367 Borrelli, N.F. and Luong, J.C. (1988) Semiconductor microcrystals in porous glass. *Proc. SPIE*, **866**, 104–109.

368 Nogami, M. *et al.* (1990) CdS microcrystal-doped silica glass prepared by the sol-gel process. *J. Non-Cryst. Solids*, **122**, 101–106.

369 Borrelli, N.F. and Smith, D.W. (1994) Quantum confinement of PbS microcrystals in glass. *J. Non-Cryst. Solids*, **180**, 25–31.

370 Reynoso, V.C.S. *et al.* (1995) PbTe quantum-dot doped glasses with absorption edge in the 1.5-μm wavelength region. *Electron. Lett.*, **31**, 1013–1015.

371 Lipovskii, A. *et al.* (1997) Synthesis and characterization of PbSe quantum dots in phosphate glass. *Appl. Phys. Lett.*, **71**, 3406–3408.

372 Lipovskii, A.A. *et al.* (1997) PbSe quantum dot doped phosphate glass. *Electron. Lett.*, **33**, 101–102.

373 Andreev, A.D. and Lipovskii, A.A. (1999) Anisotropy-induced optical transitions in PbSe and PbS spherical quantum dots. *Phys. Rev. B*, **59**, 15402–15404.

374 Dantas, N.O. *et al.* (2002) Optical properties of PbSe and PbS quantum dots embedded in oxide glass. *Phys. Status Solidi B*, **232**, 177–181.

375 Kang, I. and Wise, F.W. (1997) Electronic structure and optical properties of PbS and PbSe quantum dots. *J. Opt. Soc. Am. B*, **14**, 1632–1646.

376 Kolobkova, E.V. *et al.* (2002) Fluorophosphate glasses with quantum dots based on lead sulfide. *Glass Phys. Chem.*, **28**, 251–255.

377 Malyarevich, A.M. *et al.* (2008) Semiconductor-doped glass saturable absorbers for near-infrared solid-state lasers. *J. Appl. Phys.*, **103**, 081301.

378 Okuno, T. *et al.* (2000) Strong confinement of PbSe and PbS quantum dots. *J. Lumin.*, **87–89**, 491–493.

379 Reynoso, V.C.S. *et al.* (2002) Effect of PbS impurity on crystallization mechanism of phosphate glasses studied by differential scanning calorimetry. *Mater. Lett.*, **56**, 424–428.

380 Guerreiro, P.T. *et al.* (1997) PbS quantum-dot doped grasses as saturable absorbers for mode locking of a Cr:forsterite laser. *Appl. Phys. Lett.*, **71**, 1595–1597.

381 Henneberger, F. *et al.* (1990) Nonlinear optical properties of wide gap II–VI bulk semiconductors and microcrystallites. *J. Cryst. Growth*, **101**, 632–642.

382 Zhang, J.Z. *et al.* (1994) Femtosecond studies of interfacial electron–hole recombination in aqueous CdS colloids. *Appl. Phys. Lett.*, **64**, 1989–1991.

383 Klimov, V. *et al.* (1996) Ultrafast carrier dynamics in semiconductor quantum dots. *Phys. Rev. B*, **53**, 1463–1467.

384 Savitski, V.G. *et al.* (2002) PbS-doped phosphate glasses saturable absorbers for 1.3-μm neodymium lasers. *Appl. Phys. B*, **75**, 841–846.

385 Lagatsky, A.A. *et al.* (2004) Passive mode locking of a Cr^{4+}:YAG laser by PbS quantum-dot-doped glass saturable absorber. *Opt. Commun.*, **241**, 449–454.

386 Savitski, V.G. *et al.* (2005) Intensity-dependent bleaching relaxation in lead

salt quantum dots. *J. Opt. Soc. Am. B*, **22**, 1660–1666.

387 Ohno, T. *et al.* (2004) Recovery of 160 GHz optical clock from 160 Gbit/s data stream using modelocked laser diode. *Electron. Lett.*, **40**, 265–267.

388 Delfyett, P.J. *et al.* (1991) Optical clock distribution using a mode-locked semiconductor laser diode system. *J. Lightwave Technol.*, **9**, 1646–1649.

389 Vieira, A.J.C. *et al.* (2001) A mode-locked microchip laser optical transmitter for fiber radio. *IEEE Trans. Microwave Theory Tech.*, **49**, 1882–1887.

390 Takara, H. (2001) High-speed optical time-division-multiplexed signal generation. *Opt. Quantum Electron.*, **33**, 795–810.

391 Reithmaier, J.P. *et al.* (2005) InP based lasers and optical amplifiers with wire-/dot-like active regions. *J. Phys. D: Appl. Phys.*, **38**, 2088–2102.

392 Liu, J. *et al.* (2007) Uniform 90-channel multiwavelength InAs/InGaAsP quantum dot laser. *Electron. Lett.*, **43**, 458–460.

393 Guffarth, F. *et al.* (2003) Radiation hardness of InGaAs/GaAs quantum dots. *Appl. Phys. Lett.*, **82**, 1941–1943.

394 Marcinkevicius, S. *et al.* (2002) Changes in luminescence intensities and carrier dynamics induced by proton irradiation in $In_xGa_{1-x}As$/GaAs quantum dots. *Phys. Rev. B*, **66**, 235314.

395 Tang, X.F. *et al.* (2008) 40-Gb/s polarization-insensitive all-optical clock recovery using a quantum-dot Fabry–Perot laser assisted by an SOA and bandpass filtering. *IEEE Photon. Technol. Lett.*, **20**, 2051–2053.

396 Tang, X.F. *et al.* (2009) Low-timing-jitter all-optical clock recovery for 40 Gbits/s RZ-DPSK and NRZ-DPSK signals using a passively mode-locked quantum-dot Fabry–Perot semiconductor laser. *Opt. Lett.*, **34**, 899–901.

397 Lin, C.Y. *et al.* (2009) Compact optical generation of microwave signals using a monolithic quantum dot passively mode-locked laser. *IEEE Photon. J.*, **1**, 236–244.

398 Kim, J. *et al.* (2009) Hybrid integration of a bowtie slot antenna and a quantum dot mode-locked laser. *IEEE Antennas Wireless Propag. Lett.*, **8**, 1337–1340.

399 Camacho, F. *et al.* (1997) Improvements in mode-locked semiconductor diode lasers using monolithically integrated passive waveguides made by quantum-well intermixing. *IEEE Photon. Technol. Lett.*, **9**, 1208–1210.

400 Gubenko, A. *et al.* (2005) High-power monolithic passively modelocked quantum-dot laser. *Electron. Lett.*, **41**, 1124–1125.

401 Wang, H. *et al.* (2007) Dynamic switching of a 10 GHz quantum dot mode-locked laser using an integrated quantum dot switch. International Conference on Photonics in Switching, pp. 113–114.

402 Brezinski, M.E. and Fujimoto, J.G. (1999) Optical coherence tomography: high-resolution imaging in nontransparent tissue. *IEEE J. Sel. Top. Quantum Electron.*, **5**, 1185–1192.

403 Fischer, P. *et al.* (2005) Deep tissue penetration of radiation: 3D modelling and experiments. European Conference on Lasers and Electro-Optics, Munich, Germany, p. 641.

404 Brezinski, M.E. *et al.* (1996) Optical coherence tomography for optical biopsy: properties and demonstration of vascular pathology. *Circulation*, **93**, 1206–1213.

Index

a

absorber compression coefficients 130
absorption recovery time 78, 83, 97, 102, 105, 109, 126, 127, 129, 131, 136, 151, 156, 159, 175, 181, 185, 193
absorber saturation factor 131
absorption coefficient 14, 81, 83, 86, 121, 197, 208, 213, 215
ambipolar diffusion coefficient 124, 147
amplified spontaneous emission (ASE) 40, 57, 71, 155, 157, 168, 199, 201
Andronov–Hopf bifurcation 115, 138
ASE. *See* amplified spontaneous emission (ASE)
Auger process 32, 215

b

bleaching relaxation kinetics 10
Bohr radius 207, 209
Bragg reflector 77, 123, 186, 188, 194, 197, 200
bulk amplifier 56, 58, 60, 71

c

carrier density 17, 22, 25–28, 33, 34, 36, 47– 49, 51, 53, 55, 58, 60, 61, 62, 64
– dynamics in 56, 81
– hole saturation 69
– rate equation 106
– single pulse leads to modulation of 65
– in wetting layer (QW) on 137
chirp measurement 161, 163, 164
colliding pulse mode (CPM) locking 101
combined mode-locking/Q-switching regime 117
complex dielectric permittivity 142–144
Coulomb electron–hole interaction 96

Coulomb interactions 14, 26, 81, 94
Cr:forsterite laser 193–196, 198, 199, 210
– average output powers 196
– mode-locked pulse durations 196
– schematic presentation 194
cross-phase modulation factors 145
Cr:YAG laser 212, 213

d

datacoms 219, 220
delay differential equation (DDE) model 110, 112, 114, 125, 126, 132–141, 146, 149
– direct numerical implementation 116
– limitations to 118
– steady-state solutions, bifurcation analysis 115
– version 137
dimensionality 18
– reduced 8, 94
– role in semiconductor materials 1–4
diode-pumped semiconductor disk lasers (SDLs) 201
dispersion curve 111, 149
dispersive correction 123
dispersive element 105, 119
distributed Bragg reflector (DBR) 77, 123, 188, 194, 197
– GaAs/AlAs-based
– – bandwidths of 186
– structure 196
distributed time domain models, limitation 125
dots-in-a-well (DWELL) system 11
double heterostructure 1, 2
dual-wavelength mode-locking regime 178
dynamic modal analysis 125, 126, 138, 153

Ultrafast Lasers Based on Quantum Dot Structures: Physics and Devices.
Edik U. Rafailov, Maria Ana Cataluna, and Eugene A. Avrutin
Copyright © 2011 WILEY-VCH Verlag GmbH & Co. KGaA, Weinheim
ISBN: 978-3-527-40928-0

e

edge-emitting lasers   120, 164
Einstein relations   15
electron–hole energy separation   16
electron-hole energy states   207
electron–hole interaction   10
electron-hole pair   14, 15, 18, 31, 41, 95, 215
electron–hole transition   14
energy-resolved formula   19
energy structure   6, 11–14
– excited level (EL)   11
– ground level (GL)   11
III–V epitaxially grown quantum dots   4–6
erbium-doped fiber amplifiers (EDFAs)   57
excited-state absorption (ESA)   184, 213
excitonic approximation   17, 27, 36, 43, 44, 66, 67, 69, 72, 73, 92, 140, 146, 147, 148, 149, 152, 153

f

Fabry–Perot cavity   188
Fabry–Perot resonators   119
fiber-based laser systems   199
Franz–Keldysh effect   82
frequency-resolved Mach–Zehnder gating   164
frequency resolved optical gating (FROG) systems
– characterization   164
– SHG FROG trace   163
frequency stabilization techniques   172

g

GaAs-based Bragg mirror   193
GaAs substrate   5, 6, 24, 25, 178, 186, 200, 217
gain coefficient   15, 18, 25, 46, 47, 140
gain/group velocity dispersion (GVD)   104, 123
Gaussian distribution   16
generic laser theory   131

h

Haus's model   113, 114, 118
Haus's mode-locking theory   107
Henry linewidth enhancement factor   39, 54, 58, 80, 104, 131
high-speed communications   175
homogeneous broadening energy value   22
hybrid mode locking technique   101

i

II–VI semiconductor nanocrystals in glass   207–209
InAs/GaAs submonolayer (SML)   204
InAs QD layers   167
InAs QD technology   217
InAs quantum dashes   7
infinite response digital filter   122
InGaAs/InAs dots   6
InGaAs QD absorber mirror   200
InGaAs QD laser   165
inhomogeneous broadening   7–9

k

Kerr-lens mode-locking (KLM) technique   184, 209
– Ti:sapphire laser   99
Kramers–Kronig relations   39, 41, 142

l

large-signal iterative model   109
laser diodes
– mode-locking techniques   100, 101
– monolithic   82
laser operating parameters   173
lattice constant   5, 186
LEF. See linewidth enhancement factor (LEF)
light–matter interactions, in quantum dots   37
– active layer confinement factor   39
– for bulk materials at laser and amplifier carrier densities   39
– calculating resonant refractive index contribution   40
– characteristic electron and hole capture times   49
– characteristic relaxation time for the QD–WL hole burning   50
– choice of QD model, hierarchy of approaches   45–48
– complex operator describing gain and contribution to   44, 45
– decay of wings of inhomogeneous broadening function   41
– distribution function of electrons in quantum dot, evolution of   44
– distribution of holes, expression   43
– dynamic capture and escape rates   49
– dynamic equation, for electron contribution to   50
– effective
– – fraction of empty WL states, calculations   48, 49

– – relaxation rate, for QD electron 49
– – transverse width of waveguide mode, to introduce 40
– electron contribution to total quasi-equilibrium gain 50
– electron–hole pair energy separation 41
– excited-state transitions affecting refractive index 41
– full three-dimensional distribution 38
– ground-state electron distribution function, modified in presence of 42
– hole contribution, defined 51
– hole recombination 42
– introducing complementary homogeneous broadening function 41
– Kramers–Kronig relations 39
– local photon density and light signal intensity 40
– longitudinal wave vector determined by 38
– Lorentzian homogeneous transition broadening 45
– modal gain
– – for ground-state transitions expression 40
– – and resonant (active layer) contributions to 39
– – using expression for the material gain 40
– nonequilibrium corrections calculation 48
– nonlinear stimulated recombination term 51
– nonresonant dielectric properties 37
– optical signal in device 42
– quasi-equilibrium stimulated recombination rate for 49
– in QW and bulk lasers and amplifiers, the modal gain calculation 39
– rate of stimulated recombination of electron 42
– rotating wave notation 38
– slow-wave equations, for amplitudes of left- and right-propagating waves 38
– spontaneous recombination rate, calculation 48
– stimulated recombination
– – in presence of light signal 42
– – rate 45
– total nonequilibrium ground-state carrier density, equation for 49
– two-dimensional densities, of electrons and holes 40
– use of
– – Kramers–Kronig transformation 40, 41
– – Lorentzian homogeneous broadening function 40

– wave and slow-amplitude approximations, introducing through 38
– wave equation for light propagation 37
linewidth enhancement factor (LEF) 39, 80, 100, 157, 158, 176, 181
– in amplifier 104
– decrease, carrier pileup on ES 179
– distortions 59
– of gain and absorber sections 112
– lower diffusion of carriers, associated with 156
– in QDlasers 41, 179
– spectral dependence 158
Lorentzian homogeneous broadening function 20, 21, 40, 53, 144
Lorentzian spectrum 110

m
Mach–Zehnder modulator 164
Markoffian relaxation approximation 18
Matrix elements 11–14
Maxwell–Bloch equations 63
MBE. *See* molecular beam epitaxy (MBE)
metal organic chemical vapor deposition (MOCVD) 5
ML laser 118
mode-locked integrated external cavity surface emitting laser (MIXSEL) 202
mode-locked lasers 63, 92, 99, 103, 110, 118, 120, 122, 124, 126, 130, 131, 141
– dynamics model 103
– parameters affecting ML properties 129
– – compression coefficients 130
– – gain and group velocity dispersion 130
– – linewidth enhancement factors 130, 131
– – s-factor 129
– quantum dot lasers
– – noise performance, different configurations 171
– – performance 162
– repetition rate 166
– semiconductor lasers 119, 169, 219
mode locking 15, 42, 77, 100, 101
– equation 106
– – master equation 107
– master equation 113
– regime 167
– techniques 100, 183
– theories 126
molecular beam epitaxy (MBE) 5, 187, 190, 203
monolithic mode-locked quantum dot lasers 99, 158–161, 180, 181

- future directions 181, 182
- generic mode-locked laser models, predictions 126–131
-- laser performance depending on operating point 126–129
-- mode-locked laser behavior, affecting parameters 129–131
- noise characteristics 167–174
-- performance under optical injection 172–174
-- pulse repetition rate stability and optical feedback 170–172
-- timing jitter 167–170
- performance at elevated temperature 174–176
-- p-doping, use 176
-- pulse duration trends, at higher temperatures 175
-- stable mode locking 174, 175
- pulse generation, exploiting different transitions 176–180
-- excited-state transition 179, 180
-- mode locking via ground and excited states 176–179
- quantum dot materials, advantages 154–158
-- amplified spontaneous emission, lower level 157
-- broad gain bandwidth 154, 155
-- linewidth enhancement factor 157, 158
-- low temperature sensitivity 155, 156
-- low threshold current 155
-- QD saturable absorbers 154
-- suppressed carrier diffusion 156
- quantum dot mode-locked lasers, features 131–154
-- delay differential equation model 132–141
-- modal analysis for 153, 154
-- traveling wave modeling 141–153
- semiconductor lasers, mode locking theoretical models 103–126
-- frequency and time–frequency treatment 125, 126
-- large-signal time domain approach 109–120
-- small-signal time domain models 103–109
-- traveling wave models 120–125
- semiconductor mode-locked lasers 99–103
-- mode-locking techniques in laser diodes 100, 101
-- passive mode locking 101–103
-- place 99, 100
- ultrashort pulse generation 158–167
-- chirp measurement and pulse compression 161–164
-- external cavity QD mode-locked lasers 166, 167
-- monolithic mode-locked quantum dot lasers 158–161
-- higher power 164, 165
-- higher repetition rates 165, 166
Moss–Burstein effect 14, 81
multipopulation traveling wave models 149, 153. See also traveling wave models

n

Nd:YVO$_4$ laser
- passive mode locking 197
non-Gaussian distribution 16
nonlinear algebraic equation system 112
nonlinearity coefficient 51–54
- accurate evaluation of nonlinearity coefficient value and 53
- approximate Lorentzian homogeneous broadening function 53
- effect of
-- gain saturation due to hole burning 54
-- TPA at modest optical powers 54
- electron contribution
-- to gain compression coefficient 52
-- to nonlinearity 53
- heuristic expression, used for bulk and QW materials 52
- hole contribution, defined 52
- homogeneous and inhomogeneous broadening 53
- nonlinear dependence, of quasi-linear gain with 53
- total gain compression coefficient 52
- two-photon absorption coefficient, of laser waveguide 54
- at very high light intensity values 54
non-Lorentzian homogeneous broadening function 21
non-Markoffian relaxation of polarization 21

o

Optical amplifier, numerical parameters assessment 55
- acceptable margin DG of pulse patterning and 60
- amplified spontaneous emission noise 57
- degree of distortion, and degree of patterning 57
- disadvantages of SOAs 58
- dynamics of carrier density in SOA 56

– estimated maximum patterning-free bit rate 60
– first approximation, effect of pulse nonlinearity, used in 61, 62
– formula for critical pulse duration at 61
– gain bandwidth 57
– Henry factor α_H 58
– input light pulse energies 58, 59
– linear gain–carrier density relationship 60
– linewidth enhancement factor, presence of 59
– optical gain, determined from 55
– photon density, equation for 56
– polarization sensitivity 57
– pulse saturation energy 58
– saturated gain in presence of light 56
– saturation power 55
– semianalytical SOA theory, to evaluate 59
– slow saturation due to carrier density depletion by 60
– theory of pulse amplification in (bulk) SOA and 58
optical coherence tomography (OCT) 220
optical communications 6, 55, 60, 61, 151, 155, 160, 170, 217–219
optical frequency 21, 38, 42, 107, 113, 143, 170
optical injection 172, 173
optical Kerr effect 184
optically pumped semiconductor disk lasers (OP-SDLs)
– output power *vs.* pump power characteristics 204
– power scaling 205
optical time-division multiplexing systems (OTDMs) 167, 217

p

passive mode locking 93, 101–103, 110, 150, 153, 158, 166, 169, 183, 187, 193, 199, 212, 214
patterning-free amplification 63, 68, 69, 71–73
PbS QDs 94, 214
– bleaching decay of 211
– ESA spectra for 216
– in phosphate glass 211
p-doped lasers 181
periodic carrier density modulation 124
phonon bottleneck 9, 10, 26
Planck's constant 2
pulse amplitude, simulated dependence 128
pulse broadening 100, 103, 153, 155, 161, 165, 181
pulse compression 161, 163, 164

pulse duration 59, 61, 63, 65, 71, 78
pulse generation, exploiting different transitions
– excited-state transition as tool for 179
– mode locking via ground and excited states 176–179
pulse repetition rate stability 170, 172
pulse shaping mechanisms 185

q

QDH. *See* quantum dash (QDH)
QDMLL, optical integration 219
QDs. *See* quantum dots (QDs)
QD saturable absorbers. *See also* saturable absorber (SA)
– absorption recovery time 88, 89
– advantages of 154
– Auger mechanisms 90
– capture and escape rates 90
– characteristic time and applied electric field 90
– decrease in relaxation time, with increasing pump fluence in 94
– dynamics of absorber recovery 89
– fast and slow times estimation 90–92
– – excitonic approximation 92
– – at high voltages 91
– – pulse duration 92
– – slow relaxation stage 92
– for ground- and excited-state electron occupancies 89
– kinetics of bleaching relaxation of 89
– level structure and electron kinetic processes in dynamic model of 93
– under reverse bias 90
QD SESAMs
– constructions, limitation to 88
– for efficient passive mode locking of solid-state lasers 187
– – emitting at 1 µm 187–193
– – emitting at 1.3 µm 193–199
– grown on semiconductor mirrors 187
– mode-locked semiconductor disk lasers incorporating 201–204
– for passive mode locking of fiber lasers 199–201
Q-switching oscillations 118
Q-switching regime 129
quantum confined Franz–Keldysh effect (QCFKE) 82
quantum-confined Stark effect 160
quantum confinement effects
– on absorption and photoluminescence 208

– at longer wavelengths, appropriate semiconductor materials 209
– in technologically important wavelength regions 209
quantum dash (QDH) 2, 6, 7, 17, 160, 217
quantum dot (QD)
– materials 100
quantum dot absorbers 194
quantum dot-based semiconductor saturable absorber mirror (SESAM) 187, 194, 202
– linear optical properties 188
– macroparameters 191
– multilayer structure 189
– nonlinear optical properties 202
– nonlinear reflectivity change *vs.* pulse energy density 190
– optimization 203
– passive mode locking 190
– p–n junction 195
– potential 201
– pulse shaping influence 201
– reflection intensity 189
– reflectivity dependence 195
– schematic presentation 188
– voltage–current characteristics 192
quantum dot (QD) lasers 1, 4, 15, 42, 100, 103, 123, 169, 172, 180, 181, 182, 217
– challenge to 165
– characterization 170
– diodes 155
– dynamics of 148
– features 178
– – of mode-locking behavior in 129
– generation of shorter pulses from 181
– InAs/InP 218
– on InP substrates 22
– laser emission 176
– mode-locked, performance using different configurations 162
– potential 179
– in silicon-based photonics 219
– in space applications 218
– for stable mode locking in semiconductor 160
– switch-on of mode-locked 136, 161, 169, 176
– threshold current 157
quantum dot optical amplifiers
– at high bit rates 63
– – applied current density, different values 75
– – in bulk materials, reduced (DoS) expression 66, 67
– – condition $f_{e,h} \to 1$ in bulk/QWs 66
– – conditions for low distortion propagation of 63
– – degree of trade-off between 66
– – dependence of gain G on output pulse energy E_{out} 68, 70
– – determine carrier density $N_{min}(\Delta g)$ from 64
– – electron/hole degeneracy 67
– – energy U_{nl} and fast nonlinearity 65
– – estimated maximum patterning-free bit rate 66
– – estimation for patterning-free bit rate 64
– – evaluating carrier density variation 64
– – experimental proof of patterning-free amplification at 72
– – factor that adds to high saturation energy in QDs 67, 68
– – fast saturation of gain, and expression 65
– – Fermi quasi-levels for QWs 69
– – for ground-state occupancy of dot i expression 72
– – increase in current leads to 69
– – net gain of SOA 63
– – plateau of patterning-free operation 69
– – plateaus of constant gain 69
– – in QWs under high pump 71
– – reduced DoS for electrons 67
– – reduced patterning, at high currents in QW amplifiers 71
– – relevant to communication applications and 63
– – safe operating rate, expression 65
– – saturation pulse energy for 73
– – during short pulse, recombination neglected 64, 65
– – small-signal gain 69
– – SPM due to single-pass amplification 68
– – superiority of QDSOAs over QW and 71
– – two-photon absorption (TPA) 74
– – ultrafast and distortion-free amplification 63
– – weaker condition, estimate for bit rate 65
– short-pulse operating regime 62, 63
– – InAs QDs, in tandem with aQDmode-locked laser 63
quantum dots (QDs) 2, 4, 6, 7, 11, 13, 14, 80, 89, 94, 153
– advantage 196
– crosspoint switch 219
– devices, optical gain 154
– doped glasses 6, 210
– – nonlinear response characteristics 214
– dynamics 13, 153
– – features 153

- fast time constants 215
- GaAs 203
- InAs 160, 167, 203
- InGaAs Stranski–Krastanov 204
- kinetic theory of 22–37
- laser devices 220
- lasers (see quantum dot (QD) lasers)
- light–matter interactions in 37–51
- material systems 4
- mode-locked lasers
-- broad bandwidth available in 154
-- communication rates using 164
-- external cavity 166, 167
-- modal analysis 153, 154
-- noise characteristics of 167–174
-- performance at elevated temperature 174–176
- optical amplifiers
-- estimated maximum patterning-free bit rate in 66
-- for nonlinear operation and limiting function 76
- self-assembly of 5
- Stark effect in 138
- structure 178
- waveguide device
-- carrier lifetime, pump-probe measurements 189
quantum dots, kinetic theory of 22
- absolute value of scattering rate 34
- application of MEM to ultrafast dynamics of 35
- approaches proposed in 31
-- to consider electron–hole pair localized in dot as 31
-- to treating hole kinetics is, in a way, opposite to 32
-- to treating the hold dynamics rates fully microscopically 32, 33
- asymptotic behavior
-- at both low and high N_{wl} 29
-- at low populations of WL 29
- Auger-assisted capture speed 27
- Auger-type process 26
- bimolecular nature, of spontaneous recombination 25
- calculated dependence of peak ground- and excited-state gain in 36
- carrier escape 28
-- in semiphenomenological approach 28
- carriers pumped into wetting layer by 23
- correction to the Fermi distribution 35
- direct capture of carriers from wetting layer into 24
- dots on GaAs substrates 25
- downward relaxation and upward escape rate 24
- dynamic equations 22, 23, 30
- dynamics of hole populations in QDs 31
- effective capture and escape times depend on 30
- effective complementary occupation factors f''_w in wetting layer 29
- effective degeneracy for wetting layer 24
- effective distribution function for wetting layer 24
- Einstein relations 25
- electron level energies 23
- electrons, capturing into dots 23
- fast relaxation of QD amplifier 28
- Fermi distribution in energy 28
- gain–current relation in QD devices 36, 37
- ground-state occupancies 34
- at high carrier densities 26
- InAs/GaAs dots, expressions 30
- interlevel kinetic behavior, for Pauli blockage 34
- kinetics of quantum dots, in terms of occupation functions f_E and f_G 34
- master equations model (MEM) for occupancies 34
- microscopically calculated scattering rates
-- in and out of dot as functions of 33
- microscopic approach, for calculation of rate 31
- net stimulated recombination rate 25
- nonlocal Coulomb interactions 26
- out-scattering rates, after very sharp initial increase with 34
- Pauli exclusion principle 27
- phonon bottleneck 26
- for prediction of damping of the relaxation oscillations 34
- process of relaxation, from excited to ground state 29
- pumping of QD laser or amplifier 27
- quasi-Fermi distribution in WL and 29
- rates of capture, and escape processes 23
- relation between characteristic capture and escape times 28
- spontaneous recombination rate 25
- steady-state total carrier density, dependence of 27
- thermal activation 23
- trickle-down approximation 28
- values of electron rates interprettation 33
- wetting layer carrier density 25
quantum dot saturable absorber operation

– features of 87
– – bandwidth of QD SAs 88
– – saturation fluence 94–97
– – ultrafast recovery of absorption 88–94
quantum dots (QDs)-based saturable absorbers
– bleaching relaxation dynamics 211
– simplified structure scheme 192
– structure scheme 194
quantum dots (QDs) material
– advantages, in mode-locked laser diodes 154
– – advantages of QD saturable absorbers 154
– – amplified spontaneous emission, lower level of 157
– – broad gain bandwidth 154, 155
– – linewidth enhancement factor 157, 158
– – low temperature sensitivity 155, 156
– – low threshold current 155
– – suppressed carrier diffusion 156
– energy levels, schematic 177
quantum dots (QDs) mode-locked lasers 156, 172, 217
– bifurcation diagram 136
– delay differential equation model 132–141
– light–current characteristics 180
– performance 174, 176
– – at elevated temperature 174, 175
– – pulse duration trends 175
– – use of p-doping 176
– pulse duration characterization 161
– sources 160
quantum dots (QDs) monolithic laser 154.
  See also monolithic quantum dot mode-locked lasers
quantum dot theory 11
quantum dot waveguide device 9
– measurements of carrier lifetime 9
quantum wells (QWs) 11, 184
– lasers 121
– materials 124
– mode-locked laser, regimes 127
quantum wire (QWR) 1
QW materials 5, 6, 41, 52, 53, 156, 218
QWs. See quantum wells (QWs)

r
reference optical frequency 110
refractive index 1, 5, 18, 25, 37–41, 46, 76, 100, 105, 119, 121, 122, 123, 138, 142, 144, 145, 147, 158, 179, 181, 186, 197
– calculating resonant 40
– dispersions 142
– fast nonlinear dynamics of 76
– leading to increase in LEF 158

– for low-finesse, resonant SESAM with 197
– operator 142, 145
– optical confinement, enabled through 5
repetition frequency 99, 101, 108, 123, 128, 190, 191, 210, 214, 216

s
SA. See saturable absorber (SA)
saddle-node bifurcation 138
saturable absorber (SA) 6, 9, 37, 38, 77, 78, 80, 83, 154, 160, 165, 176
– based on QD-doped glasses 207
– Cr:YAG laser passively mode locked 212, 213
– geometry of 84
– – SESAM construction 85–87
– – waveguide construction 84
– II–VI semiconductor nanocrystals in glass 207–209
– – optical properties 209
– IV–VI semiconductor QD-doped glasses 209, 210
– macroscopic parameters 184–187
– operation 77
– – absorber recovery time 78
– – for CW or quasi-CW operation 79, 80
– – Henry linewidth enhancement factor 80
– – index modulation, at transition frequency 80
– – modulation depth 77
– – and nonlinear phase modulation 80
– – of QW and bulk saturable absorbers 80
– – saturation fluence of SA 79
– – slow or fast absorber limit 79
– – in two-level system 80
– for passive mode locking around 1.3 μm 210–212
– PbS QD-doped glass saturable absorbers 214–216
– physical processes in 80–84
– – absorbing medium, placed in electric field 81, 82
– – absorption coefficient 84
– – carrier–carrier scattering time 84
– – creating screening potential ξ_s 82
– – dependence of absorption coefficient on 80, 81
– – expression for saturation fluence, due electro-optical saturation 83
– – fast absorber saturation characteristic 84
– – fast saturable absorption for pulses 83
– – Franz–Keldysh effect 82
– – mode locking of solid-state lasers 84

– – quantum confined Franz–Keldysh effect (QCFKE)   82
– – relaxation time   83
– – strong dependence of carrier sweepout on field, together with   83
– QD-doped glass saturable absorbers   210
– – Cr:YAG laser passively mode locked with   212, 213
– – for passive mode locking   210–212
– ultrafast pulse laser based on quantum dot (*see* QD SESAMs; ultrashort pulse generation)
saturable absorption coefficients   121, 213
saturation fluence   78, 79, 81, 94–97. *See also* QD saturable absorbers
– absorption saturation, formula for slow saturable absorber   95, 96
– analytical estimate for F_{sat} 95
– fully inverted QD amplifier, gain   94
– for ground-state absorption coefficient   94
– homogeneous broadening of transitions   95
– load resistance of the SA circuit   96
– physical processes contributing to   94
– and reverse bias   97
– saturation flux values   97
– Stark effect   96
Schott glass filters   208
Schroedinger equation   19
screening potential   82
self-consistent profile (SCP) model   107–109
self-electro-optic effect device (SEED)   81
self-phase modulation (SPM)   39, 40, 44–46, 57–59, 68, 80, 100, 103, 104, 121, 137
semiconductor-based absorber mirrors   187
semiconductor disk lasers (SDLs)   204
semiconductor lasers   1, 2, 38, 39, 52, 77, 99, 100, 103, 201
– mode locking theoretical models   103–126
– – frequency and time–frequency treatment   125, 126
– – large-signal time domain approach   109–120
– – small-signal time domain models   103–109
– – traveling wave models   120–125
semiconductor materials   1
– dimensionality, role of   1–4
– role of dimensionality   1–4
semiconductor mode-locked lasers   99–103
– mode-locking techniques in laser diodes   100, 101
– passive mode locking   101–103
– place   99, 100
semiconductor nanocrystal formation   207

semiconductor optical amplifiers (SOAs)   55, 62, 74, 164, 170, 217
IV–VI semiconductor QD-doped glasses   209, 210
semiconductor QD lasers   160
semiconductor quantum dot (QD) structures   187
semiconductor saturable absorber mirror (SESAM)   38, 77, 78, 119, 184, 185
– InAs   189
– modulation depth   185
– operational spectral range   186
semiconductor ultrafast lasers
– pulse generation   157
SESAM. *See* quantum dot-based semiconductor saturable absorber mirror (SESAM)
s-factor   129
signal-to-noise-ratio   167, 175
single-population models   152, 153
size selection rule, in QDs   19
slave laser, RF signals   174
small-angle X-ray scattering (SAXS) data analysis   212
SOAs. *See* semiconductor optical amplifiers (SOAs)
solid-state lasers   84, 92, 99, 100, 161, 183, 185, 186, 187, 192, 193, 199, 209
– applications   183
– based on vibronic gain materials   99
– domain of   220
– generation of shorter pulses from   188
– passive mode locking   186, 193
SPM. *See* self-phase modulation (SPM)
standing wave factor   84, 85
stationary mode-locking equation   106
Stranski–Krastanov-grown quantum dots   5, 11, 204
Stranski–Krastanov (SK) technique   203
suppressed carrier diffusion   156

t

tapered lasers   164, 165
Taylor series   113
telecoms   220
three-stacked QDs in laser structure   5
time–bandwidth product   175
time domain lumped models   103
time domain spectroscopy   178
time–frequency treatment of mode locking   125, 126
timing jitter   167–170
Ti:sapphire (Ti:Al$_2$O$_3$) laser   183
transition matrix elements   11, 13

traveling wave amplifiers (TWAs) 55, 121
traveling wave models 120–125, 141, 151.
  See also monolithic quantum dot
  mode-locked lasers; semiconductor lasers
– dispersive correction 123
– distributed Bragg reflector (DBR)
  section 123
– distributed time domain 120
– dynamic correction 122
– first-order effects 121
– hole burning effects 122
– infinite response digital filter 122
– limitation 125
– magnitude of periodic carrier density
  modulation 124
– phenomenological relaxation
  equations 121
– of quantum dot mode-locked lasers 141
– – effects of multiple levels and 141–153
– traveling wave equations are coupled
  with 124
two-photon absorption (TPA) 54, 74
two-section laser diode 103

u
ultrafast dynamics 13, 14, 35
ultrafast optoelectronics 4, 209
– applications 4
ultrafast quantum dot lasers
– applications 217
– – biophotonics 220
– – datacoms 219, 220
– – medical 220
ultrafast solid-state lasers 99
ultrashort pulse generation 158–167
– chirp measurement and pulse
  compression 161–164
– external cavity QD mode-locked lasers
  166, 167
– monolithic mode-locked quantum dot
  lasers 158–161
– toward higher power 164, 165
– toward higher repetition rates 165, 166
ultrashort pulses 8, 62, 63, 100, 155, 157, 161,
  164, 165, 167, 179, 181, 183, 184, 189, 196
– generation 183, 184, 187, 192, 193, 199,
  201, 202, 209, 212
– laser technology 183
– – development 184
– solid-state laser 187, 201

v
vertical cavity surface emitting lasers 38, 94
vertical external cavity surface emitting lasers
  (VECSELs) 119, 201
– VECSEL chips 120
vibronic gain materials 99
vibronic lasers 100, 154

w
waveguide absorber 77–79, 84, 87
waveguide saturable absorber 77
wavelength-division multiplexing systems
  (WDMs) 218
wetting layer (WL) 11, 14, 137
– analyzing pulsations of dot and 153
– densities interact with 140
– distribution function and effective
  degeneracy for 24
– distribution over dot energy levels 48
– dynamics 146

y
Yb–fiber laser, setup to test QD absorber
  200
Yb:KYW lasers 187, 190, 191, 214,
  215, 216